Linux Administration Best Practices

Practical solutions to approaching the design and management of Linux systems

Scott Alan Miller

BIRMINGHAM—MUMBAI

Linux Administration Best Practices

Group Product Manager: Rahul Nair
Publishing Product Manager: Rahul Nair
Senior Editor: Shazeen Iqbal
Content Development Editor: Rafiaa Khan
Technical Editor: Arjun Varma
Copy Editor: Safis Editing
Project Coordinator: Shagun Saini
Proofreader: Safis Editing
Indexer: Manju Arasan
Production Designer: Aparna Bhagat
Marketing Coordinator: Hemangi Lotlikar

First published: February 2022

Production reference: 1120122

Published by Packt Publishing Ltd.
Livery Place
35 Livery Street
Birmingham
B3 2PB, UK.

ISBN 978-1-80056-879-2

www.packt.com

To my father, who had the wherewithal and foresight to introduce me to programming and computers at a very young age and taught me to see technology as a business tool. And to my wife, Dominica, and my daughters, Liesl and Luciana, for suffering through the writing of a book on top of all of the normal craziness that life is always throwing at us. My team makes this all possible.

– Scott Alan Miller

Contributors

About the author

Scott Alan Miller is an information technology and software engineering industry veteran of 30+ years, with more than a quarter of a century on UNIX and Linux. His experience has included companies of every size, in every region of the world, in nearly every industry. Scott has been a technician, lead, manager, educator, consultant, writer, author, speaker, and mentor. Today, and for more than the last 20 years, Scott has led the IT consulting team at NTG. He now lives in Nicaragua.

About the reviewer

René Jensen has 21 years of professional experience with UNIX/Linux administration, both as an employed administrator and, for the last 9 years, as a consultant. His experience ranges from branches such as medical, banking, tax, and mobile business, to working in areas such as CI/CD, container deployment, architecting server clusters, daily operations, and many other areas.

I would like to thank my family for being patient, since my work started as a hobby and I spend a lot of time going in depth with new challenges.

Table of Contents

Section 2: Best Practices for Linux Technologies

3

System Storage Best Practices

4

Designing System Deployment Architectures

5

Patch Management Strategies

6
Databases

Section 3: Approaches to Effective System Administration

7
Documentation, Monitoring, and Logging Techniques

8

Improving Administration Maturation with Automation through Scripting and DevOps

9

Backup and Disaster Recovery Approaches

10

User and Access Management Strategies

11

Troubleshooting

Index

Other Books You May Enjoy

Preface

Linux Administration Best Practices is a guide for understanding the context, decision making, and ideologies behind one of the most critical functions in business infrastructure: systems. Systems, that is, operating system level management, remains the cornerstone of communications and infrastructure. Linux remains the most popular operating system family of choice today and is only gaining more and more traction, making the need for well-trained, deeply knowledgable Linux administration teams that much more important.

Who this book is for

This book is intended for those IT professionals working with Linux or as system administrators who want to take their craft to the next level. This book is about best practices and so we approach the role and thinking of the system administrator rather than learning individual tasks. This book will challenge how you think and how you approach system administration. This book will not teach you about the tasks of system administration, but it will teach you how to think like a career administrator.

What this book covers

Chapter 1, What Is the Role of a System Administrator?, explains the actual role and mandate of the system administration function and how to apply this to your own role in your career and your organization.

Chapter 2, Choosing Your Distribution and Release Model, goes through how to choose the right Linux variation for your organization (Linux comes in many flavors and styles), understanding the importance of release models.

Chapter 3, System Storage Best Practices, attempts to take you from newbie to master regarding storage, which remains one of the least understood areas of system administration, taking a high-level approach.

Chapter 4, Designing System Deployment Architectures, breaks down assessing deployment approaches and when different models will work for you.

Systems do not exist in a vacuum; they are deployed in conjunction with other systems, often needing to interoperate for functionality or redundancy. Combining systems in meaningful ways is complex and can be counterintuitive.

Chapter 5, Patch Management Strategies, looks at patching and updates, which might sound mundane but are at the core of any system task list, and are often more complex than is realized. Good patch management will help protect you and your organization from disasters both accidental and malicious.

Chapter 6, Databases, digs into database concepts and how they apply to the systems realm so that you can provide better support and guidance to your database users. Technically not part of the operating system, database management has historically fallen to system administrators.

Chapter 7, Documentation, Monitoring, and Logging Techniques, looks at strategies for both manual and automated processes for tracking the desired state and current state of our systems.

Chapter 8, Improving Administration Maturation with Automation through Scripting and DevOps, looks at many different ways to approach automation considering practicality and real-world needs in addition to perfect, theoretical approaches. Everyone talks about the automation of system tasks, but many organizations fail to do it.

Chapter 9, Backup and Disaster Recovery Approaches, goes far beyond conventional wisdom and approaches disaster recovery with a fresh eye and modernism to provide ways to make backups easier and more effective than they normally are. The single most important task in system administration is protecting data.

Chapter 10, User and Access Management Strategies, looks at best practices for maintaining users, discusses decision making and user management approaches, and investigates architectures for remote access to the operating system. What good is a system if no one can access it?

Chapter 11, Troubleshooting, looks at how taking a planned, intentional approach to troubleshooting improves resolution speed, reduces stress, and might just find problems that would have been kept hidden otherwise. Nothing is harder than figuring out what to do when something is wrong and the pressure is on.

To get the most out of this book

You are expected to have general knowledge of Linux and operating systems. We assume experience in working with systems in a production environment and a business setting. This book covers high-level concepts rather than technical processes and so a working knowledge of Linux and operating systems is beneficial.

The software covered in this book are Linux-based operating systems such as Ubuntu, Fedora, Red Hat Enterprise Linux, and Suse.

No running system is necessary to use this book. This book focuses on high-level concepts and while knowledge of Linux and operating systems is very useful, there is no need to be hands-on with a running system while reading this book.

Download the color images

We also provide a PDF file that has color images of the screenshots/diagrams used in this book. You can download it here: `https://static.packt-cdn.com/downloads/9781800568792_ColorImages.pdf`.

Conventions used

There are a number of text conventions used throughout this book.

Code in text: Indicates code words in text, database table names, folder names, filenames, file extensions, pathnames, dummy URLs, user input, and Twitter handles. Here is an example: "**System** is a reference to **operating system** and designates the scope of the role: managing the platform on which applications run."

Any command-line input or output is written as follows:

```
$ lvcreate -l 100%FREE -n lv_data vg1
```

Bold: Indicates a new term, an important word, or words that you see onscreen. For example, words in menus or dialog boxes appear in the text like this. Here is an example: "As we progress through our exploration of **Linux System Administration**, the idea of hats and really digging into job roles and functions will become more and more clear."

> **Tips or Important Notes**
> Appear like this.

Get in touch

Feedback from our readers is always welcome.

General feedback: If you have questions about any aspect of this book, mention the book title in the subject of your message and email us at `customercare@packtpub.com`.

Errata: Although we have taken every care to ensure the accuracy of our content, mistakes do happen. If you have found a mistake in this book, we would be grateful if you would report this to us. Please visit `www.packtpub.com/support/errata`, selecting your book, clicking on the Errata Submission Form link, and entering the details.

Piracy: If you come across any illegal copies of our works in any form on the Internet, we would be grateful if you would provide us with the location address or website name. Please contact us at `copyright@packt.com` with a link to the material.

If you are interested in becoming an author: If there is a topic that you have expertise in and you are interested in either writing or contributing to a book, please visit `authors.packtpub.com`.

Share Your Thoughts

Once you've read *Linux Administration Best Practices*, we'd love to hear your thoughts! Scan the QR code below to go straight to the Amazon review page for this book and share your feedback.

https://packt.link/r/1800568797

Your review is important to us and the tech community and will help us make sure we're delivering excellent quality content.

Section 1: Understanding the Role of Linux System Administrator

The objective of Section 1 is to help you to comprehend the scope, responsibilities, role, and mandate of the System Administrator function. We take the reader past the concept of tasks and really attempt to dig into the purpose and value of the role at a much deeper level.

This part of the book comprises the following chapters:

- *Chapter 1, What Is the Role of a System Administrator?*
- *Chapter 2, Choosing Your Distribution and Release Model*

1
What Is the Role of a System Administrator?

Few things in our industry sound like they should be simpler to answer than this one, simple question: *what is a system administrator*? And yet, ask anyone and you'll get some widely differing opinions. Everyone seems to have their own take on what the title or role of **System Administrator** implies, including and possibly most varying in people who use this title for themselves or from the companies that hand it out!

Welcome to system administration and specifically **Best Practices of Linux Administration**. In this chapter we are going to dive into understanding the *job*, *role*, and *functions* of a real system administrator and try to understand how we, in that role, fit into an organization.

In tackling this book, it is necessary both for myself to have some semblance of a clear course in writing, but also for you to understand if this book is for you, or to grasp the scope that I am attempting to cover, for me to clearly define what a system administrator is to *me*.

Understanding exactly what is expected of a true system administrator will be the foundation for applying that definition of the role to the upcoming best practices that apply both to system administration generally and specifically to **Linux administration**.

In this chapter we are going to cover the following main topics:

- Where are system administrators in the real world
- Wearing the administrator and engineering hats
- Understanding systems in the business ecosystem
- Learning system administration
- Introducing the IT professional

Where are system administrators in the real world?

I think that one of the most challenging things about attempting to understand what a system administrator is comes from the fact that the *title* of *system administrator* is often given out, willy nilly, by companies with little to no understanding of **Information Technology** (**IT**), systems, or administration and treat it like a general filler for IT roles that they do not understand or know how to name. It also has a strong tendency to be given out in lieu of pay raises or promotions to entice junior staff to remain in an otherwise unrewarding job in the hopes that an impressive title will help them later in their career, so much so that in the end, the number of people working as system administrators is a very small number of people compared to the number of people with the title. In fact, it is no small stretch to guess that the average person with the title of system administrator has never thought about the meaning of the title and may have little inkling of what someone in that role would be expected to do.

If we look solely by title, system administrators are everywhere. But they exist mostly at companies too small to have plausibly employed a system administrator at all. Systems administration as a dedicated job is nearly exclusive to large companies. Most companies need someone to do the tasks of system administration, but only as a part of, and often only a small part of, their overall duties. It is the nature of IT that in small and medium sized companies you typically have generalists who *wear many hats* and do every needed IT role while having little to no time to focus on any one specific function. Whereas in large enterprises you generally get focused roles, often grouped into focused departments, that do just a single IT role: such as system administrator. But even in some enterprises you find departments organized like separate, small businesses and still having generalists doing many different tasks rather than separating out duties to lots of different people.

There is nothing wrong with this, of course. It is totally expected. It's much like how, as a homeowner, you will often do a lot of work on the house yourself, or you might hire a handyman who can do pretty much whatever is needed. You might need some plumbing, painting, carpentry, wiring, or whatever done. Whether you do it yourself, or your handyman does, you do not refer to either of yourselves as plumbers, painters, carpenters, and others. You are just a handy person, or the person that you hired is. You still recognize that a dedicated, full time, focused plumber, painter, carpenter, or electrician is a specialized role. You might do all those tasks occasionally, you might even be good at it, but it's not the same as if that was your full-time career. If you decided to claim to be these things to your friends, they would quickly call you out on the fact that you are quite obviously not those things.

System administrators are like plumbers. Everyone who owns a house does at least a little plumbing. A handyman who does *home maintenance* full time might do a fair amount of plumbing. But neither is a plumber. A very large housing development, or a construction crew might have a dedicated plumber on staff. Maybe even more than one. And nearly every homeowner must engage one from time to time. If you are me, regularly. Most plumbers either work for large companies that have need of continuous plumbing services or they work for plumbing contracting firms and have the benefits of peers and mentors to help them advance in their knowledge.

Nearly every business no matter what size we are talking about needs system administration tasks done. For very small businesses it is not uncommon for these tasks to amount to no more than a few hours per year, and when needed the scheduling is often unpredictable with many hours needed all at once and large gaps of time in which nothing is needed. In large businesses, you might need tens of thousands of hours of system administration tasks per week and require entire departments of dedicated specialists. So just like plumbers, you find small businesses either hiring IT generalists (akin to the homeowner's handyman) or outsourcing system administration tasks to an outside firm like an **Managed Service Provider** (commonly referred to as an **MSP**) or keeping a consultant on retainer; and you will find large companies typically hiring full time specialist system administrators that do nothing else and work only for that firm.

System administration tasks exist in every business, in every industry and create the foundation of what I feel is one of the most rewarding roles within the IT field. With system administration skills you can chart your own course to work in a large firm, be a consultant, join a service provider, or enhance other skills to make yourself a better and more advanced specialist. Without a firm foundation in system administration a generalist will lack one of the most core skills and have little ability to advance even in the generalist ranks. And at the top of the generalist field, true CIO roles primarily pull from those with extensive system administration comprehension.

At this point we know what a system administrator is, where you will find them in businesses, and why you might want to pursue system administration either as your career focus or as an enhancement to a career as a generalist. Now we can go into real detail about what a system administrator really does!

Wearing the administrator and engineering hats

In this section, we will explore two parts:

- How does administration and engineering differ.

- How to identify the role you are performing.

The name *system administrator* itself should clue us in as to what the role should entail. The title is not meant to be confusing or obfuscated. Yet many people believe that it is some kind of trick. If you spend enough time working in the small business arena you might even find that many people, people who are full time IT professionals, may not even believe that true system administrators exist!

System is a reference to **operating system** and designates the scope of the role: managing the platform on which applications run. This differentiates the system administrator role from, say, a database administrator (commonly called a **DBA**) who manages a database itself (which runs on top of an operating system), or an application administrator (who manages specific applications on top of an operating system), or a platform administrator (who manages the hypervisor on which the operating system runs), or a network administrator (who manages the network itself.) Being called a *system* administrator should imply that the focus primarily or nearly entirely of the person or role is centered around the care and feeding of operating systems. If your day isn't all about an operating system, you aren't really a system administrator. Maybe system administration is part of your duties but being a system administrator is not the right title for you.

Administrator tells us that this role is one that manages something. The direct alternative to an administrator is an **Engineer**. An engineer plans and designs something; an administrator runs and maintains something. I often refer to these roles as the **A&E** roles and often the titles are used loosely and meaninglessly based on how the speaker thinks that it will sound. But, when used accurately, they have very definite meanings and in each area within IT (*systems*, *platforms*, *network*, *databases*, *applications*, and others.) you have both working in concert with one another. Of course, it is exceptionally common to have one human acting as both an engineer and an administrator within an organization, the roles have extensive overlap in skills and knowledge and necessarily must work in great cooperation to be able to do what they do effectively.

The difference between the role of an administrator and the role of an engineer

There is a key difference between the two roles, however, that impacts organizations and practitioners in a very meaningful way, which is very important to discuss because otherwise, we are tempted to feel that separating the two roles is nothing more than pedantic or a game of semantics. This difference is in how we measure performance or success.

An engineering role is measured by throughput or the total quantity of work done. It's all about productivity – how many systems the engineering team can design or build in a given period of time.

The administrator role would make no sense in this same context. Administrators manage systems that are already running rather than implementing new ones. An administrator is measured on availability, rather than productivity. This may sound odd to hear, but this is mostly because the average organization has little to no understanding of administration and has never contemplated how to measure the effectiveness of an administration department.

Hats

I spend a lot of time talking about hats. Understanding hats is important.

By hats, I mean in the sense of the different job roles that we take on and understanding when we are performing the tasks of one role or another – referred to as *wearing a hat*. For example, if I work in a restaurant I might act as a waiter one day and we would say that I was wearing a waiter hat. The next day I might be working in the kitchen making salads and then I would be wearing the short order cook hat.

This may sound a bit silly, but it is important to understand. The term *hats* tends to allow us to understand the difference between *being a thing* and *performing the actions associated with that thing* better. We all know what is required to *be a plumber*, but most plumbers drive trucks to their job sites. Does this make them truck drivers? Is the skill of driving a truck actually part of the skill of plumbing? No, it is not. But it is a generally useful ancillary skill. But a large plumbing company could easily consider hiring professional drivers that drive their plumbers from job to job to keep everyone focused on what they are trained to do, lower insurance costs, increase plumbing billable hours, and even allow for hiring excellent plumbers who maybe do not even have a driver's license!

In IT, it is so common to be asked to perform the duties of myriad different roles and functions that we often forget that we are doing it. IT professionals are often seen as a jack of all trades, able to do a little of everything and often this ends up being true just from the nature of what makes people get into IT in the first place. But it really helps when we separate our intended roles, our specialties, our training, and what earns us our salaries from when we are just *helping out* by performing other duties outside of our strict arena because we happen to be skilled, flexible, or just otherwise helpful.

As we progress through our exploration of **Linux System Administration**, the idea of hats and really digging into job roles and functions will become more and more clear. Thinking of our functions as wearing a specific hat at a specific time is a powerful tool for understanding, and possibly more importantly, for communicating our roles, requirements, capabilities, expectations, and responsibilities to others and ultimately, to ourselves. In the real world, it is very rare to find a company or role where the engineering and administration aspects of systems can be truly separated. This has benefits, and it has consequences. But it is worth mentioning that most many larger companies do, in fact, keep these two roles apart – sometimes even to the extent of having them in different departments with unique management teams. There are a lot of benefits to these being separate, primarily because the soft skills needed for the two aspects of systems are generally opposing.

Engineers benefit from the strong soft skill of being good planners. Engineering is all about planning, almost exclusively. You get to take your time, think about how a system will be used, and design around that future need. Your job is to prevent emergencies rather than reacting to them.

Administrators benefit from the strong soft skill of being good perceivers – that is, responding to events in real time, rather than planning ahead. Administrators are tasked with managing live, running systems and that means that their primary challenges are presented to them in real time, and being able to triage, prioritize, and work under pressure is paramount.

Those who make good engineers rarely make good administrators and vice versa. While the technical skills are almost a total overlap, the human skills are highly opposing and very few people manage to be truly adept at both planning and triage. Those highly skilled at administration will also naturally deprioritize engineering work because they know that they can react to it effectively later, even if poorly planned for.

There is other language that can often guide us here as well. Engineers tend to work in a world of projects. There is a goal, an end point when their work is handed off to the administrators. Administration works in a world of steady, ongoing support. There is rarely a starting or ending point and the term *project* would make no sense to them. They run the systems that the company needs until those systems are replaced, generally with a new workload that the administrator runs in the same way. An administrator might need to give input to a project manager as a project might need to report what its long-term administration impact is going to be, or to get the blessings of administration that they are willing to take on the results of the project when it is handed over to them. But the project itself would always be done, by definition, by an engineering role.

Knowing when we are wearing an engineer's hat or when we are wearing an administrator's hat gives us the power to understand how we can function effectively as well as how to communicate our needs and capabilities to the rest of the organization. Most organizations are blind to these kinds of psychological and performance ramifications of the job roles and require that we provide this understanding. The better that we are at identifying our role and our needs means the better prepared we are to attempt to articulate those to management.

A common problem arising from requiring IT staff to function as both an engineer and an administrator at the same time is burnout. Being an administrator naturally leads to being very busy with requests and always being the first called when things go wrong: systems performing slowly, a computer crashing, a new bug discovered, and patches needing to be applied. Administrators tend to not only work long hours but also be on-call extensively. Getting downtime when they can get it is critical to their ability to remain in the role long term.

Engineers have no need to be on-call and are not the responsible role during an emergency or problem. Engineers, however, would not be expected to need special downtime to compensate for the extreme demands on their time and schedule. They do not get interrupted, they do not have to carry a beeper (although no one literally carries a beeper anymore), they do not miss their kids' school play or have to answer questions seconds before boarding a plane, or have to run out of a family dinner to walk someone through fixing a problem remotely. They get to live scheduled lives where they can rest like normal jobs do. The image of the IT worker running ragged and never getting a break, always being expected to drop everything for the company, and living off of promises that often never materialize of a chance to rest sometime in the future is one of system administrators in nearly all cases. No other role carries so much responsibility at the times when it is most critical.

Combining these roles together means that not only would a person wearing both hats risk being on-call and having to drop everything to respond to issues all day, every day, but then needing to fill every potential remaining moment with project work for their engineering role that expects them to have available capacity to do. Trying to stay productive on projects with deadlines while also trying to react to every question, ticket, or outage is a recipe for burnout or worse.

If we can convey to management the difference between the administration function and the engineering function, we can begin a dialogue on how to make our jobs tenable, which ultimately makes things better for both parties. Being pushed too hard leads to a loss of productivity, mistakes, oversights, and staff turnover.

The fifteen-minute rule

In the software development world, which is all engineering with no administration, the standard rule of thumb is that any interruption that requires an engineer's focus will cost that engineer the time of the interruption plus fifteen minutes additional time to get back to the point of productivity where they were before the interruption. It does not take much to see that a handful of interruptions throughout a day would reduce the effective time of an engineer to essentially zero. They might be attempting to work all day, sitting at their desk trying to focus on the task, burning through their available time, but if they keep getting interruption, they will just spin their wheels and feel more and more exhausted and disheartened as they do not get a chance to rest, but neither do they get to be productive and show something for their efforts.

This fifteen-minute rule is known as task switching overhead.

The administrator's role is one of nearly all interruptions. Monitoring systems to see what might be going wrong, being alert for new patches requiring their attention, responding to tickets or questions from other teams, and so forth. Administrators are, of course, impacted by the fifteen-minute rule just the same as an engineer is. But unlike the engineer, an administrator is normally resolving a problem, or a request and the worst is a small, atomic chunk that is completed before the next interruption takes focus. When a task is completed, or an outage resolved the next task to be tackled is a different one whether it is already lined up and ready to go or doesn't arrive for some time. The administrator must go through task switching between each issue regardless, so the overhead that this incurs is a given that cannot be avoided.

Not only does task switching overhead help us to explain and understand why administrators and engineers need to either be different people or be given extreme accommodation, but it also helps to explain the much broader need for quiet, effective work environments for all staff. Interruptions don't just come from system emergencies but from anywhere like meetings, water cooler banter, office workers who just stop by, general office noise, fire drills, you name it.

Tools to measure skills for an administrator or engineer

If we are in a larger organization where these roles can be kept separate, we can show how measuring each by their unique benefit can make each department better. If we are in a smaller organization and are forced to transition from wearing one hat to another, we need to be able to effectively work with the organization to demonstrate our needs and work with management to produce working schedules or extra resources can make us more effective.

One of the best tools that I've seen used to understand the soft skills for administrator versus engineer is the *Myers-Briggs Type Indicator* of *Judging* and *Perceiving*. Myers-Briggs is an industry standard psychological exam that, when handled well, I feel can be highly beneficial for helping us to understand our own natural strengths and weaknesses.

The following shows an overview of the different personality types and their respective cognitive functions in the Myers Briggs model:

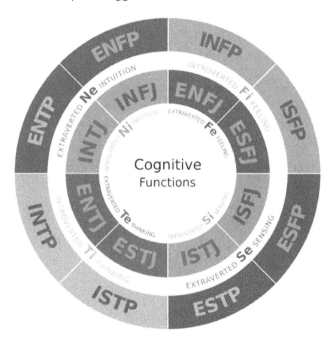

Figure 1.1 – Myers Briggs Cognitive functions of the personality types

Engineers typically require a strong leaning towards the judging profile, while administrators need a strong leaning towards the perceiving profile. Judging effectively equates to *planning*, in this case, and perceiving equates roughly to the ability to react or perform triage. Almost no one is good at both planning and responding. Knowing your strengths lets you focus on what will make you happy and what will let you excel. While knowing your weaknesses allows you (and your organization) to adapt accordingly.

The wonderous variety of the role

One of the more challenging aspects of a book like this is that the role of system administrator, even when we agree on the definition of what a system administrator is, varies so wildly that when performing the role at one company to performing the same role at another company our position can look like a completely different career path! It is easy to find system administrators who spend nearly their entire day manually deploying software and never, ever do performance tuning; another system administrator might do little other than performance tuning; another might focus on managing system applications or databases; one will manage lots of printers, but most admins will never manage even a single printer; and another managing users and userland storage concerns, while again, most, will never manage users at all. All of these system administrators may never meet another system administrator who does a similar workload to what they do, and few will ever get a second job within their field that has them do the same tasks again. Each system administration position is, effectively, unique. Almost absurdly so.

Even more extreme is that so many people think of system administration purely in terms of servers, and certainly the title tends to only be applied there, but there are careers in desktop, appliance, and even IoT realms as well. Linux, in all its forms, may remain a rather small player in the desktop and laptop space, but companies of all sizes do deploy it in production and need system administrators who understand the aspects of the operating system as they pertain to end user devices. Do not casually dismiss the desktop administration subset of systems as being all that different. While desktop administration is often less stressful or critical than server administration it is often more varied and complex and often presents some unique and extreme technical challenges.

In this book I am going to do my best to look at system administration from a high level and provide insight and guidance that applies to essentially everyone - whether you are a full-time working system administrator, a student hoping to become one someday, or an IT generalist for whom system administration is part of your regular tasks. I feel that too many guides to system administration take a myopic approach and try to look at the field as being only one or two of the myriads of aspects out there and totally forget that most of the field will never encounter the need for most of what they are taught.

Here's an image to entertain you with the jack of all trades and the master of one:

Figure 1.2 – Generalist Vs Specialist

There are three essential types of books on system administration. They are as follows:

- The first type takes a high level and attempts to present the *field* of administration.
- The second type digs in to detailed and highly specific tasks.
- And the third focuses on teaching skills required by a certification exam.

All three of these styles of books or resources are very valuable. In this book, we are tackling the first and we will leave detailed commands, syntax, tools, and other *in the weeds* implementation details up to other excellent books that already focus on those things today. I will also attempt to keep my advice as distribution agnostic as is reasonably possible. Linux is all but unique in that it is an enormous and varied ecosystem rather than a single product. When talking about best practices we are given a natural pass on needing to dig in too deeply to a specific operating system built off Linux, but it is still tempting to lean overly heavily on a certain one or two popular implementations.

It is also important to make a distinction between what is intrinsic to Linux, to an operating system built from Linux (more on that later), or intrinsic to system administration versus what is simply convention or assumption. In approaching a book like this, we can go in so many potential directions and I will, whenever possible, attempt to clarify when something is simply convention versus when something is truly how it must be done.

Great, so now we know what aptitude and psychological skills we will need to succeed at system administration, and we know about how awesome this role can be. We are getting a clear picture of what the purpose of the system administrator is. Next, we need to see how that role is going to fit into the IT department and the business itself.

Understanding systems in the business ecosystem

Systems, that is the operating system layer of IT infrastructure, plays one of the most critical roles within IT and the business. In most businesses, most of the most critical aspects of IT functionality tend to fall to the systems roles to oversee. The system administrator is often saddled with tackling security, performance, data integrity, storage, planning, access control, backups, innovation, design, consulting, and so much more. No other role commonly must address all, if even any, of those functions.

For many reasons, system administrators often form the backbone of an IT departmental infrastructure. The operating system, because of its deep roots into storage and networking, and its close association with data and applications, sits in the position with the greatest control and visibility throughout the IT organization. System administrators may have little direct contact with end users but tend to be the nexus around which nearly all the infrastructure and support departments focus and rely.

System administration is generally seen as forming the largest group of focused IT professionals, as well, at least within the scope of purely technical roles. End user touching positions such as helpdesk or deskside support may involve larger headcount, but much of those roles often involves customer service or end user training that accounts for much of their time spent during a day. Within the technical realms of networking, storage, applications, databases, and so forth, it's most likely that systems will represent the largest portion of your team and cover the broadest range of skills.

Being so central, core, and often large, you can think of systems as often functioning as the glue that holds the IT department together; or you can think of systems as being the hub onto which all other IT disciplines will tend to attach. The following diagram demonstrates this:

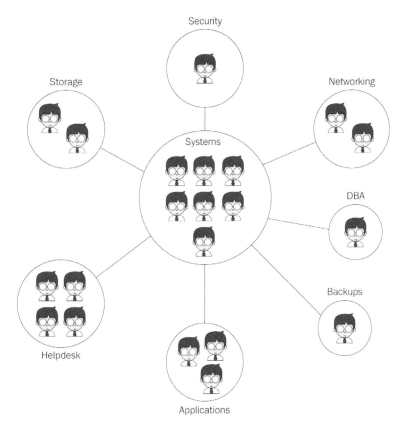

Figure 1.3 - Systems at the Core of the IT Ecosystem

Unlike networking who might know little or nothing of systems, for example, systems professionals cannot be unfamiliar with networking concepts, even extremely detailed ones. Or unlike a database administrator who can generally just assume that someone else will handle backups properly, the system administrator is often the comprehensive point of communications that must understand how the database is talking to storage, how storage is quiescing data, and how a backup is being decoupled from the system. Just as the operating system is essentially the communications hub of a workload, the system administrator naturally becomes the point of holistic understanding and management of a workload.

In my experience, systems administration often is expected to work as a full consulting peer to other teams. I've seen applications, database, and networking teams all turn to systems departments when they are looking for broader experience within their own realms. The nature of system administrators to have to deeply understand not only their own realm, but all adjacent ones, leads to a department that can often act as a consultant to others, much as how IT often does so to the business at large.

Naturally, system administration gains the greatest overall knowledge of the overall workings of a business from the IT perspective which because it touches nearly every aspect of a business, is often one of the most important views that there can be of an organization. This will lend itself to the system administrator also being in a key role to report on and potentially advise business leaders within the organization as to issues and opportunities. Of all specialized roles within IT the system administrator is the most likely to interact heavily with the business on a large scale.

Now we have a good feel for how our role is going to fit into the IT department and the business. Next, we will learn about to tackle learning this role in the first place.

Learning system administration

If you are not already working as a system administrator, you might feel that getting the necessary experience and training to become one will prove to be difficult. This is not true. In fact, moreso than nearly any other area of IT you can experiment with and practice nearly every aspect of system administration for very low cost in a home lab. System administration may have a lot of facets and favor those with deep knowledge, but it is also ultimately accessible to those willing to put in the effort.

In the following sections, I will describe some ways that you can get started on learning to become a System Administrator.

Build a home lab

Few people are as big of a proponent of building a home lab as I am. My own home lab experience has been among the most valuable of my career. And in my role as a hiring manager, no matter how much office experience someone has, I always want to know what they do at home on their own time. Home labs show dedication, interest, passion, and provide unique opportunities to really learn a process from end to end. In my decades in system administration, I have found that it is actually a rare administrator who has gained any real experience running a system from cradle to grave and understanding what a system and workload lifecycle are really like. Having that experience, even if only at home, can be a major point of differentiation in an interview or on the job. It can easily be the stand out factor that makes all of the difference.

Virtualization has brought the cost and complexity of a home lab from something that was onerous in the late 1990s to something that is almost an afterthought in the 2020s. The resource needs of a lab are often incredibly small, while the spare resources of most computers, by comparison, are large. And this is before we consider cloud computing and the ability to do lab work using on demand cloud resources today. While not free, cloud computing is often extremely affordable as another resource that a home lab can deploy expanding the student's experience even further. Leveraging both gives you even more to put on a resume and to discuss at the interview table.

Of course, to make this experience valuable, you must really dedicate a lot of effort - every bit as much, or more, than you would if it was a production system at a paying job. Take the time to automate, secure, monitor, update, and maintain real workloads even if you just use them around the house. A web server, a house VoIP phone PBX, email, instant messaging, a database, a media server, a remote access server, a DNS filter, you name it. There are many things that you can run in your own lab that are similar, or even identical, or heck even superior, to workloads that you would see in a common business environment. Do not be afraid to use your lab to make interesting and fun projects for family and friends, too.

Getting family and friends involved

Building a home lab that is only for yourself can be challenging because you never get to see your environment being used. When possible, I highly recommend recruiting friends and family to get involved as well. This lets you get end user feedback and experience systems more like they are meant to be used. Running systems only for yourself is far more difficult to get the experience that would match what happens in a business.

When I was first running my own home labs for system administration, I ran a family website, an email server, and probably the most interesting a full PBX. A **VoIP PBX** (or **Voice over IP Private Branch Exchange**) for home can be great because it tends to be complex compared to most home style workloads, can be used completely realistically as a valuable part of your family's communications platform and few experienced IT professionals have worked with them themselves.

In the PBX example, you can approach it with having extensions in multiple rooms and offering extensions to family members outside of your physical home. Using a PBX to build a free calling mechanism for family to stay in contact with each other easily can be truly utilitarian beyond the IT experience that it provides. Building something that provides real value and gets really used will not only make your projects more enjoyable but also makes the educational experience more like the real world.

Part of the secret to leveraging the home lab experience is to not only do *everything right* as you would in the best of businesses, but also to be prepared to discuss and explain in detail what you have done and how that meets and possibly exceeds the experience that someone is going to have gained from having worked in more typical work environments. Highlighting best practices, holistic management, experimentation, and the chance to have worked on the latest technologies can make all the difference between an interviewer dismissing home experience as inconsequential to them feeling that their own staff may be ill prepared to interview you due to your broader experience base and more up to date knowledgebase!

Start as a generalist and progress onto a specialist in the System Administrator field

Beyond your home lab there are other ways to gain experience before attempting to jump straight into the big business world where system administrators are most likely to be found. Of course, a common path taken is to work as a generalist in a small company where systems administration is part of the job. This approach does work but does not work often or as well as is generally expected. Making the leap from small business generalist to big business specialist can be almost as challenging as going directly from student to big business specialist in a single leap. So, we need to look at some ways to either get there directly or ways that we can assist in the transition.

Volunteer for non-profits or non-business organizations

Another option is volunteering. Non-profits and other non-business organizations often need resources that they have no means to find, attract, or afford. Volunteering in these types of organizations can be a great way to gain production experience and to demonstrate how dedicated you are to your craft. They can rarely provide any sort of mentoring or guidance, which is a significant negative, but they also generally provide little oversight or structure and will appreciate the effort that you put in, giving you more latitude to choose approaches that benefit your experience as much as benefiting them. As a volunteer, you will easily find that you can spend more time focusing on real solutions for the organization rather than needing to *play politics* or attempt to please a manager since you are working for free and have no job to defend or promotion to angle toward.

One of my better career decisions early on in my system administration years was to use my free time for a couple of years to volunteer as the sole system administrator of an all-Linux environment at a K-12 school. The school had almost no existing computers at all, and certainly no network or Internet access, and I was able to put on my engineering hat and design their entire environment including desktops, servers, telephones, networking, printing, and others. And once implemented, I was able to switch to my administrator's hat and manage the entire desktop, server, telephone, and storage environment from cradle to grave built entirely from Linux. I was able to do everything and do it all myself. I paid the price as an administrator for my oversights as an engineer. I saw the initial cost and design considerations and how they played out with end users. I deal with a workload from procuring the initial hardware, installing the software, configuring, deploying, training, and supporting. That experience didn't just catapult me forward in my career in the short term but is so significant that it is a valid talking point even in senior enterprise CIO level interviews much later in my career.

Presenting this experience in a Fortune 100 interview was very easy. Showing that I understood the processes and ramifications of decisions was great. Being able to demonstrate the ability to research, plan, and action every step from racking a server to production support of a system was unique. In the enterprise those skills rarely exist because typically different teams often handle each different portion of the process. Being able to show holistic oversight of the entire process brough much to the table. This range and type of experience was something other candidates did not have and made a huge different in my career trajectory.

Do not overlook doing small time support for friends and family. Almost everyone knows someone who could use some technical support at home and mostly what people need at home is system administration support! It goes without saying that this will be desktop support rather than support of servers. Real world support of desktops is certainly better than no experience at all. Putting together many different types of experience is best to show the most well-rounded total portfolio.

Self-study

Learning system administration has the benefit of there being so many great resources available on the market from books and certifications to online communities and videos. Both structured and unstructured materials abound, and thousands of passionate professionals are always online in communities just waiting to answer questions and help others in the field.

It is tempting to seek out classes or the newer trend of boot camps to learn IT skills in general, and system administration in specific, but in most cases I will advise against that approach. Boot camps especially are designed to teach you only very specific skills necessary to get up and running and through an interview process on a specific process and cannot take the necessary time to deeply teach topics. This is dangerous and often leads to long term career failure. More formal traditional class style education can be better, and some people learn especially well that way, but there are caveats.

Typically, classroom learning happens at a very slow pace and is not flexible to your schedule. This is only so much of a problem, but you lose the opportunity to move as fast as possible while focusing on what matters to you most. Very rarely can you find classroom education that is current and relevant. While possible, it is almost unheard of and there are seldom checks and balances to verify that what is being taught is useful or even accurate. Most importantly is that classroom learning is not a sustainable approach. Once you are working in the field you will not have the time to continuously take classes to keep your knowledge up to date, nor will you be able to find classes that provide for the immediate needs of the next project or challenge. Classroom learning is essentially a special case that is only available to people before they enter their careers. It is not a useful approach in an ongoing way. Because of this, learning through a classroom setting does not demonstrate a practical skill for the workplace.

Instead, teaching yourself through books, articles, experimentation, online communities, and other self-starting means that show that you are able to learn without someone else's assistance are crucial both because you will almost never have someone available to know everything that you need to know before you do and be prepared to teach it to you, and because you will be spending your entire career constantly needing to stay current and learn new technologies and techniques. Any serious profession involves lifelong learning, but very few demands it to the degree that is required in IT. Showing the ability to teach yourself whatever is needed is a very big deal to a potential employer.

Learning to teach yourself gives you a broader range of potential learning opportunities, more flexibility in how and when you learn, the potential to start learning at any time, and the ability to move at the fastest possible pace. Of course, as a reader of this book, you have already taken at least this step in seeking out formal, self-paced learning.

Age does not matter

You might be wondering when you should consider tackling the beginning of studies into system administration. Is this something you begin at university? Do you need several years in the field working in networking or perhaps in helpdesk before you can move over to systems? Do you start running desktop support before you can switch to servers?

Great questions, I am glad that you asked. The answer is actually: *You can start at any age!* For real, I mean it. While landing a big corporate job might be out of reach for almost any high schooler getting hands on experience and education in systems is anything but. Between books, online videos, articles, home labs, volunteer opportunities, and sometimes even actual paying work with smaller companies a high school student has so many potential avenues to build a resume, get an interview, and get their foot in the proverbial door that the best time for anyone to start studying to be a system administrator is always *today*.

A big advantage of IT compared to most other fields is that we have a lot of industry-backed certifications that can do quite a lot for building a resume early on in your career and almost none of them require previous industry experience or that you be of any specific age or have a degree. In fact, many high school trade programs specifically target assisting students get their first industry certifications during high school. A motivated student could go far beyond what any normal school program could provide and quite quickly building a resume of both experience and certifications.

Getting a great job might require turning eighteen before getting a chance to really apply, but there is every possibility that a high school student could use their school years to build up an impressive level of study and walk almost immediately into a great starting role and move up very quickly. If a student started studying towards an IT career in middle school and stuck with it for four to six years, it would not be unreasonable to have gained more hands-on experience than many mid-career working adults. Not every business will be keen to hire someone so young regardless of what they can bring to the business, but we don't want *every job*, we want that *one* job and a business that is really serious about their IT is going to potentially view that candidate as being the most impressive.

Internships

In traditional trade situations the standard means of gaining experience was through an internship. Today this is much less common, yet it still exists. Most internships will be unpaid. Internships can be extremely valuable, especially for younger learners generally of high school or university age, because they can provide access to traditional businesses, rather than non-profits or charities, and should provide a mentor to guide you while learning. Even paid internships are assumed to be quite poorly paid making them quite difficult for someone beyond student ages to typically consider.

Be wary when approaching a company to discuss an internship. Most companies offering internships do so believing that they are getting low-cost labor when the purpose of an internship is supposed to be to give back to the industry and community. In doing this we often see no mentoring or guidance. The intern is just used to do jobs that require no skill and teach them nothing. In companies that do at least attempt to treat an internship properly it is quite common for the company to lack the skill or resources to mentor and simply not be able to do so. Potentially the worst result is being mentored by someone who is not well trained themselves and the intern being taught incorrectly and coming out with fewer skills than when going in!

Do not let this dissuade you from investigating the option of finding an internship. Internships look great on a resume and even poor experiences can be used to make good talking points in an interview. If you were taught everything incorrectly, perhaps you can present a *what not to do* that was taught to you by example! Just be wary and do not think that all internships should be bad or involve doing menial labour or not being mentored. A true internship is about providing you an education via a mentor who is able to show you how the job is done in the real world. It is not an excuse to get cheap labor. That does not mean that you will never do any work in an internship. You very well may and getting your hands dirty in a real environment is part of the goal. But a true intern is there to learn and not to work *per se*.

Now we have a good idea of how to approach positioning ourselves to moving into a system administration role as well as how to tackle the problem of continuous life-long learning. Now we can look at the IT professional in the context of the business as a whole.

Introducing the IT Professional

I would be remiss to go this far talking about systems administration within IT if I did not step back for a moment and talk about the role of the **IT professional** in the broader sense, as well. It's all too easy to assume that you've picked up this book because you are an experienced IT professional looking to hone their craft, tweak their skills, or maybe make some adjustments to get yourself into the **Linux Administration** path and probably many of you are doing just that. But some of you might be new to the field and wondering if Linux Administration specifically, or systems administration more generally, are the area in which you are wanting to focus your attentions.

First, I have to say, that after well more than thirty years in the field very little else is as generally rewarding as working in IT. IT isn't just an enormous field with countless opportunities, but it is one that gives you geographic opportunity, the chance to explore any business market (finance, manufacturing, insurance, healthcare, veterinary, hospitality, research, government, military, journalist, media, software, tourism, marketing, and so on), and myriad different roles within the field. Everyone's IT journey is a unique one, and the chances for your career to be rewarding and exciting are higher than in nearly any other field. IT is uniquely positioned as not only a technical field, but also as a customer service field, but most importantly as a *core business function*. IT builds and maintains the business infrastructure. As such we are a key player in every aspect of the internals of an individual business, even before we consider the fact that we work in all businesses!

As we alluded to previously, system administrators are, in some ways, the *IT of IT*, the most core and broadly reaching role within the IT department acting much like a *meta-IT* role often combining or connecting all of the other roles.

One area of IT, though, that I think is worth special mention is the topic of the general titles of *Information Technology* and *IT Professional*. Let's look at both closely:

- First, Information Technology. Certainly, this is what *IT* stands for, but truly IT is not about technology. It's about the information, communication, storage, security, and the efficiency of the business - the infrastructure of the business. As such, technology naturally is assumed, but sometimes IT can be about whiteboards, notepads, and just bringing good decision making and common sense to the organization. I often compare IT to legal and accounting departments: each has a focus, but each also is just *part of the business* itself.

- Second, we call ourselves professionals because, well to be honest, because it sounds great. Everyone wants to be called a professional. This aligns us with doctors, lawyers, civil engineers, and so forth. But the truth is, none of those are good analogues to what we do in IT. All those fields start with stringent certifications, follow exacting rules, and would be *just following the script* if their approaches were applied in IT. There is a reason that IT certifications are almost exclusively about products rather than roles, and that is because the concept of certifying someone for an IT role conceptually doesn't really make sense. Why is that?

IT roles cannot really be certified in a meaningful way because if you could codify IT, you could automate IT, but you can't. IT is primarily creative and works as a business function more like the CEO than to any other department. IT's job is to maximize the profits of the business through improvements and good decision making in business infrastructure, which is an extremely broad mandate. The CEO has the same mandate, just without the limitation of being focused on the infrastructure. You would never certify someone as a *CEO Professional*, that's crazy. CEOs are totally creative, wild, unique. IT is the same way or should be. We would never accept a CEO that just did what everyone else did, there would be no value to them. Business professionals like the CEO or the IT department are there to take massive amounts of information and training; add common sense, experience, and creativity; and then apply all of that to a unique business in relationship to its customers, market, regulations, and competition. Almost nothing in these roles is repeatable on a large scale.

At the end of the day, the IT department, whether just one person or one hundred thousand people, has the singular function of helping the company to maximize profits. To do that we have to understand the business inside and out, business as a general concept, technology, decision making, risk and reward, and so on. You can't make those kinds of business decisions, take those kinds of profit risks, if you are tied to the confines associated with *professionals*.

A doctor, as a prime example, has so many strict rules to follow and everything centers around their *certification* process and their overall focus is all about *avoiding mistakes*. A doctor will prioritize any number of lives lost through inaction over being the direct cause of a death themselves. The *professional* approach is to avoid mistakes at the tactical level, while eschewing the strategic level.

IT, like any business function, is the opposite. We must look at the overall risk assessment, calculate potential reward, and make decisions that, mathematically, make sense for business profits. As IT *professionals*, it's not just okay to risk *losing a patient* from time to time, but if we never lose one, chances are pretty good that we are too risk averse to make good decisions. IT should never be about avoiding failure at all costs (or even just at irrational costs), but about choosing the level of risk that is dictated as wise based on the math and logic of the situation. For this reason, I often refer to those in our field as IT Practitioners because this better reflects the proper mindset that we should have when properly representing our field in a company that takes the role of the IT department seriously (and, by extension, takes itself seriously.)

The fallacy of success at any cost

Something that I have heard from businesses, which should instantly have set off alarms in the minds of every executive and manager there, is concepts like *we cannot have server downtime at any cost*. This comes up from the normal course of risk assessment and discovery. System engineers ask what the value of a system is so that they can gauge the needs of risk mitigation only to be told something like *downtime is not an option* or *we have to be up one hundred percent of the time, at any cost*.

Of course, if we really take time to think about it, we know that they are just avoiding the question by stating something absurd. We are left with nothing to work with, no way to know how to approach system design. No system has zero chance of failure, that is not possible. Saying that we must protect against all possible failures *at any cost* means that to even attempt to fulfill that demand that we just consume literally all resources that the organization can provide purely for risk mitigation. Any IT department attempting to follow that directive with any sincerity would bankrupt the firm. It should be obvious that no workload, in any scenario, is worth that. Yet, it is surprisingly often that company management expects IT to work with little other direction in determining which workloads get what degree of protection.

Knowing business, finance, and ITs place in the business are necessary when IT has to force the business to act rationally and effectively.

Don't be surprised to find that in your role as an IT practitioner, especially one in system administration, that you will be playing an instrumental role in guiding and advising the company in many different business capacities. While we would hope that other roles, such as CEO and CFO were more broadly trained in business practices, the harsh reality of business is that it is far easier to become one of those roles with little or no training than to be an effective system administrator without that business training. IT roles at a decision making or influence level require so much business knowledge and practical business thinking to do the job in with any semblance of success that often we must act as advisors to every level of the business as often there is little business experience elsewhere in the organization.

Summary

Hopefully at this point you have some solid understanding of what we mean by the concepts of System Administrator and Linux Administrator and how that role will potentially fit into an organization. We will treat the rest of this book as addressing both administration and engineering within the systems realm and with an eye towards the unique needs of Linux as our system family of choice.

We have looked at understanding what systems are, and how they fit into the IT departmental offerings. We have looked at the engineering and administration aspects of roles. We have broken down how a role can be dedicated or just one of many hats worn by a specialist. We have even broken into understanding more about IT as a general field and how we could go about acquiring an education to let us make the move into an IT career.

At this point, I think that we are ready to start to tackle the specific best practices of Linux Administration!

In our next chapter we are going to move on from examining what it means to be a *System Administrator* and focus in the same way on what it means for an operating system to be a *Linux Operating System* and look at how Linux fits into the technology ecosystem, who the key players are, and how we select Linux for our workloads.

2
Choosing Your Distribution and Release Model

When we talk about **Linux system administration**, we probably jump quickly to wondering what flavor of Linux we are going to be talking about. This is typically the first topic that pops into our minds when having even a casual conversation with a business owner or someone in another, non-technical department. What we rarely spend much time thinking about is release and support models and how these play into our planning, risk, and expenditure models.

A quarter century ago we used to be educated regularly about the merits, caveats, and machinations of different software licensing models. Today terms such as *open-source* are used constantly and no one is surprised to hear them but like with all things technical as the adoption rate of a term increases the general understanding of it likely decreases. Therefore, we need to investigate some nuances of licensing as this plays a role in how an operating system will interact with the outside world, at least from a legal perspective.

In this chapter we are going to cover the following main topics:

1. Understanding Linux in Production

2. Linux Licensing

3. Getting to Know Key Vendors and Products

4. Grokking Releases and Support Models: Rapid, Long Term, and Rolling

5. Choosing Your Distribution

Understanding Linux in production

Linux is used in every aspect of business and production systems today. Simply by being a Linux-based system actually tells us incredibly little about what a device might be doing or how it might be used. Unlike macOS, which essentially guarantees that the use case is either a desktop or a laptop end user device, or Windows Server, which all but assures us that a system is an infrastructure or line of business (LOB) server. Having a system be built on Linux gives us very little to go on when looking to determine the intended use of that system. Linux is used on servers, in virtualization, in desktops, laptops, tablets, routers, firewalls, phones, IoT devices, appliances, and more. Linux is everywhere. And Linux is doing just about everything that there is to do. There are almost no roles that Linux does not cover, at least some of the time.

For the context of a book on **Linux Administration**, we are going to assume that we are talking about Linux in the standard *GNU/Linux* vein using the industry standard set of baseline tools, and in multi-user environments. We will focus almost entirely on the idea of Linux on servers but will consider that end user devices such as desktops and laptops are valid hardware candidates as well. Much of what we discuss will, of course, apply to things such as Android devices or even to non-Linux systems. Linux itself is an almost unlimitedly large topic, but there is a generally accepted *this is what people mean by Linux* idea that we will work with. So, we aren't talking about Android, Chrome OS, and so on.

Linux itself is a *kernel*, not an operating system. But as a kernel, it has formed the backbone (and only a small overall common component) of several related **UNIX** operating systems.

Is Linux UNIX?

You'll likely hear people say that Linux isn't UNIX at some point in your Linux career. And to some degree, they are correct, but not in the way that they likely mean. Linux is only a kernel, one piece of a UNIX operating system. But operating systems built on Linux are, by and large, UNIX - at least according to the most definitive possible source, Dennis Ritche, one of the creators of UNIX. Linux implements both the UNIX approach and ecosystem, as well as the UNIX interfaces. It is a UNIX, just as FreeBSD and others are. UNIX is both a standard and a trademark. But the two are not necessarily maintained together. The waters are a bit muddy here. But standard Linux systems, any that we will be discussing in this book, implement the UNIX standard (known as POSIX originally and now a super set of POSIXs, known as SUS). So, they are a UNIX variant or derivative, just as Dennis Ritchie said that they were way back in 1999. He said the same thing about the BSD operating system family, as well.

But the story is a little better than that. While most operating systems built from Linux have never bothered to pay for any kind of UNIX certification, one of them recently did: EulerOS by Huawei which is built from CentOS, which in turn, is built from Fedora. Only EulerOS as a product officially carries the UNIX trademark designation, but it shows that the broader ecosystem is meeting the specifications. The nature of the certification process is that it makes it easy and obvious to certify large, proprietary, commercially backed UNIX projects such AIX or macOS. But in the Linux space where each distribution counts as a unique operating system and often defining exactly which OS is unique from another is a bit of a gray area (Is Kubuntu covered under Ubuntu? Is Ubuntu covered under Debian?) and where projects are often volunteer efforts that have no revenue source to cover a meaningless certification process it would be all but impossible to certify them through a complicated and expensive certification process, especially when that process has no value. In reality, the Linux and BSD ecosystems have demonstrated that the utility of the UNIX certification process has run its course and the process is now detrimental to the industry and serves no purpose. At the end of the day, being compatible with UNIX is worthless, it is Linux and BSD that other systems want to maintain compatibility with.

Microsoft demonstrated this last point beautifully in the last few years when their traditional UNIX compatibility layer called Windows SFU (Windows Subsystem for UNIX) was rebuilt and renamed WSL (Windows Subsystem for Linux) and intentionally made fully compatible with Linux in order to be able to run Linux-based operating systems on top of the Windows kernel. Microsoft, once the largest maker of traditional UNIX systems with their Xenix product now sees only Linux compatibility as valuable.

If there ever truly was an operating system war, UNIX won through and through. And within UNIX, Linux won.

So, we are continuously challenged to define exactly what we mean by Linux. Everyone has their own definition whether intentionally or simply by not understanding either the relationship between a kernel and an operating system, or by not understanding the role of Linux distributions. It does not take too long working with Linux in business before you will get a feel for what people mean when they are discussing it, however. Somehow while not having any formal definition we have managed to arrive at some sort of standard in the industry. When saying a system is or is not Linux, we know not to include *Android* or *Chrome OS* when talking about administration, but to include them when we are talking about Linux adoption on the desktop. There is a lot of nuance and there is no reliable way to know completely for sure when someone means one thing or another and we can assume that over time these definitions will morph as they are not actually based on anything solid and are not accurate.

Because of this strange approach to Linux naming conventions where nothing is formal and we use contextual clues to determine what is meant, we actually use phrases such as *switching from Chrome OS to a Linux desktop*, which should make no sense as Chrome OS is as much Linux as any other Linux-based operating system is.

Now, at the highest level, we should have a decent idea of what Linux is and, more importantly, what most people mean when they talk about something being Linux or even being a UNIX system. We are also ready to defend why Linux is absolutely a true member of the UNIX family at our next system administration cocktail reception! Next up is licensing, but don't worry because the Linux universe makes licensing as easy as it gets.

Linux licensing

Few discussions of Linux happen without the topic of **licensing** being mentioned. Mostly this happens for a few reasons: because **Linux licensing** is so different from nearly all of its competitors that it plays a significant role in business decisions, because it is the largest and most prominent open-source product on the market regardless of category, and because it arose in popularity in conjunction with the rise of the open-source software movement and quickly became its poster child. Most people instantly connect (and sometimes even confuse) Linux with any mention of open-source software, which leads to a lot of confusion as there are millions of other equally open-source software packages out there and when mentioning closed source software, no one jumps to any one comparable poster-child software package and assumes that that is what we are talking about. Linux, for whatever reason, gets treated differently than pretty much any other product on the market in how people talk, name, and think about it.

Linux itself is licensed under the **GNU General Public License** (aka **GPL**) also known as a **CopyLeft** which is the best known of all open-source licenses. This license provides us with many advantages in the use of Linux.

Being the best-known open-source license alone is a significant benefit. Understanding of how this license applies to organizations is extensive and licensing resources abound. Applying this license to organizations is standard industry knowledge and because essentially all companies globally use software licensed under the GPL at least a little, if not extensively, any concerns about the license must be addressed for myriad reasons. While other open-source licenses, such as BSD or MIT licenses, might be seen as superior from an end user IT department perspective, they are certainly less well known.

The GPL guarantees that access to the source code for Linux is available to everyone, always, for free. This gives organizations important protections. It increases security by allowing for code reviews by anyone who is interested, and with a product like Linux the number of governments, large businesses, security research firms, and interested developers who try to do so is extremely large. Linux is easily the most reviewed code in history. Open-source software of this nature retains every security advantage of closed source software while adding the extremely critical options for public review and, perhaps even more importantly, public accountability.

Freely available source code also guarantees companies using Linux that any vendor disruptions that might occur, which might include a bankruptcy, or a vendor suddenly deciding to change strategy and dropping a product or taking a product in an unwanted direction, can be mitigated in several ways such as another vendor taking up the product on their own and providing an alternative support and production path, or even the customer themselves doing so. Most importantly, an original vendor does not retain the power to hold customers truly hostage through a change in pricing, licensing, or product availability. The customer, and the market, always retain options should the product have value.

A great feature of open-source software of this nature is that every customer benefit from the protections, security, and flexibility of the license even if they do not do things like review the code themselves or fork the product to make their own release.

It is critical to understand, however, that open source does not mean nor imply free. Famously in the open-source community the expression *free as in freedom, not free as in beer* is used to explain this. Meaning that the code is free, but products made from the code may or may not be free. You are totally allowed under the license, as a vendor, to use the free and open-source code to compile a resulting product that is not free to purchase. There are Linux operating system vendors who take this approach. This doesn't remove the benefits of the code being open source, it simply means that you have to track licensing and costs the same as you work with many other operating systems such as Windows or AIX.

Most Linux derived operating systems are completely free, though, and this is a very big deal in the industry. Being completely free has obvious benefits, but these can be misleading as the lack of upfront cost makes for an easy target for salespeople to prey on the emotionally confusing saying *you get what you pay for*, which we all know intrinsically is untrue and has no foundation in reality or logic, but it's oft repeated and easy to make sound plausible. We can instantly counter this with the option to pay as much as you want for Linux, no vendor will turn down the option of a donation and one must wonder *does the value then increase because you voluntarily paid more for it?* Of course not, the entire notion is nonsensical. And yet this is a common argument made that simply throwing money at products is its own reward.

Beyond avoiding up front purchase costs, freely available software, especially operating system software, can do a lot for a business. Probably the biggest benefit is **flexibility**. With a free Linux option, you can deploy software as needed anywhere in the organization without managing licenses, getting budgetary approval, or trying to figure out how to coordinate with vendors to purchase. It's actually not uncommon for licensing overhead to actually exceed technology overhead in the deployment of non-free operating systems! This becomes even more exacerbated in the *cloud* and *DevOps* world when deployments may become automated, transparent, and unpredictable.

Free also means that staff, subsidiaries, employee candidates, partner companies, customers, or anyone really, is free to deploy the same systems for compatibility, education, standardization, testing, staging, and on and on. In some cases, proprietary licensing doesn't even allow for desired scenarios or, when it does, can be cumbersome or onerous. Not in all cases, of course, but in most of the key real-world scenarios this is the case. With the notable exception of the **BSD** family of operating systems which are also open source and generally free (e.g., **FreeBSD**, **NetBSD**, **OpenBSD**, and **Dragonfly BSD**) no other market competitor of the Linux distributions is available in a form that allows it to be easily used in home labs, testing, or partner environments nor on a range of varying hardware platforms. macOS, for example, is only available on the very limited range of *Apple Mac* end user hardware. Windows is limited to *AMD64 architecture hardware* in any meaningful way and licensing for it is complex, confusing, and expensive. *AIX* and *Solaris* are limited to extremely niche, high end server hardware from only a tiny selection of vendors and require expensive licensing. Linux and BSD families are unique in their ability to be deployed, potentially for free, in essentially any conceivable scenario.

And there we go, Linux licensing in a nutshell. At this point you should have the confidence to be able to deploy Linux solutions in your business and understand how licensing affects you, when you need to pay, and for what it is that you might decide to pay. You are prepared to separate concepts like licensing and support and invest where it makes sense for you.

> **Tip**
>
> If you find this topic to be really interesting, I recommend some classic books that delve into the topic of open-source licensing more thoroughly such as *Free as in Freedom* by Sam Williams (2002) and *The Cathedral & the Bazaar* by Eric S. Raymond (2001). These titles were key tomes from the height of the open-source revolution and are considered nearly canon in the Linux Administration space. Well worth reading to gain better insight into the club to which Linux Administrators belong.

Now that we have gone over concepts at the Linux level itself, we need to start coming down for the fifty thousand foot view to a closer ten thousand foot view and start talking about real world vendors and tangible products in the Linux space.

Key vendors and products

Unlike key competitors to Linux, such as Windows by *Microsoft*, macOS by *Apple*, Solaris by *Oracle*, or AIX by *IBM*, Linux has no single vendor representing it, but rather has quite a few vendors each providing their own products, support, and approach to Linux. This, of course, makes discussing Linux exceedingly difficult because Linux isn't a single thing, but more of a concept: *a family of related things that often share many commonalities, but don't necessarily have to.*

Describing the Linux family of operating systems is a rather daunting task as it is far more complex than just half a dozen sibling operating systems. In reality, Linux is a complex tree of root and derivative distributions with derivatives of derivatives and operating systems from all levels of the tree gaining and losing prominence over time. Thankfully, we can ignore the far more confusing and convoluted UNIX family tree of which Linux is just one branch! Analyzing the entire UNIX pedigree would require an entire book of its own and quite honestly a lot more research than I think would be prudent. Suffice it to say that UNIX is a huge topic and primarily historical interest that would not provide us with a lot of value to get too deeply into.

What about BSD?

Any mention of the Linux family of operating systems in the context of system administration will also require us to address the spiritual sibling family of operations systems: BSD. BSD is like Linux in many ways. Like Linux, BSD is open source. Like Linux, BSD represents a family of similar and closely related, but ultimately unique, operating systems. Like Linux, BSD is nearly always free. And like Linux, at least one member of the BSD family has been fully certified as a UNIX, while all members are UNIX in any meaningful way.

Almost everything that we will be covering in this book is equally applicable to both Linux and BSD families, and in many cases even more broadly than that. BSD is not exactly the same as Linux and does have its differences, as would any other operating system. Some key differences at a high level include licensing with BSD being licensed under the BSD license instead of under the GPL like Linux, and BSD referring to the *ecosystem around the kernel* rather than the kernel itself, and Linux referring to the kernel regardless of the ecosystem around it. So truly opposite terms in that regard. BSD is actually analogous to the term GNU rather than to Linux.

BSD truly represents a Linux alternative that shares no code and no licensing, yet is spiritually related in nearly every sense. Both were started from around the same time period, both to replicate UNIX at a time when UNIX was dominant, but expensive and very difficult to obtain access to without either academic access or large corporate access. And today both run nearly all the same applications, in extremely similar ways, and are often seen as direct alternatives to each other in nearly all arenas included cloud, servers, virtualization, desktops, laptops, mobile devices, and even IoT!

Also, like Linux, BSD splintered or *forked* into many products. Four key BSD family members exist in the server world: FreeBSD, OpenBSD, NetBSD, and DragonFly BSD with FreeBSD being by far the most dominant. In the desktop and mobile world BSD looks very different from Linux with the key products being non-free and almost entirely from a single vendor, Apple. Apple's macOS, iOS, iPadOS, and others are all derived from FreeBSD and remain commercial members of the BSD family giving BSD far more visibility and utility than people realize.

While tracking and understanding market dominance for any operating system is extremely difficult as there is no certain means of knowing what operating systems are truly deployed in the wild, let alone which ones are getting use or how new deployments compare to old ones, how to count primary devices against shelved tertiary devices and so forth, it is generally considered that Linux is the overall dominant operating system family today, and likely, but less broadly accepted, that BSD is in second place mostly because of mobile end user devices, but with a very strong server presence as well and macOS suddenly representing a major desktop presence as well. At this point, Microsoft's Windows NT family is the only serious competitor with Linux and BSD overall in the market, although it is still very much dominant on the desktop and laptop.

At the base of the **Linux family tree** there are several roots. These are operating system distributions built with a Linux kernel and all of the accouterments surrounding it to be assembled into a full operating system. There are probably hundreds of distributions at this level, but only a few that truly matter and that we will mention. These are **Debian**, **Red Hat**, and **Slackware**. Each of these represents its own base level operating system that you can deploy, each also represents a family of hundreds, or thousands of operating system derivatives based on the core work of the initial operating system.

We are only going to dig into server operating system choices. If we began listing important desktop Linux distributions our list would more than triple, and it would move from a stable list of **vendors** and **products** that changes very slowly over the years to a list that is disrupted every year or two with key products entering and exiting the market. Selecting an appropriate desktop operating system is a far more personal choice compared to the *purely business-focused decision process* of server selection. What looks and feels good to the end users or management is what often matters most, because it's an end user interface system. Servers, however, have no user interfaces so our decisions are very different.

Debian

Let us look first at Debian. This is the largest and most interesting family of **Linux distributions**, as well as one of the oldest having first released in 1993. Debian is very well known on its own and a great number of IT departments use Debian as their primary server distribution and many applications are developed with it as their primary target. Debian is famous for its openness and lack of commercial oversight. It is often considered the *most free* of the large distributions. Because of this, Debian has found a place less as a distribution to be deployed in its own right, but more so as a distribution used as a starting point for building other distributions. More of the Linux world is built from a starting point of Debian than from any other source.

Debian is completely free, to the point that they do not even offer any paid services, nor do they include any proprietary components or drivers. This makes it an ideal base or *seed* system for building another distribution, but often means that companies are wary of choosing Debian as their directly installed operating system of choice as original vendor support options do not exist. So, gauging the popularity of Debian is convoluted because as a base it is nearly everywhere, but installed natively it is rare.

Debian is famously light, stable, and conservative. It is not the most popular system to deploy but is a solid choice to consider.

Ubuntu

Debian's best known derivative distribution, and the single most deployed Linux based distribution on servers, is **Ubuntu**. Ubuntu could be described as little less than an industry juggernaut having come to the Linux world late in the game, first providing a release in 2004 long after nearly everyone else on this list was well established. Ubuntu moved to desktop dominance, practically defining what Linux on the desktop looked like, in just a few years and today has sat for some time as the most popular, most broadly deployed flavour of Linux with no signs of slowly down. If anything, Ubuntu adoption rates continue to increase both as Linux itself continues to grow in absolute terms and as Ubuntu continues to grow inside of the Linux space in relation to other Linux distributions.

Ubuntu is built from Debian, but extended with more features and polish and, most importantly, it has a commercial vendor backing it: **Canonical**. With Canonical you have options of getting primary vendor support for Ubuntu which means that many organizations requiring primary vendor support are able to choose Ubuntu where Debian was not an option for them. Ubuntu most certainly has many differences from Debian itself including some additions of non-open-source packages, but for many it is through of basically being a vendor supported, business ready version of Debian.

Ubuntu leads, at this current time, across deployment types. They lead in desktops, traditional servers, cloud servers, and more. To a limited extent, you can even find Ubuntu on some phones, tablets, single board computers like the *Raspberry Pi*, and IoT devices! Ubuntu is everywhere. Ubuntu is pretty much on everyone's short list.

In its early days, Ubuntu focused very hard on desktop features and on marketing themselves as being easier for people new to Linux. Whether or not this was true, this benefitted Ubuntu in getting a lot of developer and other non-IT adoption and in getting many IT professionals from the Windows world to be willing to test the Linux waters. It also gave Ubuntu a stigma that took some time to overcome. Today this is purely historical and discussions of easy to use or being desktop focused are long gone.

IBM Red Hat Enterprise Linux (RHEL)

Another early **Linux vendor**, Red Hat also, like Debian, entered the market in 1993 and has remained one of the largest and most influential vendors for its entire history. In 2019 Red Hat was purchased by IBM. Since 2002, Red Hat's flagship Linux distribution has gone by the moniker **Red Hat Enterprise Linux**, but as this name is too long the industry refers to it as **RHEL**.

RHEL is nearly unique in the Linux space as one of the rare distributions that is not free. Of course, being built from GPL-licensed Linux and other components requires that Red Hat make the source code of their operating system be open and free, but the GPL puts no such conditions on the compiled final product. This makes RHEL a much more limited use product on its own. It is one that always comes bundled with vendor support.

Because of this, RHEL is seen as the big business solution for Linux and for a long time was one of only two big vendors making products with an enterprise focus, the other being *SUSE*. For most of its history, Red Hat was the market leader with SUSE trailing behind. But in recent years, Ubuntu has pulled ahead in market terms, even if it hasn't in profit terms.

RHEL is seen as an extremely conservative distribution that is slightly heavier in standard installs compared to most of its competitors. RHEL often gets the most security focus and review and has the biggest support vendor behind it.

RHEL alternatives

As this book goes to press a lot is changing in the Red Hat world. For many years, CentOS was a Linux distribution built to be a RHEL-clone as closely as possible, made by recompiling the RHEL source code, while removing any trademarked names, logos, and so forth, and making a binary compatible distribution for free that was RHEL in all but name. Red Hat eventually bought CentOS but operated it for some time exactly as it had always been. But just before this book was written, Red Hat discontinued CentOS leaving the market in disarray. Until that point, CentOS was the primary product in the Red Hat ecosystem and most companies using RHEL in some fashion were actually doing so using CentOS which was free and came without support. The sudden cancelation has left that portion of the market in disarray.

A few other immediate options have presented themselves in the aftermath of this announcement. One is Rocky Linux which is essentially the CentOS project recreated by its original founder with the same mission. Another is Oracle Linux which was built the same was that CentOS was, by taking the RHEL source code and recompiling it, but by Oracle a major industry vendor (and the creator of the Solaris operating system) and offering it still for free, but also with optional Oracle vendor support and with a few extra features that RHEL itself did not have.

In reaction to the market furor over the sudden death of CentOS and the obvious market move to take their business to other non-IBM vendors, Red Hat has announced programs to provide a limited number of RHEL licenses for free for small businesses or IT professionals needing access to learn the operating system.

How this will play out in the next few years will be interesting and is difficult to predict. But needless to say the strong market position and longevity of RHEL, are not the near certainties that we would have thought that they were just a few months ago.

This also highlights a general rule of business: products from large proprietary vendors are often the riskiest, contrary to most emotional responses from business owners, because their direction is set by boards, investors, and market opportunity making it surprisingly easy for large products to be suddenly abandoned or discontinued even if they are doing well in the market. Often smaller vendors that are more focused can actually be more stable in the ways that most affect their customers. They simply have far more riding on keeping their existing products viable and cannot afford to gamble on killing off major product lines or alienating their user base.

Fedora

Fedora is an interesting player in the business Linux space. Technically Fedora is part of Red Hat, which is in turn a part of IBM. Fedora operates much like an independent nonprofit but is not a legal entity on its own. So many people believe it to be its own company but it is not. Fedora was originally its own project before it was merged with an early Red Hat Linux product in 2002.

Fedora, as a product serves two key purposes in which we are interested in the aim of this book. The first is that it produces one of the best completely free Linux distributions, simply called Fedora, which is designed to be able to be used well as both a desktop and a server distribution. The second is that Fedora is focused on innovation and their work is used as the foundation for what becomes both CentOS Stream and RHEL.

Originally, Red Hat was the seeding distribution and Fedora worked from their code and extended it. But over the years, Fedora is now the base seed distribution from which RHEL is derived. In that regard, Fedora is a bit similar to Debian in Debian's relationship with Ubuntu, but with RHEL instead. Like Debian, Fedora does not offer vendor support. And also, like Debian, Fedora's downstream child product is far better known than the parent is – that is that RHEL and CentOS are better known than Fedora on which they are based.

In other ways, however, we find in practice Fedora and Ubuntu are more alike overall both with similar foci broadly on many different use cases, both being used as the base for many other products, both sharing similar design strategies. There are no exact analogues, unfortunately.

Fedora's biggest challenges come from its lack of primary vendor support options. It also suffers from not being seen as a top tier distribution by many software makers. These vendors tend to then target Ubuntu and RHEL while Fedora often only gets support as an afterthought or because of its naturally high compatibility with RHEL.

OpenSUSE and SLES

The last of the big three *old guard* Linux vendors, **SUSE** products its first product release in 1994, but was a support vendor for older distributions like SLS and Slackware, which didn't have a support vendor of their own, as far back as 1992. Linux itself dates only back to 1991 and SUSE is often believed to be the first commercial support vendor for Linux. SUSE originally used other distributions as the basis for **Suse Linux**, but its underlying projects all faded away during the 1990s and since about 2000 SUSE Linux has been its own root distribution seeding others from itself.

SUSE has always made for confusing product and company names. SUSE makes its primary products under the **OpenSuse** name. These products are free and quite popular. Additionally, SUSE Linux **Enterprise Server** is a commercially supported copy of one version of OpenSUSE. There is no direct comparison to either the Red Hat/Fedora ecosystem nor the Debian/Ubuntu ecosystem. All three primary ecosystems operate rather differently.

OpenSUSE's Linux distributions are very popular in Europe, especially in Germany where they are headquartered. In North American markets they are far less prevalent.

Digging into distribution history

It is easy, and interesting, to get lost trying to track down the real history not only of Linux as a kernel, but also of all of the distributions that have been created from it over the years and how they relate to one another. So many distributions have come and gone. So many foundations and guiding organizations have attempted to control or oversee different aspects of Linux or Linux distributions, but most have faded into obscurity.

For even more fun, go back and try to track UNIX history starting in the 1960s and try to understand where the products on the market today have come from! At this point, UNIX is well over fifty years old and Linux turns thirty years old this year, in 2021! These are old systems with a lot of history and both UNIX and Linux being rooted at least partially in open source has made for so many different *stories* all stemming from one beginning. Unlike a product like Windows, which is complex enough as it is, UNIX isn't the product of a single vendor and is not really a product on its own. So, the story of what we have today is complex.

While this history is interesting, it is hard to say how much of it would have real value today. With the amount of history behind us and the rather extensive disassociation that we have between products today and vendors of the past, knowing the story of how that we got where we are is more of a party trick than useful IT knowledge. Knowing the basics of current vendors and distributions is very important, but knowing who came first, who is older, from whose work they built and so forth is banal. All of today's key vendors are nearing twenty years old or older and so all of the people involved in the early days are all gone. Even the vendors themselves would struggle to track down their products' own histories at this point.

Other Linux distributions

There are so many potential **Linux distributions** that you could consider, and if you are working on desktops you might reasonably go with any number of niche or unproven systems because the ability to deal with instability or support issues on desktops is generally pretty flexible. In the desktop Linux world, many the most popular alternative desktops are built on top of one of the major distributions such as **ElementaryOS**, **Zorin OS**, and **Linux Mint** which are all additional layers on top of Ubuntu (which is already built on top of Debian.) A few such as Solus start from scratch and use no one else as a seed distribution.

If we are looking at servers, we are going to struggle to justify using any distribution outside of the extreme mainstream. Mainstream distributions get more vendor support, more testing, broader third-party support for both software and hardware compatibility, and so on. But beyond the obvious, big distributions also have better community documentation, more how-tos and guides, and more professionals in the wild with specific experience on your exact operating system. Part of deciding what operating system should be installed required considering all of the possibilities and that includes finding additional system administration resources or replacement ones. If we had the ability to hire proactively with adequate time to train any good system administrator could get up to speed on even a completely obscure and unusual operating system easily as the basic concepts always remain constant. However, being able to find someone that is knowledgeable, comfortable, and available in a time of crisis means we must stick to distributions that are broadly known.

Over time, new distributions will rise, and old ones will fall. Change happens. The best practice in this case is to evaluate the industry and determine based on current factors if the advice in this book is current and still relevant, or if perhaps a distro here has lost popularity or run into support issues or if a new distribution with great features and amazing industry support has arisen and requires consideration. Nothing is static, especially not in IT. I do my best here to give you a good starting point on choosing your distribution(s), but never make the mistake of thinking that this list is gospel or unchanging. IT always requires us to take all of the factors that we can, apply what we know, and make the best determination based on current factors as they are applicable to us. My guide here hopefully helps you see how I look at these distributions and why these are the ones that I single out for key consideration at the time of this writing.

The myth of popularity

We must balance the idea of believing that just because something is popular that it will have a lot of support available in the market with the idea of practical support. Just because a product is popular or has many people offering support for it does not mean that support is better or easier to obtain in a practical sense.

For years, this is something that we have discussed about Windows and Linux operating systems. There were, and remain, far more Windows system administrators marketing themselves on the market than there are Linux administrators. But experience tells us that when hiring an administrator in the Linux world it is relatively easy to find at least reasonably qualified candidates. Run a series of interviews and almost every candidate will be able to do the job, even if not well. But run a similar interview looking for the exact same position but for Windows instead of for Linux and you will likely get twice or thrice the candidates to wade through with only a very small number able to do the tasks at the same level as the Linux candidates.

Why is this? Many reasons, we assume. For one, the majority of Windows shops allow or require work to be done from a GUI, rather than from a command line. This means that most people who have never done a task before can simply poke around, apply some basic common sense, read what is on the screen and appear to a casual observer to be able to do system administration duties, even if they have never seen a server before in their lives. But nearly all Linux shops allow no GUIs on servers and when they do have no administration tools in the GUI forcing all work to be done at the command line where it is far harder to simply look around and hope for the best. Familiarity with the commands and outcomes is very much needed. Basically, bluffing is much harder. This isn't because Linux is harder, at all. Not in the least.

This is because of a culture in the Linux world of GUIs being unacceptable on servers and the opposite culture in the Windows world. In the early days, in the early 1990s, Linux had no GUI, and everything was done at the command line and at the same time Windows was GUI only and the full command line capability wasn't added until decades later. This started a trend that has remained to this day. If you talk to someone who works on Linux as an administrator, they will simply assume that everything is command line only and have the same conversation with a Windows admin and while they will be familiar with the command line, they will almost universally assume that it is a partial tool used in conjunction with the GUI. Of course, if you wanted to administer Windows completely from PowerShell today, you can and you can do so very well. And likewise, if you really wanted to, Linux can be administrated from a GUI. No one does because that's a bad idea, but it is possible.

Another factor is that small Windows shops abound and tend to throw around the title of system administrator where little, if any, administration work has really happened. In the late 1990s and early 2000s we had a problem that Microsoft and other certification vendors were handing out certifications like candy on Halloween with many, possibly most, people obtaining certifications doing so through brain dumps or boot camps and having no experience and potentially even no knowledge of Windows or system administration. The use of these *paper administrators*, as they came to be known, threatened to break many IT shops that thought that they had an easy way to hire without doing any validation of their own and ended up with the industry developing a culture of IT often having no knowledge or training which, given the critical position that IT staff is in, putting companies at severe risk, but because IT is so complicated, companies often failed to identify the failings of the process, even after the fact as few companies do *post mortems* on things like hiring processes. Most companies would not even know how or where to begin to evaluate the efficacy of decision processes in that way, so they simple ignore it.

For whatever reason, finding someone claiming to be a Windows administrator is easy, but finding one with any real level of knowledge is quite difficult. Linux, with a much smaller pool of claimed administrators, the ability to hire one that can provide you with good results is decently easy. This is just an example case and is not caused by anything intrinsic to Windows or Linux, it is just resulting artefacts of many historical factors coming together, so do not attempt to read into it anything that isn't there. The only point of this example is to show that popularity or abundance of a resource on the market is not the same as having useful support.

Using multiple distributions

New system administrators often believe that they need to pick a single distribution for their entire organization and stick to it through thick and thin. This is called a **homogeneous environment** and has many obvious advantages. It is easier to learn, easier to automate, easier to test, easier to monitor and stay on top of the latest news and changes. Certainly, consider that a homogeneous environment might be right for your organization.

There are many reasons why we might need to consider a heterogeneous environment, that is one in which there are multiple Linux distributions (and more broadly speaking, perhaps other operating systems entirely like Solaris, AIX, macOS, FreeBSD, or even Windows!) Not many companies of any size are truly able to be homogeneous regardless of if they prefer Windows, Linux, or *other*. This goes for both the server world and the end user device space. These days most IT shops are forced to content with a mixture of systems on servers, desktops, laptops, and even cellular phones.

This variety is spreading rapidly not just in software but in hardware as well. In the last few years alternative CPU architectures have started to gain a foothold. This is how the industry was for its first several decades until a little less than fifteen years ago. In the mid-2000s the industry broadly consolidated around the AMD64 architecture (there were always niches of alternative architectures out there, but they were truly niche for many years) with all other major contenders failing completely or being relegated to very specialty roles. But around 2020 there was such a leap forward in non-AMD64 architecture popularity that we now have to consider heterogeneous hardware environments again, as well.

Hardware architecture may seem like an aside, but it is not. Every operating system and each Linux distro are made for only a limited number of hardware architectures. This means that your choice of operating system will determine your hardware options, or that your choice of hardware will determine your operating system options. Ideally decisions around this will be made holistically considering all of the needs of both software and hardware and coming to a design decision that creates the best resultant package rather than looking at the two in isolation and hoping to come to a working compromise that doesn't account for the results.

In general, my advice is to avoid unnecessary *operating system sprawl* in your organization. It is all too easy to end up with a dozen different operating systems, each doing a niche task, and each requiring a bit of special care and feeding different from all of the others. But don't take that as saying that you should be homogeneous unless you have no choice. There is a balance that will make sense for you. Attempting to be homogeneous will almost certainly be a fruitless effort that will result in unnecessary struggles. And allowing unchecked sprawl will almost certainly result in an unnecessarily complicated environment. Be thoughtful in your choice to deploy something new and different, but do not fear it.

Making the choice

Knowing the likely range of Linux distributions, and understanding our design to avoid operating system sprawl within reason, how should we approach our final operating system selection decision?

The simple answer is *take everything into account*. That is easier said than done. For me when I am looking to deploy a new workload my first question is about the workload itself. *On what distribution does its vendor support running it? On what distribution does its vendor recommend running it? What is the current support path and the future support paths by distribution?*

Working with, rather than against, the workload application vendor can be a big deal. Choosing the wrong distribution can result in poor performance or installation problems or even stability problems long term. It is common for vendors to officially support more operating systems than they regularly test against. For example, perhaps a vendor uses Debian in their development lab and tests every release against it, but only tests Fedora once a year or so, even though Fedora is on their support list. If you choose Fedora thinking that the vendor supports it, you might find deployment or stability problems with the application on Fedora long before the vendor does, and you might find them ill prepared to support something outside of their ken. This is the fault of the vendor for not making it clear enough that all of their support is not equal, but it is a very common scenario. Even a vendor who does an excellent job of supporting and testing many release targets still has one or two that they test more frequently or heavily. Generally, this is the operating system in use by their developers.

Different operating systems will have different performance characteristics as well. Generally, between different Linux distributions this difference is nominal. There can be large differences, even for very simple applications, between Linux and Windows, for example. Between Linux distributions we are most likely to see performance differences less from core features like the kernel configuration or system libraries, but much more likely to see differences caused by things like different programming language versions which can often vary widely between distributions based on several factors behind the scenes that are determined by the distribution's governance.

Once we know how our distribution will impact our workload we can work with other factors, like which distributions are already in our organization or part of our future planning, cost, current knowledge, support options, and so forth to determine what is the right distribution for us. In an ideal world, operating system and hardware considerations would also play a part in the selection of workloads in the first place, but in reality, it is shockingly rare for organizations to take a holistic view of their entire business and instead often make decisions, such as which applications to deploy, in isolation without considering how that will impact the business across the board. As system administrators we often have little say as to which workloads we will be supporting and so we must therefore react as best as we can to accommodate them.

And there we have it, a survey of popular Linux vendors and products. I know, it is a lot to take in. Working with Linux tends to result in getting to know many vendors and distributions and often just taking different ones out for a test drive on a regular basis. From a career perspective having many of these in your knowledge arsenal is important. The more you know the more value you have to an employer and the more career doors will be open to you.

In our next section we are going to look at support and release models commonly found for Linux and look at real world examples available for the vendors and distributions that we just learned about.

Releases and support: LTS, current, and rolling

Picking your vendor might seem like it gets you everything that you need to get started on your Linux deployment, but it does not. We still have to consider release management as part of our distribution decision plan.

A release model or release regime is an approach to how the distribution will update itself. There are three standard models followed by essentially all vendors. These are **rolling releases**, **short term releases**, and **long-term releases**.

Not all vendors provide all the different models. And each vendor approaches support differently. We will do our best to make useful generalizations, but when you make your decision you will need to consider the current strategies available from your prospective vendors as well as considering the quality of the support that they provide.

There is a second factor that is often confused with the release model, and that is the support period. The two are roughly related, but a vendor can make updates and support closely tied together, or potentially not at all. In the real world, almost all vendors seem to stick to just a few common models and vendors who traditionally did not follow these models, like Microsoft, have changed to match the models of the big Linux distributions.

To make things more complicated, each vendor provides different models under totally different naming conventions so that once you learn how one vendor does it, you will have no better understanding of what the next vendor might be doing.

We will start by exploring each release model before we try to decipher which vendors provide for each model. We should also talk about the differences between a patch and a release.

A release is a new version of an operating system. We often talk about major and minor releases. But like anything else in this topic this is all by convention. If you are familiar with the Windows world, major releases would be analogous to Windows 2000, XP, Vista and so forth. Minor releases were called **Service Packs**. So, *Windows XP* was a major release coming between Windows 2000 and Windows Vista, while *Windows XP SP2* was a minor release coming between *Windows XP SP1* and *Windows XP SP3*. Red Hat follows a similar and easy to understand model with major releases looking like **RHEL 7** and **RHEL 8**. Minor releases would be in format of **RHEL 7.2** and **RHEL 7.3**.

The idea behind a major release is that software in one major release may not be backward compatible with previous major releases. A major release is meant to provide a stable target for software vendors or projects to be able to write for and test against and feel confident that the software that they make will continue to work on a major release for its entire lifecycle. Therefore most software designates major releases against which it is known to work such as *Compatible with RHEL 8* or *Built for Windows 7*.

Minor releases are intended to be significant changes that maintain backwards compatibility. In theory you should be able to update a system from its initial minor release all the way until its final minor release without any compatibility issues with software as long as your major release does not change. Of course, in practice there is always a risk anytime that there is a change, no matter how minor it is supposed to be, that some compatibility will change. The purpose of minor releases is to be able to provide new functionality and bigger updates that might affect training, end users, performance, and other factors, but not stop software from working properly and not to require extra testing.

The major/minor release system is not adhered to be all vendors. In the Linux world's big commercial vendors, Red Hat and Suse, use the major/minor system, but Canonical (who makes Ubuntu) does not. Traditionally, Microsoft use a major/minor system, but this ended with the release of Windows 10. It should be noted that Red Hat uses the major/minor system with RHEL but does not do so with Fedora.

Canonical pioneered a now standard date-based naming convention with Ubuntu that approaches versioning completely different. Ubuntu releases are numbers by the year and month in which they release and also with a somewhat whimsical name to make them easier to remember (or something like that, honestly the numbers are really easy to remember and the animal-based names are not at all.) Microsoft copied this when they moved to Windows 10. Fedora does something similar but with a simple incrementing number that is not based on a date (nor on an animal.)

The extreme differences in underlying approach combined with the differences in naming conventions can make it tricky to understand the cycles from one vendor or product to another. We will cover some in this book, but of course they are subject to change and all three major vendors have changed these in the past (as has Microsoft) and are likely to change again in the future. The only large vendor that has remained steadfast with the same naming convention for decades is Apple with their macOS and iOS products and even those have had some marketing changes here and there to keep us on our toes.

Best practice here is to take some time and learn the current release patterns, conventions, names, and versions of the vendor(s) that you are working with and keep yourself current with them. This is something everyone working with software needs to know and understand.

> **Patches can have schedules too**
>
> It might seem crazy to talk about having patch schedules. If software is vulnerable and a patch is available, surely, we need access to that right away and there should be no intentional waiting or schedule for patch releases. For the most part, that is how vendors work. Patches come out continuously when they are available, and releases work on whatever scheduling regime is in place for the product.
>
> Famously Microsoft has not followed this strategy and their Patch Tuesday is well known. Except for the most critical patches Microsoft has long held to a scheduled patch release of just one time per month!

What does support mean?

If you say that your software has **available support** or is *supported*, we pretty much all immediately think of the same things. However, the term support means completely different things in different situations and never is simply saying that something is supported enough to actually mean anything.

On its own the term support means literally nothing, but it conjures up assurances in the minds of those who hear it which makes it a powerful marketing term. Both the offer of included support puts our minds at ease while something lacking support causes us to panic. No company in their right mind would ever use software without support, right? Well, yes and no. What does that even mean? Everything has and does not have support depending on what you mean at the time.

What often matters when discussing support, or what managers mean most, when it comes to operating systems is whether the primary vendor is continuing to supply security patches or stability patches when problems are discovered. Software is a living thing and regular patching is one of the most important things that we need our vendors to provide. This kind of support is where we absolutely must rely on our software vendor. There is no reasonable possibility that we can handle this through any other channel, and it is essentially the purpose of having an operating system distribution in the first place.

After patching, we tend to assume that a vendor saying that a product is supported will mean that they provide updates and consulting support should anything with the operating system fail. This is not a given in any sense and some of the biggest vendors, like Microsoft, do not actually include this when they refer to a system as being supported. In the Linux world, most major vendors do provide some level of support for users who run into problems or need help, and this can be very valuable. But this is actually an area where Linux tends to have something above and beyond other arenas and it can represent a significant selling point, but of course is only available in situations where support is paid for and so many companies opt not to have this kind of support.

We tend to refer to this top level of support where we have the right to submit a ticket for assistance to the operating system vendor and expect a quick response to help us with nearly anything from actual problems with the operating system itself to simple configuration or use issues or possibly even training as paid vendor support. Some companies completely depend on this type of support while some others have never used it, and some are even unaware that it exists.

Support can also mean third-party support. This can come from big vendors such as Oracle (who offers third-party vendor support for RHEL) or from small consulting firms such as Managed Service Providers (MSPs). Nearly any product can get some level of support via third parties, which can lead to an unclear support situation as some products expect support to be via a third party and others expect it to be first party.

And then there comes Community Support, a term used for any support from places like online forums and communities. This has a tendency to get a lot of focus in the Linux world and very little in other realms. All operating systems or IT products really have this kind of assistance. Linux gets more attention, most likely, because there tends to be more knowledge in the field and very little is secreted by the vendor, as is the nature of open source. Closed source software gets less benefit from community support because details must be either discovered through luck or released by the vendor. Community support often gets a lack of respect from management, but it is actually an important sign of what makes the Linux ecosystem strong: vendors that make everything public and communities of professionals looking to support one another. In many cases community support is so significant that it can completely change how we approach a product and can actually be better than vendor support.

Basically, support means many different things, and everyone has their own definition. We need to think broadly about what support needs our organization has and how different approaches will impact us. Many large organizations require primary vendor support with a service level agreement for internal political reasons. Just keeping a system running in a healthy way often requires nothing more than basic security patches and moderate community support. Evaluate your own needs, remove emotions, and do not assume that the term support implies more than it does.

We have now established what these terms mean and how releases and support and tied together and how they are provided. This gives us a framework for exploring real world models from actual products and vendors.

Release model: rapid release

The best known and arguably most important release model has no official or standard name. You will hear names like Rapid Release, Short Term Support, Current, and similar applied to it, but generally it is seen simply as the standard against which other models are measured against. Kind of the *happy medium* of models.

The key products are:

- Fedora
- Ubuntu

We start our discussion with **Rapid Release** models because this is what most people perceive as being the standard or baseline approach to releases. These are the most well-known and understood releases even if not the most commonly used in practice. They are quickly becoming more popular in recent years.

Rapid release (which could also be called *Standard* or *Common Release*) models have a convention of releasing new versions roughly every six months, but of course every vendor is free to use whatever time frame that they want. We would generally consider something to fall into this category if the release was at least two months apart, but no more than a year apart.

The best known and easiest to explain product in this category is Fedora, which is made by the Red Hat division of IBM. Fedora releases a new major version roughly every six months and has no minor versions. Fedora numbers each release sequentially to make tracking changes as easy as possible. There are no weird numbers to remember, no point releases, no skipping around. Just Fedora 1, 2, 3... 30, 31, 32, 33, and so forth. Support for each Fedora release is roughly thirteen months. So, at any given time at least the two latest, if not the three latest, releases are all under support.

We have to mention that Fedora does not have any paid primary vendor support. Support in Fedora terms refers to Red Hat's support for updates and security patches. Third party support for Fedora is, of course, extensive.

The mainline Ubuntu product is rapid release as well but with a much more confusing naming convention than Fedora uses. Like Red Hat's Fedora, Ubuntu is released every six months. Ubuntu's support is for nine months from initial release giving just three months of overlap from the subsequent version's release date until the end of Ubuntu's support commitment to the previous release.

Unlike Fedora which has only software level support, Canonical offers full primary vendor paid support options for Ubuntu's rapid release option.

These short support time frames may feel problematic and it is because of the emotional feeling of needing to find a new solution every nine to thirteen months to avoid the *end of support* that is always looming when choosing a rapid release product that so many companies jump to other models to avoid having to find a new solution on a regular basis. This is illogical because this is based on believing that updating the system in question is not an option.

Rapid release support may be *short term* in most cases, but this is just one perspective. In both example cases here (and all cases of which I am aware) getting *eternal support* is as simple as keeping your systems up to date within a reasonable time frame. With Ubuntu this means updating to the current release at least every nine months and with Fedora at least every thirteen months. In both of these cases, and in most normal cases, these are in place updates done with just a command or two not situations that involve replacing one solution with another. So, while we may call these short term support approaches, in practice they are quite the opposite. This is the industry standard approach to having a continuously supported system that always stays reasonably up to date.

Release model: LTS

The key LTS products are:

- Red Hat Enterprise Linux (RHEL)
- OpenSuse Leap
- Ubuntu LTS
- Debian

LTS might be standard for long term support, but most people think of LTS in terms of the associated release schedule. While there is no hard and fast rule for what constitutes an LTS outside of a vendor designating it as such we typically assume that LTS releases will be at least two years apart. Ideally vendors would separate the naming conventions of releases and support models to make it clearer, but few major vendors do.

Technically an LTS release means that that release will be supported for a *long time*, which what that exactly means being unique to each vendor. LTS is a marketing term, more than anything, but one that tends to inspire instant confidence in many customers and so is often used for that purpose.

The underlying theoretical goal of an LTS release is to provide a *stable* product that remains *supported* for a *long time* (we must use a lot of quotes when doing this) to make it easier for software vendors to make products that will work on that product and not have to continuously maintain updates. This makes them popular with software vendors because they require less work to support. They tend to be popular with IT departments because the primary risks of LTS releases can be many years or even decades down the road and so most negative ramifications of choosing an LTS release tend to end up being someone else's problem.

Speaking in generalities can only get us so far. Be cognizant that vendors can use any model that they want and may change at any time and can use the term LTS without any underlying change to their support. Do not let the emotional reaction to the term cause you to favor an LTS release if the overall support does not best meet your needs.

Red Hat's LTS product is RHEL. All versions of RHEL are LTS. Major Versions of RHEL release on a somewhat unpredictable schedule that started roughly every two years and has slowly increased to be closer to five years between releases and, we would guess, will continue to slowly expand in time between releases in the future.

Red Hat provides support for each release for approximately ten years. As you can easily see, the support cycle and the release cycle are both *long term* in terms of time but are not directly related to one another. The support timeline is two to five times the length of the release cycle, in this case. So, it is a long-term release (aka **slow release**) as well as a long-term support, model.

RHEL itself is not free except under special circumstances. A number of RHEL clones exist, most of which are free and have no official vendor support but do receive patches making them *supported* in a similar way. Traditionally CentOS was the main, free LTS release mirroring RHEL. As CentOS was discontinued projects following in its footsteps like *RockyLinux*, *Oracle Linux* and *AlmaLinux* cover the same territory. We can completely legitimately call them LTS releases because they keep patch level and update support for a long time.

Suse's LTS product is SUSE Linux Enterprise Server or **SLES**. This product uses a nearly identical major/minor and support model to Red Hat's and has always been considered the most direct competitor. SLES releases new major versions every three to five years on a rather irregular schedule as they feel that they are ready. Like Red Hat, official support for each major release is approximately ten years. So like RHEL, SLES is a slow release with a long term support model.

Like RHEL, SLES is not free on its own... *well not exactly*. Under the SLES brand name, SLES is not free. But SUSE makes a free version of SLES called **OpenSuse Leap**. This can be a bit confusing because Leap is the free version of SLES, but it causes the linguistic and logical challenge of having to say that SLES is not free, but Leap is the free version of SLES. Basically, SLES is not free when using the SLES name, but it is free when using the Leap name. Confusing? You had better believe it.

Both SLES and OpenSUSE Leap are slow releases and LTS. Both have support for the same length of time, but SLES has much more intensive support.

Canonical takes a very different approach with Ubuntu. Unlike Red Hat and SUSE which each make multiple products, Canonical makes only one: Ubuntu. Ubuntu releases like clockwork every six months as we mentioned above. Instead of making a separate product for its LTS release, Canonical simply designates every fourth release of its rapid release product as its LTS release. So, every two years a *special* Ubuntu release happens that gets approximately eight years of support. So, Ubuntu has only one product, and only one release cycle (every six months) but two support cycles. While the most complex to explain, this results in the most flexible and straight forward product development, testing, and use.

Debian takes yet another approach. Remember that Debian never offers paid vendor support options, but only security and software support. This influences their support model. Debian performs releases roughly every two years making it a slow-release product. However officially they only fully support the current release at any given time. Unofficially they often continue to support the previous version in many cases. This creates an odd situation where the release cycle is relatively slow, but the support mechanism is very aggressive expecting everyone using Debian to almost immediately update to the latest version which undermines the reasons that most companies want to work with a slow-release product in the first place. So, Debian is slow on release, but has anything but long-term support.

Release and support schedule interplay: The overlap

Release and support schedules on their own provide a limited picture. It is in putting the two together that we really see how we must view a distribution. With Ubuntu LTS, for example, a release cycle of two years means that we have the option to update our platform at one pace, and their long-term support cycle of eight years means that we have the option of not updating if we do not want to for several cycles without losing vendor support. Most people expect this behavior when talking about LTS models, but such behavior is not implied.

Debian's two-year release schedule is nearly identical to Ubuntu LTS' but with official support ending upon release of the latest version the guaranteed support is actually less than from even the Ubuntu interim release cycle which provides a three month overlap between the time that a new release comes out and when the support for the previous release ends.

This overlap, a number that no one ever talks about, is easily the most important factor to be discussed - if we were to believe that we could single out any one factor on its own in a meaningful way. The overlap tells us our window of opportunity not only for we as system administrators to update our systems, but also how long our third-party software vendors will have to test and release their new products while maintaining support from the operating system vendor. This overlap number varies from a low of 0 to a high of over 10 years. No one number is better or worse. Too large of an overlap encourages big problems. Too small can make maintaining systems difficult.

To make matters more confusing, there is a volunteer team of developers that work with Debian to provide unofficial long-term support for every Debian release. So, while officially the Debian team provides extremely limited support, in practice and unofficially the support for previous versions is extensive and may easily challenge any other support vendor. This has created the situation where Debian is seen as stable, conservative, well supports, and rock solid while also lacking any paid primary vendor support or official LTS options!

Release model: Rolling

The key Rolling products are:

- Fedora Rawhide
- OpenSUSE Tumbleweed

A **rolling release** means that system updates are made available immediately. We typically assume that this means that they will release *as available* which could mean that they release at any time, day or night, and potentially multiple times per day. Updates are not released on a pre-determined schedule. As soon as they are ready, they release. Of course, as the system administrator it is up to you to control when they are applied. Just because they release does not mean that they apply automatically.

Rolling releases are the rarest and as you can imagine, can have a bit of difficulty coordinating with application vendors. The product is a constantly moving target and there is little, if any, ability to designate compatibility.

Rolling releases have some interesting artefacts to consider. First, they are great because you get absolutely all the latest features as quickly as possible. No waiting for a release cycle to come around. If a new feature that you want from a component product in the operating system releases the day after your rapid release prepares to go to market, you will have to wait at least six months for any major product to get that through to you. And if you are dealing with a slow-release distribution you have to wait two to five years! That can be quite a long wait if it is an important feature for you.

A famous example of waiting on features happened when *PHP 7* released and promised to completely revitalize the PHP landscape with far better performance and features. Companies stuck on slow-release products were waiting years to get the performance upgrade while rapid and rolling release distributions were running circles around them in benchmarks. In a world where features often include improved security capabilities waiting on new features can include security risks, as well.

Why not just update the packages manually

Proponents of long-term support and slow-release distributions will often make the claim that to solve the obvious short comings of the LTS ecosystem that we can just acquire the packages that we need in some other way and keep them up to date individually. A great example is the PHP case that I mentioned.

This is entirely possible, and people do this all of the time. But just because it can be done, does not mean that it makes logical sense. The point, literally the entire point, of getting a slow release, long term support operating system is that you want there to be a predictable, supported target on which to run applications. Outside of that, everything about an LTS release is a negative to the customer organization.

If we approach an LTS release with the assumption or willingness to simply start pulling apart the operating system and modifying pieces of it to get newer features, we undermine its function. The operating system vendor will no longer provide support, or have insight, into the full stack that they did before, and the application vendor will have an unknown target product. At this point we effectively take on all of the complications, overhead, and risk of a slow-release distribution while also taking on the problems and risks of a fast or rolling release distribution while additionally adding some overhead and risk of doing it ourselves outside of the purview of the vendor.

All totally doable and it will easily work. But that it can work should never cause us to perceive this as being a good idea. Instead, it should expose how fundamentally foolish the idea of clinging to LTS distributions is if we do not actually believe in their value!

A better approach would be to select a faster release distribution that properly meets our needs while also providing whatever level of support we require and getting a more complete support and update experience, less work on our end to recreate work that the vendor has already done, and not leaving the world of tested application targets to build something ourselves for no reason.

Best Practice: Use the tools as they are designed, do not try to work around the design to satisfy an emotional or political desire to use a long-term support distribution when it is not able to meet the needs.

The underlying mistake we normally see here is losing sight of the goal. Presumably, the goal in a business is always to maximize profits. For practical reasons we often have to use intermediary goals to refine achieving our primary goal. In doing so we might distill part of our goal down to maximizing vendor support, which is already a dangerous intermediary goal. What is then happening in many organizations is taking that questionable intermediary goal and attempting to make a yet more intermediary goal of using LTS releases to meet the first intermediary goal and forgetting that just because a release is designated LTS does not mean that it provides better support necessarily or meets our needs and totally misses the point that if we do not use it as designed it is not really an LTS release any longer. At least not in practical terms.

Rolling releases are less common than other types both in that fewer vendors make them and that fewer companies deploy them. The best-known rolling release is OpenSUSE Tumbleweed. Tumbleweed is given so much attention that OpenSUSE actually promotes it most of the time as their primary product. It is anything but a second-class citizen. Red Hat's rolling release is Fedora Rawhide.

Support of rolling releases creates interesting conversation because with other releases we think of support and support time frames in relation to *time since the initial release*. But with a rolling release there are no release versions and so in some ways support is generally considered to only be for one day, or it is for forever, all depending on how you look at it.

The reason that it is hard to understand or express how support works with rolling releases is not because rolling releases are complex, but rather because they are so simple that they do not hide the real problems that we are attempting to address with other release models: the desire to maintain vendor support while avoiding updating.

With any release model if we keep our system continuously up to date, we are always covered by vendor support. What changes is not whether support is available, but what we are required to do to qualify for said support. With LTS releases we accept that we have to update (or just replace) our systems every five to ten years. With rapid release we accept that we have to update our systems every six to thirteen months. And with rolling releases we just have to accept that updating is a daily process. As long as we accept these things, all approaches have *eternal support* as long as the vendor supports the products in any capacity.

Choosing the release model for our workloads

Now that we understand the release and support models, we need to think about applying them to our own decision making. Primarily our needs are going to be defined by our workload, as is always the case. But when we have options we need to be prepared to make well thought out and strategic decisions.

In general, most people lean heavily towards LTS releases, often irrationally so unable to state any benefit to the approach. Much of this emotional baggage likely stems from earlier days of IT when operating systems only had long term releases and updates were problematic and fragile often causing breaks. This was so significant that we often think of updating an operating system as something we assume cannot be done. This was (and still is) heavily prevalent in the Windows world were updates between operating system major releases truly did cause breaks of great significance with nearly every update and rarely was able to safely update a system *in situ*. The assumption became, and remains, that to upgrade the operating system under a workload that a new system will need to be deployed and the workload migrated over to it.

This was mostly true in the UNIX world in the 1990s and earlier as well, but never to the same degree due to the prevalence of interface standards and the lack of GUI-based applications. The types of workloads typical on UNIX systems were generally able to work across many disparate systems already so version differences of a single operating system were trivialized by comparison. This fear of older systems, and especially Windows systems, not handling updates gracefully has unfortunately infected many managers and Linux decisions are often made on misinformation because of it.

The problem is exacerbated by two key things as well. One is that slow-release systems such as RHEL and Ubuntu LTS go years between updates during which time many individual changes accumulate in the underlying components. So LTS updates often do present some complications because of this. Add to this the general trend of IT departments that choose LTS releases to also avoid other forms of updates or to not update LTS versions promptly (sometimes resulting in updates of more than one version at once) and updates can often truly become quite scary.

Of course, we can tackle some of this by treating LTS releases properly and updating them as quickly as is appropriate, but we can never eliminate the additional overhead that they create entirely. By comparison we should look at a rapid release product like Fedora or Ubuntu that releases something like four to six times during the time that its associated LTS release updates once. On the surface, if you fear *updating* as a singular terrifying experience then doing so several times instead of once sounds downright dreadful. This is a misapplication of the concept, however.

For the most part, interim releases like Ubuntu happening in between LTS releases are essentially incremental releases. The LTS release will get all of the same changes, but it will get many all at once. This makes updates much scarier and more complicated. Splitting up those same changes into four smaller releases generally makes it far easier to deal with any breaking compatibility issues sooner and more gracefully.

Rolling releases, logically, extend these benefits even further and reduce system changes to the smallest possible amount per update. They do create a new problem in doing so which is that we lose the ability to identify a specific target release for testing. We then run the risk that when a vendor is testing in software will never exist on our own systems and we lose a bit of predictability. In practice rolling releases tend to work quite well and are completely valid in production, but they do add some risk through a loss of predictability and focused testing of third-party products.

If what we need is a system essentially only comprised of operating system components then a rolling release easily can pull into the lead in terms of the best support, security, and features. If we need to support third party applications on top of the operating system then we will mostly favor rapid releases, instead.

Nothing but the operating system

It was not so long ago that we would laugh at the idea that we could deploy an operating system for a desktop or a server, and never deploy another application on top of it but run the system only for components that are included in the operating system itself. If you come from the Windows world this is likely extra strange to contemplate as the operating system itself is so devoid of extra components.

In the Linux world operating systems tend to be rather full of additional components. On my desktop devices nearly every tool that I use comes included with my operating system. In fact, the Chromebook ecosystem is designed completely around this concept. In Linux servers we see this happen a lot as well with many components from web servers to databases all included in the operating system.

The overall amount and general fragility of third-party applications being run today is very different than it was just ten, let alone twenty, years ago. Another factor lessening the benefits of slow-release distributions.

Understanding support and release models is not a trivial task and surprisingly few system administrators ever take the time to really understand what it means. Now that you understand what each model means you will be able to apply future offerings from potentially new vendors and use that to make decisions for your organization. You already have a jump on many in the industry who tend to gloss over or ignore these types of decisions as if they are trivial when they can result in significant implications if chosen incorrectly.

Now that we have gone into so many different types of details about Linux distributions we are going to take all of these different aspects and concerns and attempt to combine them into a holistic decision-making strategy to help us pick our distribution for deployment for any workload that we may be considering.

Choosing your distribution

Surprisingly, picking which **distribution**, or **distro** as it is commonly called in the Linux world, can be far more of a challenge than it seems like it should be. You might be *lucky* and work for a company that has a pre-determined Linux distro standard that you have to follow, and this question is already answered for you. This is becoming an increasingly rare scenario, though, as companies begin to realize the benefits of using the right distro for the right use case, and as it becomes better known that the idea that skills standardization just doesn't benefit from keeping systems identical as much as it was commonly assumed. But the practice still exists.

At the end of the day, it is essentially nonsensical to lead with operating system over workload choices. There is relatively little value in forcing a specific operating system choice and making application choices based on that. Of course, in an ideal world, all factors are considered and weighed and licensing, support knowledge, standardization and other factors heavily impacting operating system choice will play a role, they tend to be pretty small in comparison to application choice factors. And that's assuming that we are considering Windows, macOS, and some *Linux variant*; if the decision is just about one Linux option or another then the differences between distributions drops considerably.

Of course, if you are in an environment with thousands of **RHEL servers** and are tempted to install a single Ubuntu server for a specific workload, the value of figuring out how to get that workload working on RHEL or considering another workload might be reasonable given the scale of standardization that we are talking about. But it's very rare that an IT shop will ever be in a position to have hundreds or thousands of operating system images that are homogenous. Very much an edge case scenario.

Using our workload to tell us what distributions realistic options are, and then applying any homogeneity factors, we then can factor in release and support models that best suit the needs of the application and the business. Risk aversion and affinity is, of course, a major factor.

Do not fear risk

Every business, every decision has risks. In IT one of our most important jobs is evaluating risks and balancing those risks against opportunity and reward. Reducing risk factors to pure math can be difficult to do, often impossible, but even if we cannot simply create numbers to show us what risk and reward levels are in any given decision this should not change the fundamental fact that our goal is to apply math and logic to a situation to decide the best course of action.

A real threat to businesses and one that causes businesses to fail with great frequency is risk aversion. This might sound counter intuitive, but it is not. If risk aversion, or truly conservative business practices, are based in reason and logic then this should increase a business' chances of survival. Rarely is this the case, however. Most risk aversion is emotional and illogical and emotional decisions, especially when made around risk, almost always result in more risk rather than less, regardless of what the intended outcome was to be. Much as how a panic response will rarely save you in a dangerous situation, an emotional panic-like response to business decisions will almost entirely remove your ability to make good decisions.

As IT professionals, working with our businesses to understand risk assessments, factors, and affinity levels is a large part of our value to the organization. Choosing an operating system might be a tiny individual component taken on its own, but it is a common view into the workings of a business' decision-making machinery. If a business requires expensive vendor support or risk slow-release cycles out of fear rather than out of calculated risk assessment, this tells us a lot about how decision making is viewed as a process.

Best Practice: A business should hold their staff responsible not for the outcome of decisions, but rather for the quality of the decision-making process.

This best practice applies to absolutely every facet of an organization and is in no way specific to choosing a Linux distribution or workload. That is how best practices really are, very general and widely applicable.

Summary

We covered a lot in this chapter and a lot of seemingly unimportant factors that, I believe, tend to actually be pretty important in our roles in system administration. We learned how Linux fits into modern organizations and where it comes from historically which might give us some view into where it is going, as well. We dove into open-source licensing and how software licensing plays a role in how we work with our operating systems.

We took a look at who the big vendors are today and what key products they have on the market. Then we investigated release and support patterns both in general and specifically from the major vendors on the market today. And then we looked at putting all of those factors together to try to make a decision process for picking the best distribution for our workload.

In our next chapter we are going to be leaving behind the ten-thousand-foot views and we will be digging into the very technical world of *System Storage Best Practices*. This is one of my favorite areas of system **administration**!

Section 2: Best Practices for Linux Technologies

The objective of Section 2 is to introduce you to technical best practices and a high-level understanding of the workings of difficult technological concepts that are core to the system administration function. We will investigate design elements and architectures rarely covered but critical to Linux.

This part of the book comprises the following chapters:

- *Chapter 3, System Storage Best Practices*
- *Chapter 4, Designing System Deployment Architectures*
- *Chapter 5, Patch Management Strategies*
- *Chapter 6, Databases*

3
System Storage Best Practices

Probably the most complicated and least understood components of **System Administration** involve the area of **storage**. Storage tends to be poorly covered, rarely taught, and often treated as myth rather than science. Storage also involves the most fear because it is in storage that our mistakes risk losing data, and nothing tends to be a bigger failure than data loss.

Storage decisions impact performance, capacity, longevity, and most importantly, *durability*. Storage is where we have the smallest margin of error as well as where we can make the biggest impact. In other areas of planning and design we often get the benefit of quite a bit of *fudge factor*, mistakes are often graceful such as a system that is not quite as fast as it needs to be or is somewhat more costly than necessary, but in storage overbuilding might double total costs and mistakes will quite easily result in non-functional systems. Failure tends to be anything but graceful.

We are going to address how we look at and understand storage in Linux systems and demystify storage so that you can approach it methodically and empirically. By the end of this chapter, you should be prepared to decide on the best storage products and designs for your workload taking into account all of the needs.

In this chapter we will look at the following key topics:

- Exploring key factors in storage

- Understanding block storage: Local and SAN

- Surveying filesystems and network filesystems

- Getting to know **logical volume management** (**LVM**)

- Utilizing RAID and RAIN

- Learning about replicated local storage

- Analyzing storage architectures and risk

Exploring key factors in storage

When thinking about storage for systems administration we are concerned with **cost**, **durability**, **availability**, **performance**, **scalability**, **accessibility**, and **capacity**. It is easy to get overwhelmed with so many moving parts when it comes to storage and that makes it a risk that we may lose track of what it is that we want to accomplish. In every storage decision, we need to remain focused on these factors. Most importantly, on all of these factors. It is extremely tempting to focus on just a few causing us to lose our grasp of the complete picture.

In most cases, if you study postmortems of storage systems that have failed to meet business needs, you will almost always find that one or more of these factors was forgotten during the design phase. It is very tempting to become focused on one or two key factors and ignore the others, but we really have to maintain a focus on all of them to ensure storage success.

We should begin by breaking down each factor individually.

Cost

It might seem that no one could forget about cost as a factor in storage but believe me it happens, and it happens often. As a rule, IT is a business function and all businesses, by definition, are about making money, and the cost of providing an infrastructure need always must factor in profits. So, because of this, no decision in IT (or anywhere in a business) should happen without cost as a consideration. We should never allow cost to be forgotten, or just as bad allowing someone to state that *cost is no object* because that can never be true and makes absolutely no sense. Cost may not be the primary concern, and the budgetary limits may be flexibility, but cost always matters.

Storage is generally one of the costliest components of a production system so we tend to see costs be more sensitive when dealing with storage than when dealing with other parts of physical system design such as CPU and RAM. Storage is also often easiest to solve by simply throwing more money at it and so many people when planning for hardware err on the side of overbuilding because it is easy. Of course, we can always do this and as long as we understand enough of our storage needs and how storage works it will *work* outside of being overly expensive. But, of course, it is difficult to be effective system administrators if we are not cost effective - the two things go together.

Durability

Nothing is more important when it comes to storage than durability: the ability of the storage mechanism to resist data loss. Durability is one of two aspects of reliability. For most workloads and most system scenarios, durability is what trumps all else. It is very rare that we want to store something that we cannot reliably retrieve even if that retrieval is slow, delayed, or expensive. Concepts such as data availability or performance mean nothing if the data is lost.

Durability also refers to data that resists corruption or decay. In storage we have to worry about the potential of a portion of our data set losing integrity which may or may not be something that we can detect. Just because we can retrieve data alone does not tell us that the data that we are retrieving is exactly what it is supposed to be. Data corruption can mean a file that we can no longer read, a database that we can no longer access, an operating system that no longer boots, or worse, it can even mean a number in an accounting application changing to a different, but valid, number which is all but impossible to detect.

Availability

Traditionally, we thought about data reliability mostly in terms of how available our data was when it came time to retrieve it. Availability is often referred to as *uptime* and if your storage is not available, neither is your workload. So, while availability generally takes a back seat to durability, it is still extremely important and one of the two key aspects of overall storage reliability.

There are times when availability and performance become intertwined. There can be situations where storage performance drops so significantly that data becomes effectively unavailable. Consider a shower that just drips every few seconds, technically there is still water, but it is not coming through the pipes fast enough to be able to use it.

We will be talking about RAID in depth in just a little bit, but availability and performance are good real-world examples. A famous situation can arise with large RAID 6 arrays when a drive or two have failed and have been replaced and the array is online and in the process of actively rebuilding (a process by which missing data is recalculated from metadata.) It is quite common for the RAID system to be overwhelmed due to the amount of data being processed and written that the resulting array, while technically online and available, is so slow that it cannot be used in any meaningful way and operating systems or applications attempting to use it will not just be useless from the extreme slowness but may even error out reporting that the storage is offline due to the overly long response times. *Available* can become a murky concept if we are not careful.

Performance

When it comes to computers in the twenty first century storage is almost always the most significant performance bottleneck in our systems. CPU and RAM almost always have to wait on storage rather than the other way around. Modern storage using solid state technologies has done much to close the performance gap between storage systems and other components, but the gap remains rather large.

Performance can be difficult to measure as there are many ways of looking at it, and different types of storage media tend to have very different performance characteristics. There are concepts such as latency (time before data retrieval begins), throughput (also known as *bandwidth*, measuring the rate at which data can be streamed), and input/output operations per second or IOPS (the number of storage related activities that can be performed in each amount of time.) Most people think of storage only in terms of throughput, but traditionally IOPS have been the most useful measurement of performance for most workloads.

It is always tempting to reduce factors to something simple to understand and compare. But if we think about cars, we could compare three vehicles: one with a fast acceleration but a low top speed, one with slow acceleration and a high-top speed, and a tractor trailer that is slow to accelerate and has a low top speed but can haul a lot of stuff at once. The first car would shine if we only cared about latency: the time for the first packet to arrive. The second car would shine if we cared about how quickly a small workload could be taken from place to place. This is most like measuring IOPS. The tractor trailer will be unbeatable if our concern is how much total data can be hauled between systems over the duration of the system. That's our throughput or bandwidth. With cars, most people think of a *fast car* as the one with the best top speed, but with storage most people think about the tractor trailer example as what they want to measure, but not what *feels* fast when they use it. In reality, performance is a matter of perspective. Different workloads perceive performance differently.

For example, a backup deals in steady, linear data and will benefit most from storage systems designed around throughput. Therefore, tape works so well for backup performance and why old optical media such as CD and DVD were acceptable. But other workloads, like databases, depend heavily on IOPS and low latency and benefit little from total throughput and so really benefit from solid state storage. Other workloads like file servers often need a blend of performance and work just fine with spinning hard drives. You have to know your workload in order to design a proper storage system to support it.

Performance is even more complex when we start thinking in terms of burstable versus sustainable rates. There is just a lot to consider, and you cannot short circuit this process.

Scalability

A typical physical system deployment is expected to see four to eight years in production today and it is not uncommon to hear of systems staying in use far longer. Spend any amount of time working in IT and you are likely to encounter systems still powered on and completely critical to a company's success that have been in continuous use for twenty years or even more! Because a storage system is expected to have such a long lifespan, we have to consider how that system might be able to grow or change over that potential time period.

Most workloads experience capacity growth needs over time and a storage design that can expand capacity as needed can be beneficial both for just protecting against the unknown but also by allowing us to invest minimally up front and spending more only *if* and *when* additional capacity becomes needed. Some storage systems may also be able to scale in terms of performance as well. This is less common and less commonly considered critical, yet even if a workload only increases capacity needs and not performance needs *per se*, larger capacity alone can warrant a need for increases performance just to handle tasks such as backups since large capacities mean larger time to backup and restore.

In theory, you could also have a situation where the needs for reliability (durability, availability, or both) may need to increase over time. This, too, can be possible, but is likely to be much more complex.

Storage is an area in which flexibility to adjust configuration over time is often the hardest, but also the most important. We cannot always foresee what future needs will be. We need to plan our best to allow for flexibility to adjust whenever possible.

Capacity

Finally, we look at capacity, the amount of data that can be storage on a system. Capacity might seem straightforward, but it can be confusing. Even in simple disk-based arrays we have to think in terms of raw capacity (the sum of the capacities of all devices) and in terms of the resultant capacity (the usable capacity of the system that can be accessed for storage purposes. Many storage systems have redundancies to provide for reliability and performance and this comes at the cost of consumed raw capacity. So, we have to be aware of how our configuration of our storage will affect the final outcome. Storage admins will talk in terms of both raw and usable capacity.

Now that we have a good handle on the aspects of storage that we need to keep in mind we can dive into learning more about how storage components are put together to build **enterprise storage subsystems**.

Understanding block storage: Local and SAN

At the root of any standard storage mechanism that we will encounter today is the concept of **block devices**. Block devices are storage devices that allow for non-volatile data storage that can be stored and retrieved in arbitrary order. In a practical sense, think of the *standard* block device as being the hard drive. Hard drives are the prototypical block device, and we can think of any other block device as behaving like a hard drive. We can also refer to this as implementing a drive interface or *appearance*.

Many things are block devices. Traditional spinning hard drives, solid state drives (SSD), floppy disks, CD-ROM, DVD-ROM, tape drives, RAM disks, RAID arrays and more are all block devices. As far as a computer is concerned, all of these devices are the same. This makes things simple as a system administrator: everything is built on block devices.

From a system administrator perspective, we often simple refer to block devices as *disks* because from the perspective of the operating system we cannot tell much about the devices and only know that we are getting block storage. That block storage might be a physical disk, a logical device built on top of multiple disks, an abstraction built on top of memory, a tape drive, or remote system, you name it. We cannot really tell. To us it is just a block device and since block devices generally represent disks, we call them disks. It is not necessarily accurate, but it is useful.

Locally attached block storage

The simplest type of block storage devices is those that are physically attached to our system. We are familiar with this in the form of standard internal hard drives, for example. Local block devices commonly attach by way of SAS, SATA, and NVMe connections today. In the recent past, **Parallel SCSI** (just called **SCSI** at the time), and **Parallel ATA** (aka **PATA**) just called **ATA** or **IDE** at the time, were standards. All of these technologies, as well as some more obscure, allow physical block devices to attach directly to a computer system.

It is locally attached storage that we will work with most of the time. And all block devices have to be locally attached somewhere in order to be used. So this technology is always relevant.

Locally attached block devices come with a lot of inherent advantages over alternatives. Being locally attached there is a natural performance and reliability advantage: the system is as simple as it gets and that means that there is less to go wrong. All other things being equal, simple trumps complex. Storage is a great example of this. Fewer moving parts and shorter connection paths means we get the lowest possible latency, highest possible throughput, and highest reliability at the lowest cost!

Of course, locally attached storage comes with caveats or else no one would even make another option. The negative of locally attached storage is flexibility. There are simply some scenarios that locally attached storage cannot accommodate and so we must sometimes opt for alternative approaches.

Storage Area Networks (SAN)

The logical alternative to a locally attached device is a remotely attached device and while one would think that we would simply refer to these types of block devices in this manner, we do not. A remote attached device uses a network protocol to implement the concept of *remoteness* into the storage and the network on which a remote device is communicating is called a **Storage Area Network** and because of this, common vernacular simply refers to all remote block storage as being a **SAN**.

The terrible terminology of SAN

Technically speaking, a Storage Area Network should be a reference to a dedicated network that is used to carry block device traffic and in very technical circles this is how the term is used. Devices on a SAN can be direct block devices, disk arrays, and other similar *block over network* devices. The SAN is the network, not a *thing* that you can buy.

In the common parlance, however, it is standard to refer to any device that provides storage, implements a block device interface, and connects to a network rather than directly to a computer as a SAN. You hear this every day in phrases like *did you buy a SAN?*, *we need a SAN engineer*, *I spoke to our SAN vendor*, *should we upgrade the SAN?*, and *where is our SAN?* Go to your nearest IT hardware vendor and ask them to sell you a SAN and they will without hesitation, the terminology is so standard that dollars to donuts says that they will be completely confused if you try to act like a SAN is anything but a hardware device into which you place hard drives and is connected to a network via some sort of cable.

Because storage is complex, confusing, and scary and because storage area networks add additional layers of complexity on top of the basics this entire arena became treated as black boxes full of magic and terminology quickly deteriorated and most beliefs around SAN became based on misconceptions and myth. Common myths include impossible ideas such that SANs cannot fail, that SANs are faster than the same technology without the networking layer, that SANs are a requirement of other technologies, and others.

We can only be effective system administrators if we understand how the technology works and avoid giving in to myths (and marketing.) For example, we cannot make meaningful risk analysis or performance decisions if we believe that a device is magic and do not consider its actual risk profile.

In theory, a SAN is an extremely simple concept. We take any block device whether it is a physical device like an actual hard drive, or some more complicated concept like an array of drives, and encapsulate the standard block device protocol (such as SCSI or ATA) and send that over a network protocol (such as TCP/IP, Ethernet, or *FiberChannel*.) The network protocol acts as a simple tunnel, in a way, to get the block protocol over a long distance. That's all that there is to it. At the end of the day, it is still just a SCSI or ATA based device, but now able to be used over a long distance.

Of course, what we just added is a bit of complexity, so SANs are automatically more fragile than local storage. Any and all risk and complication of local storage remains plus any complications and risk of the networking layer and equipment. The risk is cumulative. Plus, the extra networking layer, processing, and distance all must add additional latency to the storage transaction.

Because of these factors, SAN based storage is always slower and more fragile than otherwise identical local storage. The very factors that most myths have used to promote SAN are exactly their weaknesses.

The SAN approach does have its strengths, of course, or else it would serve no purpose. A SAN allows for three critical features: distance, consolidation, and shared connections.

Distance can mean anything from a few extra feet to across the world. Of course, with longer distances come higher latencies, and typically storage is very sensitive to latency, so it is pretty rare that remote block storage is useful from outside of the range of local connection technologies. If you have to pull block storage data over the WAN, you will likely experience at very least latencies that cause severe performance issues and will typically see untenable bandwidth constraints. Typical production block storage is assumed to be many GB/s (that big B, not little b) of throughput and sub-millisecond latency, but WAN connections rarely hit even a single Gb/s and even the best latencies are normally a few milliseconds if not scores or more!

Consolidation was traditionally the driving value of a SAN. Because many systems can physically connect to a single storage array over a single network it became easy, for the first time, to invest in a single, expensive storage system that could be used by many physically separate computer systems at once. The storage on the device would be *sliced* and every device that attaches to it sees its own unique portion of the storage.

When local storage isn't local

With all of the interfaces, abstractions, and incorrect terminology that often exists in IT, it can be really easy to lose track of exactly what is happening much of the time. SANs are one of these places were getting confused is par for the course. It is the nature of a SAN to take a block device that is far away and make it seem, to the computer using it, as if it were local. But it can also take something that is local, and make it seem local, when it is really remote. Did I just say that?

The best example is that of the external USB hard drive. We all use them; they are super common. Go to any local department store and pick one up. Order one online. You probably have five on a shelf that you have forgotten about. A USB drive, while external, is obviously still local, right?

Well, it isn't all that easy to say. Sure, it is physically close. But in technology terms remote means that something is *over a network* and local is *not over a network*. It does not matter how far away something is, it is the network aspect that determines local and remote devices. Otherwise, my desktop in Texas is physically attached to my dad's desktop in New York because there is a series of cables the entire way in between them.

This presents an interesting challenge because, you see, USB is actually a very simple networking protocol, as are IEEE 1394 and Thunderbolt. If you physically dissect an external drive you can see this at work, to some degree. They are made from standard hard drives, generally with SATA interfaces, and a tiny network adapter that encapsulates the SATA protocol into the USB network protocol to be sent over the network (often just two feet total distance.)

USB and its ilk might not feel like a network protocol, but it really is. It is a layer two network protocol that competes with Ethernet and can attach multiple devices, to multiple computers, and can even use things similar to switches. It is a real networking platform and that means that external hard drives attached via USB are, in fact, tiny SANs! Hard to believe, but it is true. Consider your mind blown.

Storage, being the largest cost of most systems, being able to be shared and sliced more efficiently lowered cost to deploy new physical computer systems. Hard drives, for example, might come in 1TB sizes, but a single system might need only 80GB or 300GB or whatever and with a shared SAN hundreds of computers systems might share a single storage array and each use only what they need. Today we gain most of this efficiency through local storage with virtualization, but before virtualization was broadly available only systems like a SAN were able to address this cost savings. So, in the early days of SAN, the focus was cost savings. Other features really came later. This value has mostly inverted today and so is generally more expensive than having overprovisioned local storage but can still exist in some cases.

The last value is shared connections. This is where two or more computers access the same portion of the storage on the same device - seeing the same data. This might sound a bit like traditional file sharing, but it is anything but that.

In file sharing we are used to computers having a *smart* gatekeeping device that arbitrates access to files. With a SAN, we must remember that this is a *dumb* block device that has no logic of its own. Using a SAN to attach two or more computer systems to a single logical block device means that each computer thinks of the storage as being its own, private, fully isolated system and has no knowledge of other systems that might be also attached to it. This can lead to all kinds of problems from lost changes to corrupt files, to destroyed file systems. Of course, there are mechanisms that can be used to make shared storage spaces possible, but by definition they are not implemented by the SAN and have to be provided at a higher level on the computer systems themselves.

Shared SCSI connections

In the days before SAN, or before SANs were popular and widely available, there was another technique allowing two computers to share a single pool of hard drives: shared SCSI.

With this technique, a single SCSI ribbon cable (typically able to connect to eight, sixteen, or even thirty-two devices. One device would need to be a controller, presumably on the motherboard of a computer. The other connections were open for connecting hard drives. But another connector could be connected to another controller on a separate computer and the two computers could each see and access the same drives.

Due to the limitations of needing to share a single ribbon cable between two physical computers made this technique outrageously limited and awkward, but feasible. The primary value to a setup of this nature was allowing one computer system to fail and the other to take over, or to double the CPU and RAM resources assigned to a single data set beyond what could fit in a single server chassis. But the reliability and performance limits of the storage component left the system generally less than practical and so this technique was rarely implemented in the real world. But historically it is very important because it is the foundation of modern shared block storage, it was standard knowledge expected in late 1990s systems training, and it helps to visualize how SAN works today - more elegant, more flexible, but fundamentally the same thing.

Today, the biggest use cases for shared block storage connections is for clustered systems that are designed to use this kind of storage as shared backing for virtualization. This was the height of fashion around 2010 but has since given way to other approaches to tackle this kind of need. This would now be a rather special case system design. But the technologies that are used here will be co-opted for other storage models as we will soon see.

The world of SAN has many popular connection technologies. There are super simple SAN transports that are so simple that no one recognizes them as being such including USB, Thunderbolt, and IEEE1394/Firewire. Then there are a range of common enterprise class SAN protocols such as iSCSI (SCSI over IP), FibreChannel, FCoE (Fibre Channel over Ethernet), *FC-NVMe* (NVMe over Fiber Channel), and so on. Each SAN protocol presents its own advantages and challenges, and typically vendors only offer a small selection from their own equipment so choosing a vendor will typically limit your SAN options and picking a SAN option will limit your vendor selection choices. Understanding all of these protocols moves us from the systems world into the networking one. It is rare that as a system administrator you will be in a position to choose or even influence the choices in SAN design, typically this will be chosen for you by the storage, networking, and/or platform teams. If you do get to have influence in this area then significant study of these technologies, their benefits, and their applicability to your workload(s) will be necessary but is far outside of the scope of this tome.

Block storage is not going anywhere. As much as we get excited about new storage technologies, such as object storage, block storage remains the underpinning of all other storage types. We have to understand block devices both physically and logically as we will use them in myriad ways as the building blocks of our storage platforms. Block storage is powerful and ubiquitous. It represents the majority of storage that we interact with during our engineering phases and is expected to remain at the core of everything that we do for decades to come.

When deciding between local and remote block storage there is a useful rule of thumb: *You always want to use local storage until you have a need that local storage cannot fulfill. Or you never want to use remote storage until you have no other choice.*

Surveying filesystems and network filesystems

Sitting on top of block storage we typically find a **filesystem**. Filesystems are the primary (and by primary, I mean like they make up something like 99.999% or more of use cases) manner of final data storage on computer systems. Filesystems are what hold files, as we know them, on our computer storage.

A filesystem is a data organization format that sits on top of block storage and provides a mechanism for organizing, identifying, storing, and retrieving data using the file analogy. You use filesystems every day on everything. They are used even when you cannot see them whether it is on your desktop, cell phone, or even on your VoIP phone, or microwave oven! Filesystems are everywhere.

Filesystems are really databases

If you want to get a little geeky with me for a moment and be honest you are reading a book on system administration best practices so we both know you are loving getting into some serious details, we can look at what a filesystem really is. At its core a filesystem is a NoSQL database, specifically a file database (essentially a specialized document database), that uses a raw block device as its storage mechanism and is only able to store and retrieve files.

There are other specialty databases that use block devices directly (often called raw storage when dealing with database lingo), but they are rare. Filesystems are a database type that is so common, so much more common than all other database types combined, that no one ever talks about or thinks about them being databases at all. But under the hood, they are truly a database in every sense.

To show direct comparisons, a standard database regardless of type has a standardized storage format, a retrieval format, a database engine (driver), and in some cases a database management layer (that can often allow for the use of multiple database engineers within a single system interface), and a query interface for accessing the data. Whether you compare MongoDB or MS SQL Server you will find that filesystems behave identically. The chosen filesystem on disk format is the storage format, the retrieval format is the *file*, the database engine is the filesystem drive, the database management system in Linux is the Virtual File System (which we will discuss later), and the query language is a list of underlying POSIX commands implemented in C (with simple shell-based abstractions that we can use for convenience.) Compare to standard databases and there is no way to tell them apart! Very cool stuff.

After a computer system is deployed, nearly everything that we do with it from a storage perspective involves working on the filesystem. We tend to focus heavily on block storage during engineering phases, and filesystems during administration phases. But certainly, we have to plan our filesystems properly prior to deploying a system to production. Proper filesystem planning is actually something that is heavily overlooked with most people simply accepting defaults and rarely thinking filesystem design at all.

Most operating systems have native support for a few different filesystems. In most cases an operating system has one obviously standard and primary filesystem and a handful of special case filesystems that are relegated to use on niche hardware devices or for compatibility with other systems. For example, Apple macOS uses APFS (Apple File System) for all normal functions but can use ISO 9660 when working with optical disks or FAT32 and **exFAT** used for compatibility with Windows storage devices (such as USB memory sticks or external hard drives.) Windows is similar but with NTFS instead of APFS. Windows recently has added one alternative filesystem, ReFS, for special needs, but it is not commonly used or understood.

In Linux, however, we have several primary filesystem options and scores of specialty filesystem options. We have no way to go through them all here, but we will talk about several of the most important as understanding why we have them, and when to choose them is very important. Thankfully in production systems we really only have to concern ourselves with a few key products. If you find filesystems interesting, you can research the many Linux filesystem options to learn more about filesystem design and history and you might even find one that you want to use somewhere special!

These are the key Linux filesystems today with which we need to be concerned for everyday purposes: XFS, EXT4, ZFS, BtrFS. Nearly everything that we do will involve one of those four. There are loads of less popular filesystems that are well integrated and work perfectly well like JFS and ReiserFS but are almost never seen in production. There are older formats like EXT2 and EXT3 that have been superseded by more recent updates. There are loads and loads of filesystems that are standard on other systems that can be used on Linux like NTFS from Windows or UFS from the BSD family. There are the standard niche filesystems like ISO 9660 and FAT32 that we mentioned earlier. Linux gives you options at every turn and filesystem selection is a great example of just how extreme it can get.

EXT: The Linux filesystem family

Nearly every operating system has its own, special-sauce filesystem that it uses natively or by default and is tightly associated with it and Linux is no exception.... just kidding, Linux is absolutely the exception which is amazing considering how much more robust Linux is in its filesystem options than any other operating system. Illumos has ZFS, FreeBSD has UFS, Windows has NTFS, macOS has APFS, AIX has JFS, IRIX had XFS and on, and on. Linux truly has no filesystem of its own, yet it has nearly everyone elses.

Most people talk about the EXT filesystem family as being the Linux native filesystem and certainly nothing else comes close to matching that description. When Linux was first being developed, long before anyone had actually run it, the MINIX filesystem was ported to it and became the default filesystem as the new operating system began to take off. But as the name suggests, the MINIX Filesystem was native to MINIX and predated Linux altogether.

Just one year after Linux was first announced, the EXT filesystem (or MINIX Extended File System) was created taking the MINIX Filesystem and, you guessed it, extending it with new features mostly around timestamping.

As Linux began to grow, EXT grew with it and just one year after EXT was first released its successor EXT2 was released as a dramatic upgrade taking the Linux filesystem ecosystem from a hobby system to a serious enterprise system. EXT2 ruled the Linux ecosystem almost exclusively from its introduction in 1993 until 2001 when Linux went through a bit of a filesystem revolution. EXT2 was such a major leap forward that it was backported to MINIX itself and had drivers appear on other operating systems like Windows and macOS. Possibly no filesystem is more *iconically* identified with Linux than EXT2.

By 2001 many operating systems were looking to more advanced filesystem technologies to give them a competitive advantage against the market and Linux did so both by introducing more filesystem options and by adding journaling functionality to EXT2 to increment its version to EXT3. This gave the EXT family some much needed stability.

Seven more years and we received one additional major upgrade to Linux' quasi-native filesystem with EXT4. Surprisingly, the primary developer on EXT3 and EXT4 stated that while EXT4 was a large step forward that it was essentially a stopgap measure of adding improvements to what is very much a 1980s technology. Filesystem design principals leaped forward especially in the early 2000s and the EXT family is likely at the end of the road, but still has a lot of useful life left in it.

I am going to delve into a little bit of detail for each of the main filesystem options, but to be clear this is a cursory look. Filesystem details can change quickly and can vary between versions or implementations so for really specific details such as maximum file size, file count, filesystem size and so forth please look to Wikipedia or filesystem documentation. You will not need to memorize these details and rarely will you even need to know them. In the 1990s filesystem limitations were so dramatic that you had to be acutely aware of them and work around them at every turn. Today any filesystem we are going to use is able to handle almost anything that we throw at it, so we really want to understand where different products shine or falter and when to consider which ones at a high level.

EXT4

Linux, as a category, has no default filesystem in the way that other operating systems do, but if you were to attempt to make the claim that any filesystem deserves this title today that honour would have to go to **EXT4**. More deployed Linux-based operating systems today choose EXT4 as their default filesystem than any other. But this is beginning to change so it seems unlikely that EXT4 will remain dominant for more than a couple years yet.

EXT4 is reasonably fast and robust, quite flexible, well known, and meets the needs of nearly any deployment. It is the jack of all trades of Linux filesystems. For a typical deployment, it is going to work quite well.

EXT4 is what we call a *pure filesystem*, which means it is just a filesystem and does not do anything else. This makes it easier to understand and use, but also makes it more limited.

XFS

Like the EXT family, **XFS** dates back to the early 1990s, and comes from a Linux competitor, in this case SGI's IRIX UNIX system. It is venerable and robust and was ported to Linux in 2001, the same year that EXT3 released. For twenty years now EXT3/4 and XFS have competed for the hearts and souls of Linux Administrators (and Linux Distro creators to choose them as the default filesystem.)

XFS is also a *pure filesystem* and is very commonly used. XFS is famous for its extremely high performance and reliability. XFS is sometimes specifically recommended by high performance applications like databases to keep them running at their peak.

XFS is probably the most deployed filesystem when the system administrator is deliberately choosing the filesystem rather than simply taking the default, is probably the most recommended by application vendors, and is my own personal choice for most workloads where storage needs are non-trivial.

Over the years, EXT4 and XFS have gone back and forth in popularity. My own observations say that XFS has slowly been edging ahead over the years.

The one commonly cited caveat of XFS compared to EXT4 is that EXT4 is able to either shrink *or* grow a volume once it has been deployed. XFS can grow, but cannot shrink, a volume that has been deployed. However, it is almost unheard of for a properly deployed production system to shrink a filesystem, so this is generally seen as trivia and not relevant to filesystem decision making (especially with the advent of thin provisioned block storage.)

ZFS

Releasing to Solaris in 2006, **ZFS** is generally considered to be the foundation of truly modern filesystem design. By the time that work on ZFS had begun in 2001 the industry was already beginning to take filesystem design very seriously and many new concepts were being introduced regularly, but ZFS really took these design paradigms to a new level and ZFS remains a significant leader in many areas still today.

ZFS really has three high level areas in which it attempted to totally disrupt the filesystem industry. First in size: ZFS was able to address multiple orders of magnitude more storage capacity than any filesystem before it. Second in reliability: ZFS introduced more robust data protection mechanisms than other filesystems had allowing it to protect against data loss in significant ways. And third, in integration: ZFS was the first real *non-pure* filesystem where ZFS represented a filesystem, a RAID system, and a logical volume manager all built into a single filesystem driver. We will go into depth about RAID and LVMs later in this chapter. This integration was significant as it allows the storage layers to communicate and coordinate like never before. Pure filesystems like EXT4 and XFS can and do use these technologies but do so through external components rather than integrated ones.

While ZFS is not new, having been in production systems for at least fifteen years, it is quite new for release on Linux. It took many years before a port of ZFS was made available for Linux, and then there were many years during which licensing concerns kept it from being released in a consumable format for Linux. Today the only major Linux distribution that officially supports and packages ZFS is Ubuntu, but Ubuntu's dominant market position makes ZFS automatically widely available. At this time, it is less than two years since ZFS was able to be used for the bootable root filesystem on Ubuntu. So ZFS is quite new in the production Linux space in any widely accessible way. Its use appears to be growing rapidly now that it is available.

ZFS represents probably the most advanced, reliable, and scalable filesystem available on Linux as of the time of this writing. It should be noted that from a purely filesystem-based performance perspective that ZFS is not known to shine. It is rare that storage performance tweaking at the filesystem level is considered valuable but when it is modern filesystems with all of their extra reliability typically cannot compete with older, more basic filesystems. This has to be noted as it is so often simply assumed that more modern systems are going to also be automatically faster, as well. This is not the case here.

BtrFS

Pronounced *Butter-F-S*, **BtrFS** is the current significant attempt to make a Linux native filesystem (there was a previous attempt called ReiserFS in the early 2000s that got some traction but ended badly for non-technical reasons.) BtrFS is intended to mimic the work of ZFS, but native to Linux and with a compatible license.

BtrFS trails ZFS significantly with many features still not implemented, but with work ongoing. BtrFS is very much alive and increasingly more Linux distributions are supporting it and even choosing it as a default filesystem. BtrFS feels as if it might be the most likely long-term future for Linux.

Like ZFS, BtrFS is a modern, heavily integrated filesystem that is beginning to include functionality from RAID and LVM layers of the storage stack. Performance is the weakest point for BtrFS today.

> **Stratis**
>
> Given its industry support, we need to make mention of Stratis. Stratis is not a filesystem itself *per se* but act much like one. Stratis is an attempt to build the functionality of integrated (or *volume-managing file systems*) like ZFS and BtrFS using the existing components of XFS and the standard Linux LVM layer.
>
> In its early days on IRIX, XFS was designed to be used with IRIX's native LVM and the two integrated naturally providing something not entirely unlike ZFS or BtrFS today. When XFS was ported to Linux its associated LVM layer was not ported, but the native Linux LVM was made to work with it instead. XFS + LVM has long been an industry standard approach and Stratis mearly is attempting to provide a more accessible means of doing so while integrating best practices and simplified management.

This sums up the four current production filesystem options that you will likely encounter or be responsible for choosing between. Remember that you can mix and match filesystems on a single system. It is very common, in fact, to use EXT4 as a boot filesystem for basic operating system functionality while then relying on XFS for a high-performance database storage filesystem or BtrFS for a large file server filesystem. Use what makes sense for the workload at an individual filesystem layer. Do not feel that you have to be stuck using only one filesystem across all systems, let alone within a single system!

Most of the real intense technical aspects of filesystems are in the algorithms that deal with searching for and storing the bits onto the block devices. The details of these algorithms are way beyond the scope of not only this book but systems administration in general. If you are interested in filesystems, learning how data is stored, protected, and retrieved from the disk can be truly fascinating. For systems administration tasks it is enough to understand filesystems at a high level.

Sadly, there is no way to provide a real best practice around filesystem selection. It is unlikely that you are going to have reason to seriously consider the use of any rare filesystem not mentioned here in a production setting, but all four that are listed here have valuable use cases and all should be considered. Often the choice of filesystem is not made in isolation unless you are working with a very specific product that requires or recommends a specific one for the purpose of some feature. Instead, the choice of filesystem is normally going to be dependent on many other storage decisions including RAID, LVM, physical support needs, drive media, and so forth.

Clustered file systems

All of the filesystems that we have discussed thus far, regardless of how modern they are, are *standard* filesystems or *non-shared* filesystems. They are only viable when access is guaranteed to be from only a single operating system. In almost all cases, this is just fine.

If you recall from our discussion on SANs, however, we mentioned that there are use cases where we may want multiple computer systems to be able to read and write for the same storage area at the same time. Clustered or *shared storage* filesystems are the mechanism that can allow that to work.

Clustered filesystems work just like traditional filesystems do, but with the added features by which they will write locking and sharing information to the filesystem so that multiple computer systems are able to coordinate their use of the filesystem between attached nodes. In a standard filesystem there is only one computer accessing the filesystem at a time so knowing what file is open, when a file has been updated, when a write is cached and so forth are all handled in memory. If two or more computers try to share data from a traditional filesystem, they cannot share this data in memory and so will inevitably create data corruption as they overwrite each other's changes, fail to detect updated files, and do all sorts of nasty things from outdated write caches!

Since the only shared component of these systems is the filesystem, all communications between nodes accessing a file system have to happen in the filesystem itself. There is literally no other possible way without going to a mechanism that is no longer shared storage but shared compute which is much more complicated and expensive.

To describe how clustered filesystems work in the simplest terms we can think of each computer knowing, from the filesystem, that a specific section of the block device (disks) is set aside in an extremely rigid format and size to be an area where the nodes read and write their current status of interaction with the filesystem. If node A needs to open File X, it will put in a note that it is holding that file open. If node B deletes a file, it will put in a note that it is going to delete and update it once the file is deleted. node C can tell what activity is going on just by reading this one small piece of the filesystem. All of the nodes connected know not to cache the data in this area, to state any action that they plan to take, and to log anything that they have done. If any node misbehaves, the whole system corrupts, and data is lost.

Of course, as you can tell, this creates a lot of performance overhead at a minimum. And this system necessarily requires absolute trust between all connected nodes as the access and data integrity controls are left up to the individual nodes. There is not and there cannot be any mechanism to force the nodes to behave correctly. The nodes have to do so voluntarily. This means that any bug in the code, any failure of memory, any admin with root access, any malware that gains access to a single node, and others. can bypass *any* and *all* controls and read, modify, destroy, encrypt, and others, to any degree that it wishes and all of the kinds of security and controls that we normally assume protect us do not exist. Shared storage is extremely simple, but we are so used to storage abstractions that it becomes complex to try to think about how any storage system could be so simple.

Like with regular filesystems, Linux has multiple clustered filesystems that we will commonly see in use. The most common one is GFS2 followed by OCFS2.

As with SANs in general, the same rule will apply to clustered file systems: you do not want to use them, until you have to.

Network filesystems

Network filesystems are always a little bit hard to describe but benefit well from the *you will know it when you see it* phenomenon. Unlike regular filesystems that sit on top of a block device and provide a way to access storage in the form of files, network filesystems take a filesystem and extend it over a network. This might sound a lot like a SAN, but it is very different. A SAN shares a set of block devices over a network. Network filesystems share filesystems over a network.

Network filesystems are quite common, and you probably see them every day, but we often do not think about them being what they really are. We often refer to network filesystems as *shares* or *mapped drives* and the standard protocols used are NFS and SMB (sometimes called CIFS, which is not really accurate.) Servers that implement network filesystems are called file servers and if you make a file server into an appliance, it is called a NAS (for Network Attached Storage.) Network filesystems are also often thought of as *NAS protocols* for this reason, just as block over network protocols are thought of as *SAN protocols*.

Unlike shared block protocols, network filesystems are *smart* with the machine that shares out the storage having a local filesystem that it understands and intelligence about the files involved so concepts like file locking, caching, file updates and so forth can be handled through a single gatekeeper that can enforce security and integrity and there is no need to trust the accessing nodes. The key difference is that a SAN is just storage blindly attached to a network, it can be as simple as a network adapter bolted onto a hard drive (and it often is, actually.) A device implementing a network filesystem, on the other hand, is a server and required a CPU, RAM, and an operating system to function. Shared block storage is almost exclusively used in very limited deployments with carefully controlled servers. Network filesystems can be used almost anywhere that a SAN can be used but are also commonly used to share storage directly to end user devices as their robust security, ease of use and lack of needed end point trust make them highly useful where SANs would be impossible to deploy.

Network filesystems run as an additional network-enabled layer on top of traditional filesystems and do not replace the *on disk* filesystems that we already have. In speaking of interfaces, we would describe network filesystems as *consuming a filesystem interface* and also *presenting a filesystem interface*. Basically, it is filesystem in, filesystem out.

Like with traditional filesystems, Linux actually offers a large range of network filesystem options, many of which are historical in nature or extremely niche. A common example is the **Apple Filing Protcol** or **AFP** (aka *AppleTalk*) which Linux offers, but is not used on any production operating system today. Today only NFS and SMB really see any real work usage in any way.

NFS

The original *network file system* in wide use and literally the source of the name, **NFS** dates back to 1984! NFS cannot be native to Linux as it predates Linux by seven years, but NFS has been the default network file system across all UNIX-based or inspired operating systems since its inception and represents a rather significant standard because of this. Because Linux is so prominent today, most people think of NFS as being *Linux' protocol*.

NFS is available on essentially any system. Any UNIX system, even macOS, offers NFS as does Windows Server! NFS is an open standard and all but universal. NFS maintains popularity by being simple to use, robust on the network and generally performs well. NFS remains heavily used on servers wherever direct filesharing between systems is required, especially on backup systems.

SMB

The **Server Message Block (SMB)** protocol predates NFS and was originally available in 1983. These are very old protocols indeed. SMB did not really find much widespread usage until Microsoft really began to promote it around 1990 and with the rise of the Windows NT platform throughout the 1990s SMB began to become quite popular along with it.

SMB really benefited from Microsoft's heavy use of mapped drives between their servers and workstations which made the SMB protocol very visible to many users both traditional and technical.

In Linux, support for the SMB protocol is provided by the Samba package (Samba is a joke on the SMB letters.) Linux has good support for SMB, but using it is more complex than working with NFS.

Choosing between NFS and SMB for file sharing needs on Linux generally comes down to the use case. If working with predominantly UNIX systems, generally NFS makes the most sense. If working predominantly with Windows systems, then SMB generally makes the most sense. Both are powerful and robust and can service a wide variety of needs.

Where decision making can get extremely hard is in cases where we can provide for the same need it totally different ways. For example, if you need to provide shared storage backends for virtualization you might have the option of a network filesystem like NFS or a clustered filesystem on a SAN like GFS2. These two approaches cannot be easily compared as every aspect of the two systems is likely to be different including the vendors and hardware and so comparisons typically must be done at a full stack level and not at the network technology level.

Now we have explored file systems technologies and seen a broad scope of real-world file system options for Linux systems and looked at how filesystems can be local or remote, single access, or clustered to allow for multiple access vectors. At the same time, we have a good idea of how to approach decisions involving filesystem selection and configuration. We know when choose different filesystem technologies as well as what to look for in new or alternative systems that we may not have looked at specifically here. Filesystems need not be scary or confusing but can be valuable tools in our bag of tricks that we can use to fine tune our systems for safety, scalability, access, or performance. Next, we will look at one of the least understood areas of storage, the *logical volume*.

Getting to know logical volume management (LVM)

I hate to apply terms like *new* to technology that was in use by the late 1980s but compared to most concepts in computer storage **logical volume management** (**LVM**) is pretty new and is far less known than most other standard storage technologies to the majority of system administrators. LVMs were relegated to extremely high-end server systems prior to Linux introducing the first widely available product in 1998 and Microsoft following suit in 2000. Today LVMs are ubiquitous and available, often natively and by default, on most operating systems.

An LVM is the primary storage virtualization technology in use today. An LVM allows us to take an arbitrary number of block devices (meaning one or more, generally called *physical volumes*) and combine, split, or otherwise modify them and present them as an arbitrary number of block devices (generally called *logical volumes*) to the system. This might sound complex, but it really is not. A practical example can make it seem quite easy.

An example is when we have a computer system that has three hard drives attached to it. They can be all the same, or they can be different. In fact, one could be a traditional spinning hard drive, one a modern SSD, and one an external USB drive (or a RAID array, SAN, you name it). We can add all three to our LVM as physical volumes. An LVM will allow us to treat this as a single storage pool and turn it into any configuration that we want. We might turn it all into a single logical volume so that we simply get the combined storage of the three. Or maybe we will create a dozen logical volumes and use each for a different purpose. We can have as many physical volumes as we want and create as many logical volumes as we want. Logical volumes can be any size that we want (in some cases even bigger than the total physical size!) We are not limited to traditional disk sizes. With logical volumes we often find it useful to make more, smaller volumes so that we can have more management and isolation.

With an LVM, we can think of the system as consuming a block device and presenting a block device. Because LVMs use and provide block devices (aka disk appearances) they are *stackable*, meaning if you wanted to, you could have a block device on which there is an LVM which makes a logical volume, that is used by another LVM, that makes another logical volume, that is used by another LVM, and so forth. This is in no way practical, but it helps to visualize how an LVM sits in a *storage stack*. It is always in the middle, somewhere, but other than being in the middle it is very flexible.

LVMs only need to provide this basic *block in, block out* functionality to be an LVM, but there are other features that are commonly added to LVMs to really make them incredibly useful. Some of the most standard features that we tend to expect to be found in an LVM include live resizing of logical volumes, *hot plugging* of physical devices, snapshot functionality, cache options, and thin provisioning.

On Linux, as with most things, we have not one, but multiple logical volume managers! This is becoming more common as creating integrated filesystems with their own LVMs has become the trend in recent years. In production on Linux today we have LVM2, ZFS, and BtrFS. You will, of course, recognize the latter two as being filesystems that we mentioned earlier. When most people are talking about a logical volume manager on Linux, they mean LVM2, generally just called LVM. But ZFS and BtrFS' integrated logical volume managers are becoming increasingly popular approaches as well.

Because of the *stackable* nature of an LVM, that is consuming and providing block devices, we are able to use LVM2 in conjunction with ZFS or BtrFS if we so choose and can either disable their integrated LVM layers as being unnecessary, or we can use them if they have features that we want to take advantage of! Talk about flexibility.

Whatever happen to partitions

If you recall working in IT in the 1990s, we used to talk about disk partitioning quite incessantly. It was a regular topic. How do you set the partitions, how many do you make, basic and extended partitions, what partitioning software to use, and so on and so forth. To be sure, partitions still exist, we just have not needed them for a very long time now (not since Windows 2000 or Linux 2.4, for example.)

Partitions are a very rigid *on disk* system for slicing a physical disk into separate areas that can each be presented to the system as an individual block device (aka drive.) In this way, partitions are like a super basic LVM without the flexibility. Partitions are limited to existing as a part of a single block device and the mapping of which areas of the block device belong to which partition are kept in a simple partition table at the start of the device.

Partitions were the forerunner of logical volumes and some people still use them (but only because they are not familiar with logical volumes.) Partitions are not flexible and lack important options like thin provisioning and snapshots, that logical volumes can offer, and while resizing a partition is technically possible it is inflexible, hard, and extremely risky.

Everything that partitions offered LVMs offer too, plus lots more, without giving anything up. The need for partitioning (the act of creating multiple filesystems out of a block device) has decreased significantly over the years. In the late 1990s it was standard and all but required for even the simplest server and often even a desktop to have good reason to be divided up into different filesystems. Today it is far more common to merge many block devices into a single filesystem. Mostly this is because filesystem performance and reliability have totally changed and the driving factors for partitioning have eroded. There are good reasons to still divide filesystems today. We simply need to do so far less of the time.

Many mechanisms today, such as backup utilities, leverage the power of the LVM layer to do tasks like freezing the state of the block device so that a complete backup can be taken. Because an LVM operates beneath the final filesystem layer it has certain capabilities lacking at other layers. LVM is the storage layer where we get the least critical features, but it tends to be where all of the magic happens. LVMs give us flexibility to modify storage layouts after initial deployment and to interact with those storage systems at a block level. An LVM is a core technology component in providing the feel of a modern twenty-first century operating system.

Of course, any new technology layer will have caveats. An LVM adds another layer of complication and more pieces for you, as the system administrator to understand. Learning to manage the LVM is hardly a huge undertaking, but it is quite a bit more to have to learn than if you do not have one. LVMs also introduce a small amount of performance overhead as they translate between the physical devices and the logical ones. Typically, this is a truly tiny amount of overhead, but it is overhead nonetheless.

In general the benefits of an LVM dramatically outweigh the cost and more and more systems are starting to simply deploy an LVM layer without asking the end user whether or not they want it because increasingly functionality that customers simply expect an operating system to have depend on the LVM layer and allowing systems to be deployed without one often leaves customers stranded unclear why their systems do not live up to expectations and often in a way that they do not realize for months or years after initial deployment.

Like other forms of virtualization, storage virtualization and LVMs are most important for *protecting against the unknown.* If we knew everything about how a system would be used for its entire lifespan, things like resizing, backups, consolidation and so forth would have little value, but this is not how the real-world works.

Best practices when it comes to an LVM is generally accepted to be: Unless you can positively provide solid technical reasons why the overhead of the LVM is going to be impactful and that that impact outweighs the protections that an LVM provides, always deploy an LVM.

Logical volume managers provide a critical building block to robust storage solutions and are, in many ways, what separates modern storage from classical computing systems. Understanding how logical volumes abstract storage concepts and let us manipulate and build storage to act like we want gives us many options and lead us to additional concepts such as RAID and RAIN, which we will discuss next that use LVM to construct data protection, expansion, and performance capabilities.

Utilizing RAID and RAIN

We have looked at so many ways of interfacing with our storage. But probably the most exciting is when we start to deal with **RAID (Redundant Array of Inexpensive Disks)** and by extension its descendant, **RAIN (Redundant Array of Independent Nodes)**. Before we go too far, it must be noted that RAID is a huge topic that would require a book of its own to truly address in a meaningful way. Understanding how RAID works and all of the calculations necessary to understand the nuances of its performance and risk is a major subject all on its own. My goal here is to introduce the concept, explain how it fits into a design, expose the best practices around it, and prepare you for further research.

RAID and RAIN are mechanisms for taking many *storage devices* (block devices) and using the natural device multiplicity (often misstated as redundancy) to provide some combination of improved performance, reliability, or scalability over what possibility with only an individual drive is. Like an LVM, RAID and RAIN are *mid-stack* technologies that consumer block device interfaces and provide block device interfaces and therefore can be *stacked* on top of an LVM, below an LVM, on top of another RAID, on top of a mixture of hardware devices, and so forth. Very flexible.

In fact, RAID and RAIN are actually each a specialized form of LVM! No one, ever, talks about these technologies in this way and you will get some strange looks at the company Christmas party if you start discussing RAID as a specialized LVM, but it actually is. RAID and RAIN are extreme subsets of LVM functionality with a very tight focus. It is not actually uncommon for general purpose LVMs to have RAID functionality built into them and the trend with the integrated filesystems is to have the LVM and RAID layers both integrated with the filesystem.

RAID

RAID standard for *Redundant Array of Inexpensive Disks* and was originally introduced as a set of technologies that work at the block device level to turn multiple devices into one. RAID-like technologies go way back to even the 1960s and the term and modern definitions are from 1988, which means that RAID actually pre-dates more general purpose LVM.

RAID essentially takes an array of block devices and puts them into lockstep with each other under one of many different data storage regimes, called *levels*. Each RAID level acts different and uses a different mechanism to merge the underlying physical disks into a single virtual disk. By making multiple drives act as if they were a single drive, we can extend different aspects of the storage as needed but gaining in one area normally comes at a cost in another so understanding the way that RAID works is important.

RAID is an area where there is a very high importance on the system administrator having a deep understanding of the inner workings of the storage subsystem. And surprisingly it is an area where very few system administrators truly know how their system works.

While RAID comes defined as a series of *levels*, but do not be fooled. The levels are simply different types of storage that share the underlying RAID basics. RAID levels do not really build on one another and a higher number does not represent some intrinsically superior product.

Because RAID is really a form of an LVM, it can sit anywhere in the storage stack and in the wild can be found almost anywhere. Some very popular RAID levels are actually *stacked RAID* leveraging this innate artefact of its design, most notably RAID 10.

RAID also comes in both hardware and software variants. Hardware RAID is much like a graphics card that connects directly to a monitor and offloads work from the main computer systems and talks to the hardware directly. A hardware RAID card does exactly this reducing load on the main computer system, interfacing to hardware storage devices directly, and potentially offering special features (like a cache) in the hardware. Software RAID, instead, leverages the generally much more powerful system CPU and RAM and has more flexible configurations. Both approaches are completely viable.

Each RAID level has a unique set of properties and makes sense to be used at a different time. RAID is a complex topic deserving of its own tome to tackle properly. RAID is not a topic that we can look at too quickly, which has been a danger with storage in the past. RAID risk factors are often distilled into meaningless statements such as stating numerically *how many drives can a RAID array of level X recover from?* This means nothing and is meant as a way of simplifying something very complex into something that can simply be memorized or thrown onto a chart. RAID does not work this way. Each RAID level has a complex story around performance, reliability, cost, scalability, real world implementations, and so forth.

RAIN

Over time as systems became larger and more complex, the limitations with the RAID approach began to become apparent. RAID is simple and easy to implement. But RAID is inflexible and there are some key features like simple resizing, automated rebalancing, and flexible node sizing that it handles poorly. A new family of technologies were needed.

RAIN eschews the *full block device* approach to arrays that RAID was based on and instead breaks up storage by smaller chunks, often by blocks, and handles replication at that level. In order to do this effectively, RAIN must not just understand the concept of these blocks, but also the block devices (or *disks*) that they are on, but also the nodes in which they exist. This nodal awareness lends RAIN its moniker: *Redundant Array of Independent Nodes*.

Oddly, in RAIN it is not really the nodes that are necessarily redundant but really the blocks and you can actually implement RAIN on a single physical device to directly compete with traditional RAID in its simplest form, but this is rarely done.

Because RAIN handles block level replication it can have many advantages over RAID. For example, it can use different sizes devices rather fluidly. Drives of varying sizes can be thrown *willy nilly* into servers and tied together with great efficiency.

Under RAID if a drive fails, we need a replacement drive that is capable of taking over its place in the array. This is often problematic, especially with older arrays. RAIN can avoid this issue by allowing any combination of available capacity across the array absorb the lost capacity of a failed hard drive.

RAIN comes in such a variety of implementations, each one essentially unique, that we really cannot talk about it in any standard way. Most solutions are proprietary today and while a few well known open products have been made and have made their way into being standard components of the Linux ecosystem generally they are external to the distributions they behave much like proprietary products in how we must approach them.

In the future we may see significant consolidation or at least standardization within the RAIN market as these technologies become more available and well understood. Until then we need to approach RAIN with the understanding of how block replication *could work* and know that each implementation may make drastically different design choices. RAIN might be built into the kernel or might exist as an application running higher in the stack, in some cases it could even run in a virtual machine virtualized on top of a hypervisor! How RAIN will react to a lost drive, to load balancing, locational affinity, rebalancing during loss, rebuild after repair, and so forth are not defined by any standards. To use RAIN, you must research and learn in depth about any solution that you will be considering and think critically about how its artefacts will impact your workloads over time.

RAIN is almost guaranteed to be the future of system storage. As we move more and more towards clusters, hyperconvergence, cloud and cloud-like designs RAIN feels more and more natural. And adoption of RAIN will only increase as understanding increases. This simply takes time even though the technology itself is not new.

Nearly every production system that we will ever design or support, will use some form of RAID or RAIN whether locally or remotely. By now, we are prepared to think about how the decision of what RAID level or configuration or what RAIN implementation is chosen will impact our systems. Taking the time to deeply understand storage factors in these multi-device aggregation frameworks interact is one of the most valuable high level knowledge areas for system administrators across the board. In our next section, we will build on these technologies to see how local storage can be made redundant with external systems, or nodes.

Learning about replicated local storage

Possibly the most critical storage type, and the least understood, is **replicated local storage** or **RLS**. RLS is not a difficult concept, it is quite simple. But there are many myths surrounding other concepts, such as SAN, that they have clouded the functionality of RLS. For example, many people have started using the term *shared storage* as a proxy for *external storage* or possible for *SAN*. But external storage does not mean that it is or can be shared, and local storage does not mean that it is not or cannot be shared.

The term replicated local storage refers to two or more computer systems which have local storage that is replicated between them. From the perspective of each computer system, the storage is locally attached, just like normal. But there is a process that replicates the data from one system to another, so that changes made on one appear on the other.

Replicated local storage can be achieved in multiple ways. The simplest, and earliest for was to use **Network RAID**, that is RAID technology simply used over a network. **Mirrored RAID** (aka **RAID 1**) is the simplest technology for this and makes for the best example.

There are two ways to handle this scenario, one is a hot/cold pair where one node is *hot* and has access to write to the storage and the other node(s) can read from it and, presumably, take over as the hot writeable node should the original hot node fail or relinquish control. This model is easy and is similar to many traditional models for shared storage on a SAN as well. This approach allows the use of regular (non-clustered) filesystems such as XFS or ZFS.

The other approach is a live/live system where all nodes replicating the storage can read and write at any time. This requires the same clustered filesystem support that we would need with any shared block storage. Just like with a SAN being used at the same time by two nodes, the nodes in an RLS cluster will need to communicate by storing their activity data in a special area of the clustered file system.

Replicated local storage can give us many of the benefits typically associated with a SAN or other external storage, namely the ability for multiple nodes to access data at once. While also having the benefits of locality including improved performance and resilience because there are fewer dependencies. Of course, the replication traffic has its own overhead, and this has to be considered. There are many ways that replication can be configured, some with very little overhead, some with a great deal.

It is common to feel that replicated local storage is new or novel or in some way unusual. Nothing could be further from reality. In fact, what is rarely understood, is that for high reliability storage systems RLS is always used. Whether it is used locally (that is, directly attached to the compute systems) or if it is used remotely (meaning that the remote storage uses RLS to make itself more reliable), RLS is at the core of nearly any true high availability storage layer.

RLS comes in multiple flavours, primarily Network RAID and RAIN. We could call it Network RAIN when used in this situation, but we do not. Unlike RAID, which is nearly always local only, RAIN is nearly always used in an RLS situation and so the network nature of it is nearly assumed, or at least the network option of it.

RLS in so many forms on Linux that we cannot really talk about all of the options. We will have to focus on a few more common ones. RLS is an area where there are many open sources as well as commercial and proprietary solutions with a wide variety of performance, reliability, and features; and implemented in often very different ways. RLS can add rather a new level of complexity to any storage situation because you have to consider the local storage communication, the replication communication, any potential network communication between nodes and remote storage (that is local to another node), and how the algorithms and protocols interact with each other.

DRBD

The first and simplest RLS technology on Linux is **DRBD** or the **Distributed Replicated Storage System**, which is a Network RAID layer baked right into the Linux kernel. Wikipedia states that DRBD is not *Network RAID*, but then describes it as exactly Network RAID. Whether it is or not might be little more than semantics, in practice it is indistinguishable from Network RAID in use, in practice, by description, and even under the hood. Like all RAID, DRBD consumes block devices and appears as a block device allowing it to be stacked anywhere in the middle of the storage stack just like regular RAID and LVMs.

DRBD is built on a RAID 1 (mirrored RAID) mechanism and so allows for two to many nodes with each node getting a copy of the data.

DRBD is very flexible and reliable, and because of its simplicity it is far easier for most system administrators to understand clearly as to how it works and fits into the storage infrastructure. But because DRBD is limited to full block device replication by way of mirroring, as is RAID 1, the ability to scale is quite limited. On its own, DRBD is very much focused on classic two node clusters or very niche use cases with a large number of compute nodes needing to share a small amount of identical data.

Making DRBD flexible

Because DRBD is just a software RAID tool, in effect, and because you have complete management of it, and because RAID acts as an LVM with total flexibility to sit anywhere in a stack, you can take DRBD and turn it into something far more scalable than it might first appear. But currently this process is all manual, although in theory you could script it or create tools to otherwise automate these kinds of procedures.

One powerful technique that we can use is the concept of *staggering* our RAID 1 with extra logical block devices to mimic RAID 1E which operates essentially like RAID 1 but is scalable. This technique works by taking the physical storage space on an individual node and logically breaking it into two (or theoretically more) sections with an LVM technology. In a standard Network RAID setup, the entire space of the storage on node 1 is mirrored to the entire storage space on node 2. But now that we have split storage on each node, we mirror the first portion of node 1 to a portion of node 2; and node 2 does the same but with node 3; and node 3 does the same but with node 4; and this same process carries on indefinitely until whatever the terminal node number is does this with node 1 completing the *circle* and every machine has RAID 1 for its data, split between two other nodes as its mirrored pair. In this way, we can make a Network RAID 1 ring indefinitely large.

This technique is powerful, to be sure. But it is extremely cumbersome to document and maintain. If you have a static cluster that never changes, it can work very well. But if you regularly grow or modify the cluster, it can be quite problematic.

DRBD, and most Network RAID technologies, are typically blessed with good overall performance and, perhaps more importantly, rather predictable performance. DRBD, by its nature of presenting a final block device, is inherently local. In order to access DRBD resources remotely it would be necessary to use DRBD as a building block to a SAN device which would then be shared remotely. This is, of course, semantics only. DRBD is always local because to DRBD the SAN is the local compute node, the SAN interface is another layer higher up the proverbial stack and so while the SAN would be remote, DRBD would be local!

Gluster and CEPH

While two different technologies entirely, **Gluster** and **CEPH** are both free, open source, modern RAIN solutions designed for Linux that allow for high levels of reliability and high degrees of scalability. Both of these solutions at least offer the option of having storage be local to the compute node in question. Both are very complex solutions with many deployment options. We cannot simply assume that the use of either technology tells us that storage is going to be local or remote. Local is the more common application, by far, but both have options to directly build a remotely accessible and separate storage tier that is accessed over a network if designed to do so.

These technologies are far too complex and full of options for us to dig into here. We will necessarily have to treat them at a very high level only, although this should be more than adequate for our needs.

RAIN storage of this nature is the most common approach to handling large pools of servers (compute nodes) that will share a pool of storage. This technique gives the storage the opportunity to be local, to rebalance itself automatically in the event of a failure, but rarely will guarantee data locality. The storage across the group is a pool, only. So there can be an affinity for locality with data, but there is not the strict enforcement of locality as there is with DRBD. This gives more control to DRBD, but far more flexibility and better utilization with Gluster or CEPH.

Proprietary and third-party open-source solutions

In addition to what comes baked in or potentially included with a Linux distribution are many third party components that you can install. Nearly all of these products will fall into the RAIN category and very in price, support, and capabilities. A few worth knowing the names of include *LizardFS*, *MooseFS*, and *Lustre*.

It is impossible to cover the potential range of commercial products that may be or may become available. RAIN storage is an area of current development and still many vendors make products in this space but do not make them widely available. In some cases, you can find commercial RAID or RAIN systems that are only available when included with appliances of one type or another or when included in some other project. But all of these storage systems follow the same basic concepts and once you know what those are and how they can work, you can make good decisions about your storage systems even if you have not necessarily worked with a specific implementation previously.

Virtualization abstraction of storage

It is all too easy to get lost when talking about storage and forget that most of the time, storage is not even something that we have to worry about as system administrators! At least not in the way that we have been approaching it.

The storage administrator

It is not uncommon in larger organizations to decide to separate storage and systems duties as there are so many complexities and nuances to storage that having a team that is dedicated to understanding them can make sense. If you have worked in a Fortune 500 environment you have probably witnessed this.

Some of the biggest problems with this come from separating the people who deeply understand the workloads from some of the most important factors that determine the performance and reliability of those workloads. Separation often also requires that core architectural decisions be made politically rather than technically. If you use local storage, you cannot separate the storage and systems teams in any realistic way. Because of this, many organizations have used often terrible technical design decisions to create skill silos within their organizations without considering how this would impact workloads. The deployment of SAN technology is quite often done for this purpose.

Irrespective of the efficacy of this approach, when in use this generally means that storage is taken out of the hands of systems administrators. This simplifies our role dramatically while simultaneously cutting us off at the knees when it comes to being able to provide ultimate value. We can request certain levels of performance or reliability and must trust that our needs will be met or that we, at the very least, will not be held accountable for their failures.

Similarly, it is common to separate systems and platform teams. In this case we see the same effect. The platform team, which manages the hypervisor underneath the systems, will provide storage capacity to the systems team and systems must simple consume what is made available to them.

In both of these cases storage is abstracted from the system and provided simple as a *blind block device(s)* to us on the systems team. When this happens, we still have to understand how underlying components might work, which questions to ask, and at the end of the day still have to manage file systems on top of the provided block devices. The block device interface remains the universal *hand off* interface from a storage or platform team to the systems team.

An additional aside: the same thing will often happen to the platform team. They might have to take blind storage from a storage team, apply it to the hypervisor layer, then carve up that block device into smaller block devices to give to the systems team!

In most cases today, our Linux systems will be virtualized in some manner. We have to understand storage all of the way down the stack because Linux itself may be the hypervisor (such as in the case of KVM) or be used to control the hypervisor (as is the case with Xen) or provide storage to a higher-level hypervisor (like VirtualBox) and in all these cases it is Linux managing every aspect of the storage experience potentially. Linux may also be being used to create a SAN device or storage layer in some other form. We have to understand storage inside and out, but the majority use cases will be that our Linux systems will be getting their storage from their hypervisor even if we are the managers of that hypervisor.

While they can choose to behave in many different ways, most people set up hypervisors to act like an LVM layer for storage. They are a bit of a special case because they convert from block to filesystem, then back to block for the handoff to the virtual machine, but the concept remains the same. Some hypervisor setups will simply pass through a direct block connection to underlying storage whether a local disk, a SAN, or a logical volume from an LVM. These are all valid approaches and leave more options for the virtual machine to dictate how it will interact with the storage. But by and large having the block layer of storage terminate with the hypervisor, be turned into a filesystem, creating *block device containers* on top of the filesystem and allowing the virtual machines to consume those devices as regular block devices is what is expected from virtualization that many people actually refer to this artefact of virtualization approaches as being intrinsic to virtualization itself, which it is not.

You can use this technique inside of a system as well. Examples of this include mounting file system file types like *qcow2*, *vhdx*, or *iso* files! Something that we do every day, but rarely think about or realize what we are actually doing.

Obviously when getting our storage from the hypervisor, concerns about standard (non-replicated) local storage, replicated local storage, standard (non-replicated) remote storage, or replicated remote storage are all made at a different layer than the system, but the decisions are still made, and those decisions completely impact how our systems will ultimately run.

We have learned about a lot of storage abstraction approaches and paradigms now with LVMs, RAID, and RAIN. Now we need to start to think about how we will put these technologies together to build our own storage solutions.

Analyzing storage architectures and risk

Nothing creates more risk for our systems than our storage. That should go without saying, but it has to be said. Storage is where we, as system administrators, have our greatest opportunity to make a difference, and it is the place where we are mostly likely to fail and fail spectacularly.

In order to address risks and opportunities in regard to storage, we must understand our entire storage stack and how every layer and component interact with each other. Storage can feel overwhelming, there are so many moving pieces and optional components.

We can mitigate some of the overwhelming feelings by providing design patterns for success and understanding when different patterns should be considered.

General storage architectures

There are two truly high-level axis in **storage architecture**: *local* versus *remote*, and *standard* versus *replicated*.

Of course, the natural assumption for most people is to jump immediately to believing that replicated and remote are the obvious starting point. This is actually not true. This is probably the least sensible starting point for storage as it has the least likely to be useful combination of factors.

Simple local storage: The brick

Believe it or not, the most commonly appropriate storage design for companies of all sizes is local, **non-replicated storage**! Bear in mind that *replicated* when we speak of storage architectures does *not* reference a lack of backups nor a local of local replication (such as RAID mirroring) but only refers to whether or not storage is replicated, in real or near-real time, to a second totally separate system.

We will cover this again, and slightly differently, when we look at total system design rather than looking at storage in isolation. Like most things in life, keeping things simple generally makes the most sense. Replication sounds amazing, a must have feature, but replication costs money and often a lot of it and to do replication well often impacts performance, potentially dramatically.

Replicating disaster

A common mistake made in storage design is getting an emotional feeling that the more replication that we have, the more that we are shielded from disaster. To some degree this is true, replicated some files locally with RAID 1 does a lot to protect against an individual hard drive failing and remote replication can protect against an entire node failing, but neither does anything to protect against much more common problems like accidental file deletion, malicious file destruction, file corruption, or ransomware.

If we do something simple, like delete a file that we shouldn't, then instantly our high-power replication mechanism will ensure that our deletion ripples through the entire system in a millisecond or two. Instead of protecting us, it might be the mechanism that replicates our mistake faster than we can react. Overbuilding replication mechanisms typically protects only against hardware failure and can quickly turn into a situation of diminishing returns.

That first level of RAID might be highly valuable because hard drive failure remains a very real risk and even the tiniest drive hiccup can cause significant data loss. But replicating between nodes will only protect against entire system loss which is quite a bit less common. RAID protection is relatively cheap, often costing only a few hundred dollars to implement. Nodal replication, however, will require dramatically more hardware to achieve replication generally costing thousands or tens of thousands of dollars for a fraction of the protection that RAID is already providing.

Mechanisms like RAID, especially RAID 1 (mirroring) are also extremely simple to implement and very straightforward. It is quite uncommon to encounter data loss caused by human error in mirrored RAID. The same cannot be said for replicated storage between nodes. There is far more to go wrong, and the chances of human error causes data loss is much higher. We do not simply mitigate risks by choosing to go this expensive route, we introduce other risks that we have to mitigate for as well.

Many system administrators feel that they cannot use simple, local, non-replicated storage, and problems with company politics cannot be overlooked. If your company is going to *play politics* and blame you, as the system administrator, even when the mistake is not yours and your decision was the best one for the business, then you are forced to make dangerous decisions that are not in the interest of the profitability of the business. That is not something that a system administrator can control.

As a system administrator, we can manage this political problem *in some cases* by presenting (and documenting well) risk and financial decisions to demonstrate why a decision that may ultimately have led to data loss to have still been the correct decision. No decision can every eliminate all risks, as IT professionals and especially as system administrators we are always making the decision as to how much risk to attempt to mitigate and at what financial cost should we do so.

Risk assessments

One of the hardest, yet most important, aspects of IT and especially systems administration is doing risk assessments to allow for proper planning. Risk is rarely taught either formally or organically. This is an area where nearly all businesses fail spectacularly and IT, where risk is absolutely key to everything that we do, is generally left with no training, no resources, and no support.

Teaching risk is a career in and of itself, but a few techniques that we should be using all of the time can be covered here. At its core, risk is about assigning a cost that we can apply against projected profits.

We have two key aspects of risk. One is the chance that something bad will happen, the second is the impact of that event. The first we can express as a matter of *happens X times per year* if you find that handy. The second can be expressed in monetary terms such as *it will cost $5,000*. If something will happen once a decade then you could say it is .1x per year. So, something that impacts us for five grands would have an annual cost of $500. This is ridiculously simplistic and not really how risk works. But it's an unbelievable useful tool in expressing risk decisions to management where they want millions of factors distilled to a single bottom line number.

Now we have to take our numbers that show the cost of a risk mitigation strategy. For example, if we are going to implement a replication technology that costs $300/year in licensing and requires ten hours of system administration time per year at $120/hour as can project a cost of mitigation to be $1500/year.

Next weeks need a mitigation effectiveness. Nothing is really one hundred percent. But a good replication strategy might be 95% or a typically one might be around 65% effective. With these numbers we can do some hard math.

We know that we are at risk of losing roughly $500 per year. If we spend $1500 per year, we can 95% surely stop the $500 loss. $500 * .95 = $475. So, take $1500-$475=$1025 of loss per year caused by the risk mitigation strategy. These are numbers you can take to a CFO. Do this math, you should be able to show savings or protection, not a loss. If you show a loss then you really, really need to avoid that plan. It means that the risk protection mechanism is, for all intents and purposes, representative of a *disaster* simply by implementing it.

Math, it might sound trite to say, but the average system administration and even the average CFO will often run from using basic math to show if an idea is good or bad and go purely on emotion. Using math will protect you. It means you can go to the CFO and CEO and stand your ground. You cannot argue with math. Show them the math, if they decide that the math is wrong, have them work the numbers. If they decide to ignore the math, well, you know what kind of organization you work for and you should really think long and hard about what kind of future there is at a company that thinks that profits are not their driving factor. And if something goes wrong in the future and you get blamed, you can pull out the math and ask, *if this was not the right decision, why did we not see it in the math when we made the decision?*

Nothing feels better that defending successfully a seemingly crazy decision that has been backed by math. Show that you are doing the best job possible. Do not just say it, do not make unsubstantiated claims. Use math and prove why you are making decisions. Elevate the state of decision making from guesswork to science.

Not all workloads can simple be treated this simply. But vastly more than are normally assumed can. This should be your standard assumption unless you have solid math to show otherwise. Or you are dealing with a situation that goes beyond math, such as life support systems where uptime and reliability are more important than money can every describe. Otherwise, use math.

This simplest of all storage approaches is easily thought of as *the brick*. It is simple, it is stable, it is reliable, it is easy to backup and restore, it is easy to understand and support. This solution is so effective today, especially with modern storage technologies and hardware, that I will state that it is appropriate for 85-90% of all production workloads, regardless of company size, and at least 99% of small business workloads.

RLS: The ultra-high reliability solution

What workloads and situations do not make sense for the simple architecture above almost always fall into this category: **replicated local storage** (**RLS**). RLS allows for highly available storage with great performance at reasonable cost. Nothing will match the performance and cost of straight local storage that we mentioned first, but if you need higher availability and better reliability than that solution can provide, this is almost certainly the solution for you.

RLS provides the highest level of reliability available because it has fewer moving parts. A remote high availability solution must necessarily deal with distance, cabling, and network protocols as additional components at a minimum, and typically with networking equipment (like switches) additionally, all above and beyond the risks of RLS. Remember that remote storage solutions wanting to accomplish true high availability are going to have to do so by using RLS locally in their own cluster, and then making that cluster of storage available remotely over a network so you have all of the complications and any potential problems with RLS, plus the overhead and risks of the remote access technology.

If straight local storage with no remote replication takes ninety percent of all workloads, standard RLS much take ninety percent of what remains (the two should take about ninety nine percent together.) When doing proper planning, these two options are so simple, cost effective, and safe that they are just impossible to beat under normal circumstances. These are your break and butter options.

Alternative storage reliability

While RLS might feel like the absolute end all, be all of storage protection, it really is not. It is great when you have to rely on the storage layer to handle reliability. But in that regards, it is a kludge or a band aid, not the ultimate solution.

In a perfect world we have storage mechanisms that are a layer higher than our actual storage layer with things like databases. A database management system is able to do a much better job at maintaining availability and coordinating data replication between nodes than a blind block replication mechanism can ever do. Putting replication where the application intelligence is just makes sense.

Ideally, applications would use their databases as their layer for storage and state reliability and let intelligence systems that know how the data is used replicate what matters. Databases are one of the most ideal mechanisms for replication because they know the data that they have and are able to make good decisions about it.

Because of this, many enterprise applications do not use any form of storage or even systems replication whatsoever. Using *unreliable* systems and highly reliable *applications* is a solid strategy and offers benefits that you can get in no other way. Because of this, we can sometimes ignore high availability needs at the raw storage layer and just focus on performance.

The lab environment: Remote shared standard storage

This architecture is very popular because it checks all of the boxes for salespeople: *expensive*, *risky*, and *confusing*. It is true, salespeople push this design more than any other because it generates so many ways to bill the customer for additional services while passing all responsibility on to the customer.

All architectures have a legitimate place, more or less, in an ecosystem and this one is not an exception. But before we state a use case we should state where its benefits lie: non-replicated remote storage is *slower*, *riskier*, and more *expensive* (on the surface) than local storage mechanisms. Where do we find value for such a design?

Primary the benefits are in lab, testing, and other environments where the value to data durability is negligible, but we can benefit from large environments where storage can be carved out judiciously to create maximum cost savings at a large scale.

Nearly every large environment has a need for this kind of storage at some point in their collection of workloads. The key is to identify where this mix of needs is sensible and not to attempt to apply it where it is inappropriate. Because this kind of storage is cheap at large scale (but outrageous expensive at small scale) and because managers so often mistake remote shared storage for the *magic box that cannot fail* it is often used where it is least applicable. There is no magic, it is a poor technology for most production workloads and should be chosen only with extreme caution.

The rule of thumb here is that you should only use this type of storage if you can decisively prove that it makes sense - that even standard reliability and performance are not valuable. If you have any doubts, or if the organization has any doubts, then choose a different storage architecture before making a mistake with this one means maximum risk. Performance is a *soft failure*, it is easy to correct after the fact and it normally has marginal impact if you get it wrong. Data loss is a *hard failure* where getting it wrong is not a graceful failure but a catastrophic failure and there is little ability to correct it later.

Fail gracefully

We cannot always avoid failure. In fact, much of the time skirting failure is a critical part of our job. In order to make this work we have to understand how to fail well and a key component of that is the idea of *failing gracefully*. And in storage, this is an area where this concept is more pronounced than in other areas.

The idea with failing gracefully is that if we fail, we want it to be in a small, incremental way rather than in a tragic, total disaster kind of way. Many storage decisions that we make are designed around this. We know that we might get things wrong. So, we want to make sure that we are as close to correct as possible, but also with taking into consideration *what if* we are wrong.

In this way we tend heavily towards RAID 10 in RAID and towards local storage in architecture. We want solutions that, if we are wrong, result in us being too safe rather than losing the data because we thought that it would not matter.

While it takes a lot more than just bad storage decisions to determine an entire architecture, remote non-replicated storage is the foundation point of the popular design used by vendors and resellers which they typically call a **3-2-1 architecture** and what IT practitioners called the **Inverted Pyramid of Doom**.

This architecture is traditionally deployed broadly but is by no small margin the least likely architecture to ever me appropriate in a production environment. It is slow, it is complex, it is costly, and it carries the highest risks. It makes sense primarily in lab environments were recreating the data being stored is, at most, time consuming. This is an architecture truly designed around the needs of typically non-production workloads.

The giant scale: Remote replicated storage

Our last main architecture to consider is the *biggest* of them all. Remote replicated storage. It might see like this would be the storage architecture that you would see in every enterprise, and while not exactly rare, it is seen far less commonly than you would guess.

Remote replicated storage is the costliest to implement at small scale but can become quite affordable at large scale. It suffers from extra complexity over RLS (which means less reliable) and lower performance than either RLS or straight local storage and so only makes sense when cost savings is at a premium, but a certain degree of reliability is still warranted. A bit of a niche.

Considering that the two local storage architectures were granted ninety nine percent of workload deployments (by me, of course) between them, this architecture gets most of the last percentage that is left, at least in production.

The safest system is only so safe

One of my more dramatic stories from my decades as a system administrator comes from a time that I was working at a university library as a consultant. I was brought in to work on some large-scale databases in this two-admin shop. The senior admin was out on vacation and only the junior admin was still around. The environment used a highly redundant and high availability SAN system on which all systems depended. There was an extreme amount of protection of this system from UPS to generators to hardware redundancy all at great expense.

While I was there, the junior admin decided to do some maintenance on the SAN and, for whatever reason, accidentally clicked to delete all of the volumes on the SAN. This was, of course, an accident but a very careless one by someone who was assuming that every possible failure scenario was carefully guarded against. But all of the high availability, all of the redundancy, all of the special technology did nothing to address simple human error.

With one click of her mouse, the entire library's computer systems were gone. The applications, the databases, the users, the logs. All of it. Everything was dependent on a single storage system and that system could be turned off or, in this case, deleted with essentially no effort by someone with access. All of the eggs were in a single backet that, while made very sturdily, had a big opening and could easily be turned upside down.

To make matters far, far worse the junior system administrator was not emotionally prepared for the event and was so terrified of losing her job that she had to be hospitalized and all of the potential IT staff that might have been able to have stepped in to assist were instead engaged in getting her an ambulance. Through poor technological planning, and through poor human planning, an easily avoidable disaster that should have only turned into a minor recovery disaster turned into a huge outage. Luckily there were good backups, and the system was able to be restored relatively quickly. But it highlighted well just how much we often invest in protecting against mechanical failure and how little we address human frailty and, in reality, it is human error that is far more likely to be the cause of a disaster than machines failing.

Storage architectures and *risk* is the hardest part of storage design. Drilling down into filesystem details is generally fun and carries extremely little risk to us as system administrators. If we pick EXT4 when BtrFS would have been best, the penalty is probably nominal, so much so that we would never expect anyone to ever find out that we did not make the perfect decision. But choosing the wrong storage architecture could result in massive cost, large downtime, or big time data loss.

We really have to take the time to understand the needs of our business and workloads. How much risk is okay, what performance do we need, how do we meet all needs at the optimum cost. If we do not know the answers then we need to find out.

Best practice is, as always, to take the time to learn all business needs, learn all of the available factors and apply. But that is very hard to do in practice. Some rules of thumb to assist us will come in very handy.

Storage best practices

Attempting to distill storage into **best practices** is rather hard. At the highest level, the fundamental rule to storage is that there are no shortcuts, you need to understand all aspects of the storage infrastructure, understand the workloads, and apply that combined knowledge allow with an understanding of institutional risk aversion to determine workload ideals.

Going further, best practices include:

- RAID: If data is worth storing, it is worth having on RAID (or RAIN.) If you are questioning the value of having RAID (at a minimum) on your servers, then reconsider storing the data at all.

- RAID Physicality: Both hardware and software RAID implementations are equally viable. Determine what factors apply best to your needs.

- LVM: Like general virtualization which we will touch on in future chapters, storage virtualization is not about providing a concrete feature set that we need on day one. It is about providing mechanisms to protect against the unknown and to be flexible for whatever happens in the future. Unless you can present an incredible strong argument for what LVM is not needed, use it.

- Filesystem: Do not be caught up in hype or trends. Research the features that matter to you today and that are likely to protect you against the unknown in the future and use a filesystem that is reliable, robust, and tested where you can feel confident that your filesystem choice is not going to hamper you long term.

- Storage Architecture: Unless you can prove that you require or financially benefit significantly from remote storage, keep your storage access local. And unless you can demonstrate clear benefit from nodal replication, do not replicate between nodes. Simple trumps complex.

As a system administrator you might deal with storage design decisions infrequently. But no matter how infrequent, storage design decisions have some of the most dramatic impacts on our long-term system compared to any other decisions that we make. Take your time to really determine what is needed for every workload.

Storage example

We should step back and put together an example of how these pieces might fit together in a real life scenario. We cannot reasonably make examples for every common, let alone plausible, storage scenario but hopefully we can give a taste of what we are talking about in this chapter to make it all come together for you.

To keep things reasonably simple, I am going to work as generically as possible with the absolutely most common setup found in small and medium businesses. Or at least what probably should be the most common setup for them.

Smaller businesses generally benefit from keeping their designs quite simple. Lacking large, experienced staff and often at high risk from turnover, small businesses need systems that require less maintenance and those that can easily be maintained by consultants or staff that may not possess tribal knowledge of the environment.

For these kinds of environments, and also for many larger ones, hardware RAID with hot and blind swappable drives are important. They allow hardware maintenance to be done by bench professionals without a necessity to engage systems administration tasks. This becomes extremely critical when dealing with colocation or other distant facilities. In these cases, someone from the IT department may have no way to be physically involved at all.

So, we start with some general case assumptions. At the physical layer we have eight hard drives. These can be spinning drives, SSD, NVMe, any block device. It does not really matter. What matters is that we have multiple, but we want them to all act as a single unit.

So, we add to this a hardware RAID controller. This RAID controller we use to attach all of the drives and set it to put them into an appropriate RAID level. This could be any number of options, but for this example we will say that they are in RAID 10.

From the very beginning of our example, without having even installed Linux or anything of the sort, we have used the hardware of our system to implement our first two layers of storage! The physical devices, and the first abstraction layer.

We will not show actually inserting the drives here, that is purely a manual process and chances are your server vendor has done this already for you, anyway.

As for setting up the RAID itself, every controller and vendor is slightly different, but the basics are the same and the task is always extremely simple. That is much of the point of the RAID controller - to reduce the amount of knowledge and planning necessary around RAID operations to the bare minimum both up front and during operational phases. For our example here, to make things easier to demonstrate, we are going to assume that we are dealing with a single array that is *not* the array from which our operating system is running so that we can show some of the steps more easily from the command line. But this is just an example.

Remember that hardware RAID is different for every vendor and potentially every product by a vendor. So you always need to be aware of exactly how your specific product works.

Also, we should note that to the RAID controller, each drive attached to it is a block device. This is the unique case where the block device interface, pretending to be a physical hard drive, is actually and truly a hard drive! In every subsequent case we will be using software to implement the interface of a hard drive but the drive that we represent will be logical, not physical:

```
=>ctrl slot=0 create type=ld drives=2I:1:5,2I:1:6,2I:1:7,2I:1:8
raid=1+0
```

This is a real syntax for a real-world RAID controller. Typically, you will do this task graphically from a GUI. But sometimes you will want to use a command line utility. When possible, I work from the command line, it is more repeatable and far easier to document.

Once we are past this phase of initial hardware configuration then we can proceed with the Linux specific portions of the example.

Hardware RAID controllers typically create their own naming conventions in the /dev filesystem. In our example case, the controller uses the cciss syntax and created device c0d1 under that system. All of these will vary depending on the control and configuration that you use.

Next, we are going to create a logical volume layer on top of the RAID layer to allow us more flexibility in our storage system. To do so we must start by adding the newly created device to Linux' LVM system as *a physical device*. We do this with the `pcvreate` command and the path to our new device:

```
# pvcreate /dev/cciss/c0d1
```

Very fast and easy. Now the LVM subsystem is aware of our new RAID array device. But of course, all that LVM knows and cares about is that it is a block device. That it is a RAID array specifically is actually not something that LVM can detect, nor does it matter. The point here is that it is abstracted and can be utilized the same no matter what it is. The speed, capacity, and safety characteristics are encapsulated in the RAID layer and now we can think of it purely as a hard drive as we move forward.

Another interesting point here is that when using a hardware RAID controller this abstract and virtualized hard drive representation is only a logical hard drive, but as a block device it is actually physical! Mind blowing, I know. It's really hardware, it just is not a hardware hard drive. Ponder on that for a moment.

How that our RAID array is under LVM's management we can add the drive to a volume group. In this example we will be adding it to a new volume group that we are creating just for the purposes of this example. We name this new group `vg1`. Here is an example command doing just this:

```
# vgcreate vg1 /dev/cciss/c0d1
```

Okay, now we are getting somewhere. The capacity of the individual physical hard drives being combined by the RAID controller into a single logical drive under LVM control is now in a capacity pool or *volume group* where we can start to carve out actually useful subsets of that capacity to use on our server.

With the volume group created, all that is left is to make the actual *logical volumes*. Remember that logical volumes have replaced *partitions* as the primary means of dividing a block device into consumable portions that we can use. For our example we are going to do the absolute simplest thing and tell the LVM system to make just one logical volume that is as large as possible; that is, using 100% of the available capacity of the volume group (which is currently at 0% utilization as this is the very first thing that we will have done with it.):

```
# lvcreate -l 100%FREE -n lv_data vg1
```

This command tells LVM to create a new logical volume, using 100% of the free space that is available, in volume group `vg1`, and name the new logical volume `lv_data`. That is it. We now have a logical volume that we can use! It should be obvious that we could have made a smaller logical volume, say of 50% of the available space, and then made a second one, also of 100% of what was remaining after that to give us two equal sized logical volumes.

Remember that an LVM system like the one found here in Linux, gives us flexibility that we would often lack if we were to apply a filesystem directly to the physical drives or even to the virtual drive presented by the RAID controller hardware. The LVM system lets us add more physical devices to the volume group, for example, which gives us more capacity for making logical volumes. It will let us resize the individual logical volumes, both growing or shrinking them. LVM will also allow us to snapshot a logical volume which is very useful for building a backup mechanism or preparing to do a risky system modification so that we can revert quickly. LVM does very important things.

Now that `lv_data` has been created we will need, in most cases, to format it with a filesystem in order to make it truly useful. We will format with XFS. In the real world today, XFS would be the most likely to be recommended filesystem for general purpose needs:

```
# mkfs.xfs /dev/vg1/lv_data
```

Very simple. In a few seconds we should have a fully formatted logical volume. In applying the filesystem format we stop the chain of block device interfaces and now present a filesystem interface which is the change that allows applications to use the storage in a standard way instead of using block devices as are used by storage system components.

Of course, one last step is necessary, we have to mount the new filesystem to a folder to make it usable:

```
# mkdir /data
# mount /dev/vg1/lv_data /data
```

That is it! We just implemented a multi-layer abstraction based storage system for one of the most common system scenarios. We have built an XFS file system on top of LVM logical volume management on top of a hardware RAID controller on top of multiple individual physical hard drives.

Because each layer uses the block device interface, we could have mixed and matched so many more additional features. Like using two RAID controllers and merging their capacity with the volume group. Or making multiple volume groups. We could have made many logical volumes. We could have used software RAID (called MD RAID in Linux) to create RAID using the output of the two RAID controllers! The sky is really the limit, but practicality keeps us grounded.

At this point if you *cd /data* you can use the new filesystem just as if it has always been there. That it is a new filesystem, that it is built on all these abstraction layers, that there are multiple physical devices making all of this magic happen is completely hidden from you at this point. It just works.

Now in the past, if this was 2004, we would generally stop here and say that we have described what a real world server is likely going to look like if it is implemented well. But this is not 2004 by a long shot and we really need to talk more about how we are likely going to see our storage used in the most common scenarios. So today we need to think about how our virtualization layers are going to use this storage, because things get even more interesting here.

We will assume that our /data filesystem that we just created will be used to store the drive images of a few virtual machines. Of course, these drive images are just individual files that we store in the filesystem. Very simple. No different than creating and storing a text file in /data (except VM drive images tend to be just a tad larger.)

What is neat about a drive imagine (this could be a QCOW, VHD, ISO, or other) is that they sit on top of a filesystem but, when opened by a special driver that is able to read them, they present a block interface again! That is right. We have gone from block to block to block to block to filesystem to block again! In some unique cases we might not even use a hypervisor but might use this new block device file somewhere in our regular server. Windows does this commonly with VHD files as a way to pass data around in certain circumstances. MacOS does this as their standard means of creating installer packages. On Linux it is far less common but just as available.

But assuming that we are doing something normal, we will assume that we are running a hypervisor, KVM almost certainly, and that the virtual machines that are going to run on KVM are going to use disk image files storage on our newly minted file system. In this case, much of what we have done here is likely to happen yet again inside of that virtual machine.

Some portions would not be very sensible to recreate. The physical drives are already managed by a physical RAID controller. The speed, capacity, and reliability of the storage is already established by that system and does not need to be duplicated here. The standard approach is for a single drive image file to be presented to the operating system running in a virtual machine as a single block device. No different than how our operating system was presented with the block device from the hardware RAID controller.

Now inside of the virtual machine we will often run through the same exercise. We add the presented block device as an LVM physical volume. Then we add that physical volume to a volume group. Then we carve out one or more logical volumes from that volume group. We then format that logical volume with our filesystem of choice and mount it. Of course, typically much of that is not done by hand as we have done here but rather than the installation process automating much of it.

We can add more steps such as using MD RAID. Or we can use fewer, such as by skipping LVM entirely. We could do all of the same steps with just a single hard drive and no RAID controller. This would be far less powerful physically, but all of the examples would work the same at a technical level. We could use VLM on the physical machine but not in the virtual machines, or vice versa! The flexibility is there to do what is needed. It is all about understanding how block devices can be layered, how a filesystem goes on a block device and how block files can turn a filesystem back into block devices!

Abstraction and encapsulation are amazingly powerful tools in our IT arsenal and rarely are they so tangible.

Summary

If you have survived to the end of this chapter and are still hanging in with me, congrats, we made it! Storage is a big deal when it comes to systems administration and likely no other area that you manage will you be able to bring as much value to your organization.

We have covered storage basics building on the concepts of block device interfaces, abstraction techniques, filesystems and their interfaces, and used these concepts to investigate multi-device redundancy and how it can be used to build complex and robust data storage, and how storage access across devices can be handled to meet any potential need. My goal here has been to give you the knowledge necessary to think carefully on your own about your storage needs for any given workload, and an understanding of availability technologies and how you can apply them to meet those goals most effectively.

Never again should you see storage as a magic black box or a daunting task that you dread to tackle. Instead, you can see storage as an opportunity to shine and to demonstrate how proper system administration best practices can be applied to maximize whatever storage factors matter most for your workload without simply throwing money at the challenge or worse, simply ignoring it and hoping that you can find another job before things fall apart.

In our next chapter, we are going to look at system architecture at an even higher level. Many of the most interesting concepts from this chapter will be recurring there. System architecture relies on storage architecture very heavily and many redundancy and system protection paradigms are shared. It can be quite excited to see how good storage design elements can lead to a truly high performance, highly available, and cost-effective final solution.

4
Designing System Deployment Architectures

How we deploy systems determines so much about how those systems will perform and how resilient they will be for years to come. A good understanding of design components and principles is necessary for us to understand in order to approach the design of the platforms that will carry our workloads. Remember, at the end of the day, only the applications running at the very top of the stack matter - everything beneath the applications, whether the operating system, hypervisor, storage, hardware, and others are just tools used to enable the final application-level workloads to do what they need to do best. It is easy to feel that these other components matter individually, but they do not. To put it another way, what matters is the results rather than the path taken to get to the results.

In this chapter, we are going to start by looking at the building blocks of systems (other than storage which we tackled extensively in our last chapter before taking all of those components as a whole and looking at them, to see how they can form robust carriers for our application workloads. Next, we will look at need analysis. Then finally we will move on to assembling those pieces into architectural designs to meet those needs.

By the end of this chapter, you should feel confident that, while Linux is potentially only one slice in the middle of our application stack, you are prepared to design the entire stack properly to meet workload goals. While technically much of this design is not strictly systems administration (or engineering) it most often falls to the system administrators to handle as only the rarest of organizations have highly skilled and end to end knowledgeable staff from other departments. The systems team sits at the nexus of all components and has the greatest single role visibility in both directions (up the stack to the applications and down the stack to hypervisors, hardware, and storage). It is natural that systems teams are tasked with the greater design tasks as there is no one else capable.

In this chapter we are going to learn about the following:

- Virtualization

- Containerization

- Cloud and **Vitual Private Server** (**VPS**)

- On premises, hosted, and hybrid hosting

- System design architecture

- Risk assessment and availability needs

- Availability strategies

Virtualization

Twenty years ago, if you asked the average system administrator what virtualization was they would look at you with a blank stare. We have had virtualization technologies in IT since 1965 when IBM first introduced them in their mainframe computer systems, but for your average company these technologies were relatively rare and out of reach until vendors like *VMware* and *Xen* brought these to the mainstream market around the turn of the millennium. The enterprise space did have many of these technologies by the 1990s, but knowledge of them did not disseminate far.

Times have changed. Since 2005, virtualization has been broadly available and widely understood, with options for every platform and at all price points, leaving no one with a need to avoid implementing the technology because it is out of technical or financial reach. At its core, virtualization is an abstraction layer that creates a computer *in software* (on top of the actual hardware) and presents a standard set of virtual hardware. Software that performs virtualization is called a **hypervisor**

In the last chapter we spoke repeatedly about interfaces and how something consumes or presents itself as a disk drive or file system, for example. A hypervisor is software that presents a *computer interface*, meaning it doesn't just present a hard drive, but it acts like an entire computer. If you have never used or thought about virtualization this might seem extraordinarily complex and confusing, but in reality, this is an abstraction that often makes computing far simpler and more reliable. Just like technologies that abstracted storage (like **Logical Volume Managers** and **RAID** systems) proved to be incredibly valuable once they were mature and understood, so has computer level virtualization.

There are two types of hypervisors that we will talk about in this chapter, and they are simply known as Type 1 and Type 2 hypervisors. All hypervisors present the same thing: a *computer*. But what makes a Type 1 and a Type 2 hypervisor different is what they consume.

Type 1 hypervisor

Sometimes called a *bare metal* hypervisor, Type 1 hypervisors are intended to run directly on the system hardware but, of course, can run on anything presenting itself as a capable piece of hardware (such as another hypervisor!) As such, a Type 1 Hypervisor is not an application and does not run on top of an operating system and so only needs to worry about hardware compatibility with the physical device on which it will be installed.

Type 1 hypervisors are generally the only type considered true ready for production because they install directly without any unnecessary software layers and so can be faster, smaller, and more reliable.

The Type 1 hypervisor was more difficult to initially engineer and so the earliest hypervisors were generally other types that could pass off work to the operating system. But effectively it was the introduction of the Type 1 hypervisor and enough vendors with disparate products to warrant a mature market designation that encouraged the extreme move to virtualization in the 2000s.

Type 2 hypervisor

Unlike the bare-metal hypervisor, a Type 2 hypervisor is an application that you install onto an operating system. This means that the hypervisor has to wait for the operating system to give it resources, competes with other applications for resources, and requires that the operating system itself is stable, in addition to the hypervisor being stable, in order to keep workloads running on top of it.

When virtualization was relatively new, especially in the microcomputer arena, Type 2 Hypervisors were much more common because they were cheaper and easier to make and have little need for hardware support to do what they do. A Type 2 Hypervisor lets the bare metal operating system do the heavy lifting of supplying drivers and hardware detection, task scheduling, and so forth. So, in much of the 2000s we saw Type 2 Hypervisors taking a principal role in driving virtualization adoption. They are easy to deploy and very easy to understand and because they are just an application that gets deployed on top of an operating system anyone can just install one on an existing desktop or even laptop to try out virtualization for themselves.

By the late 2000s, technology had changed rather significantly, the software had advanced and matured, and nearly all computers had gained some degree of hardware assistance for virtualization, allowing hypervisors to use less code while gaining much better performance. Type 1 hypervisors rapidly proliferated, and, before 2010, the idea of using a Type 2 hypervisor in production was all but unthinkable. Type 1 hypervisors provide a single, standard operating system installation target, moving the heavy lifting away from the operating system and over to the hypervisor, where it is generally accepted to be better positioned. Because the hypervisor controls the bare metal, it is able to properly schedule system resources and eek maximum performance out of a system. Hypervisors are expected to be only a small fraction of the size of an operating system. This means little more than a shim between virtualized operating systems and the physical system (a tiny layer of code doing the bare minimum, and being essentially invisible to the operating system running on top of it). This minimizes bloat and features, while operating systems need to be large, complex, and feature-rich to do their jobs well in most cases.

Type 2 Hypervisors have proven to be useful in lab environments, especially for situations where testing or learning is best done from a personal computing environment such as a desktop or laptop or can be useful for special case temporary workloads where there is a need to completely disable or possibly even to remove the hypervisor when it is no longer needed. But for production server environments only Type 1 Hypervisors are really appropriate today.

There are two best practices commonly associated with virtualization

- Virtualize every system, unless a requirement makes you unable to do so. In practical terms, you will never realistically see a valid exception to this rule.

- Always use a Type 1 (Bare Metal) Hypervisor for servers.

Hypervisor types are confusing

In the real world detecting what is and is not a Type 1 Hypervisor can be rather difficult. A hypervisor, by definition, really does not have any kind of end user interface of its own. This makes it something that we have to explain, but not something that we really see. Even a true operating system is hard to point to and say *see, there it is* because it is really a shell or a desktop environment running as an application on top of the operating system, rather than the operating system itself, that we see and touch. With a hypervisor, any interface that we see, of any sort, has to be being presented by something running on an operating system, not something running directly on the hypervisor.

Hypervisors, of course, need some sort of interface for us to interact with them. How they handle this varies wildly and not all hypervisors, even of the same type, are built the same. Under the hood, of course, they are always running on the bare metal, but they can use several different architectures to handle all of the functions that they need. Each different architecture has a different opportunity for how it will seem to appear to an end user – meaning that an end user sitting down to the system may experience wildly different interfaces that pretend to be things that they are or possibly are not.

In early Type 1 Hypervisors it was common to run a virtual machine (the name for a virtualized computer running on top of a hypervisor of any type) that had privileges to control the hypervisor given to it. This allows the hypervisor to be as lean as possible and allows big tasks like presenting a user interaction shell to be done using existing tools and no one had to reinvent the wheel. Using this approach meant that hypervisor engineering work was minimal in the early days allowing the virtualization itself to be the key focus.

As time has progressed, alternative approaches have begun to emerge. Running a full operating system in a virtualized environment on top of the hypervisor just to act as an interface for the end user felt like a waste of resources. Later hypervisors used creative ways to get around this making the hypervisor itself heavier but reducing the overall weight of the total system.

Virtualization tends to be confusing and vendors have little reason to want to expose the inner workings of their systems. It has therefore become commonplace to misuse terms or to suggest that hypervisors work differently than they do either for marketing reasons or to attempt to simplify the system for less knowledgeable customers. There are many hypervisors today and potentially more will arise in the future. In production enterprise environments, however, there are four that we expect to see with any regularity and we will briefly break down each one to explain how it works. No one approach is best, these are simply different ways to skin the same cat.

VMware ESXi

The market leader of virtualization today. VMware is one of the oldest virtualization products and has changed its design over time. Originally VMware followed the classic design of an extremely lean hypervisor and a *hidden* virtual machine that ran on top of it running a stripped down copy of Red Hat Enterprise Linux which provided the end user facing shell for interaction with the platform.

Today VMware ESXi instead builds a tiny shell into the hypervisor itself that provides only enough potential user interaction to handle the simplest of tasks such as detecting the IP address that is in use or setting the password. Everything else is handled through an API that is called from an external tool allowing for the heaviest portions of the user interface to be kept completely on the physical client workstation rather than on the hypervisor.

Microsoft Hyper-V

Although late to the enterprise Type 1 virtualization game, Microsoft opted for the classic approach and always runs a virtual machine in which is contained a stripped down copy of Windows which provides the graphical user interface that end users will see when having installed Hyper-V. This first virtual machine is installed automatically, by default requires no paid licensing, does not contain Windows branding, and is contrarily named the *physical* machine which together can make it appear that there is no VM at all, but rather a Type 2 hypervisor running on top of a Windows install, but this is not the case. It simply appears so based on tricky naming and the unnecessarily convoluted standard install processes. Doubt not, Hyper-V is a true Type 1 hypervisor running in the most classic way.

Xen

Coming from the same early era as VMware, Xen started with the classic approach, but, unlike Vmware, stuck with it over the years. However, unlike Hyper-V, Xen installations tend to be more manual, and the use of a first virtual machine for the purpose of providing end user interactions is not in any way hidden and, in fact, is completely exposed. What this means is that during the hypervisor installation process a virtual machine is created automatically (so it is always the first one) and that virtual machine is given special access to directly manipulate the console. So, what you see once it turns on is the console of the virtual machine itself as the hypervisor does not have one besides that.

You *can* even choose between different operating systems to use in the management virtual machine! In practice, however, Xen is always used with Linux as its control environment. Other operating systems are mostly theoretical. This exposure makes hidden classic systems, like Hyper-V, all the more confusing because in Xen it is so obvious how it all works.

Because Xen and Linux go together so tightly, it can be valuable for a Linux system administrator to have at least some knowledge of Xen, Xen management via Linux, and Xen architecture. This tight coupling does not ensure that systems and platform teams will become intertwined when using Xen, but it makes the likelihood higher.

KVM

Finally, we come to the **Kernel-based Virtual Machine** (**KVM**). KVM is special for a few reasons. First because it uniquely takes the approach of merging the hypervisor into the operating system itself. And second because it does so with Linux. Unlike other hypervisors where you have a clear separation between the platform administration who manages the hypervisor and the system administrator who manages the operating system level. Here, the two roles must be merged into one because the two components have been merged into one. You cannot separate the operating system and the hypervisor when using KVM. KVM bakes the hypervisor right into the Linux kernel. It is simply part of the kernel itself and always there.

This approach has clear benefits. It simplifies the entire system and provides most of the advantages of each different approach with relatively few caveats. There are caveats, of course, such as that there is a real risk of bloat in the hypervisor install which increases the potential attack surface. KVM's biggest benefit is probably that it leverages the ecosystem of Linux system administrators and existing knowledge so that some of the more complex aspects of managing a system that runs on bare metal such as filesystem and other storage architecture decisions, driver support, and hardware troubleshooting are all shared with Linux giving an enormous base platform and support network from which to begin.

Because of KVM's easy and well-known licensing (due to its inclusion in Linux), well known development potential, and broad availability of components it has become far and away the most popular means of building your own hypervisor platform when vendors want to create something of their own. Many large vendors in the cloud, hypervisor management, or hyperconvergence space have leveraged KVM as the base of their systems with customizations layered on top.

Is virtualization only for consolidation?

Ask most people why you *bother* to virtualize and consistently the same answer is repeated: *Because it allows you to run several disparate workloads on a single physical device.* There is no doubt that consolidation is a massive benefit, when it applies, but stating this as the only, or even the key, benefit means we are missing the big picture.

The enduring myth that we virtualize to save money through consolidation is one that it seems no one is going to be able to dispel. The nature of virtualization is simply too complex for the average person, even the average IT professional, and what it really provides remains broadly misunderstood. The real value of virtualization is in the abstraction layer that it creates which provides a hedge against the unknown - a way to make system deployments more flexible, and more reliable, while incurring less overall effort. Virtualization gives you more options for that unknown event happening sometime in the future that you cannot plan for.

A core challenge for virtualization is that it offers simply too many benefits that do not always relate to one another. Most people want a simple, stable answer and do not want to understand how exactly virtualization works and why adding a layer of additional code actually makes systems simpler and more reliable. It is all a bit too much. The reality is that virtualization has many benefits, each of which is typically enough to justify always using it, and essentially no caveats. Virtualization has no cost, essentially no performance overhead, does not add management complexity (it does, but only in some areas while reducing it in others resulting in an overall reduction.)

Inevitably people will ask if special case workloads exist for which virtualization is not ideal. Of course special cases exist. But the next response from every IT shop on earth is to proclaim that they and they alone are unique in their server needs and that they are the one solitary case where virtualization does not make sense - and then they reliably state a stock and absolutely ideal workload for standard virtualization that applies to nearly everyone and is as far from a special case or an exemption from best practices as can be. Trust me, you are not the exception to this rule. Virtualize every workload, every time. No exceptions.

Most examples that people give of why they avoid virtualization is normally examples of virtualization done wrong. Other, non-virtualization related mistakes, such as selecting a bad vendor, improperly sizing a server, or choosing a large overhead storage layer when something lean is needed could happen with or without virtualization. Everyone from the system administrator to the platform team to the hardware purchasers still have to do the same quality job that they would without virtualization. Virtualization is not a panacea but failing to be the silver bullet that removes the need to do our jobs well in no way excuses not using it every time. That is just bad logic.

Now we should have a good understanding of what virtualization really is, instead of simply a passing knowledge of its utility, and know why we use it for all production workloads. Virtualization should become second nature very quickly whether you are working in your home lab or running a giant production environment. Make it a foregone conclusion that you will virtualize and only worry about little details like storage provisioning and hypervisor selection or management tools. Next we look at virtualizations alternative and close cousin, containerization.

Containerization

Some people consider containers to be a form of virtualization, sometimes called Type-C virtualization or OS-level virtualization. In recent years, containers have taken on a life of their own and very specific container use cases have become such buzz-worthy topics that containers as a general concept have been all but lost. Containers, however, represent an extremely useful form of (or alternative to) traditional virtualization.

Container-based virtualization varies from traditional virtualization in that in traditional virtualization every aspect of system hardware is replicated in software by the hypervisor and exists uniquely to every instance or virtual machine (often called a Virtual Environment (VE) when talking about containers) running on top of it. There is nothing shared between the virtual machines and by definition any operating system that supports the hardware virtualized can run on it exactly as if it was running on bare metal.

Container-based virtualization does not use a hypervisor at all, but rather is built from software that heavily isolates system resources in an operating system allowing individual virtual machines to be installed and operate as if they are fully unique instances, but behind the scenes all virtual machines running as containers share the host's kernel instance. Because of this, the ability to install any arbitrary operating system is limited as only operating systems capable of sharing the same kernel can be installed on a single platform.

Because there is no hypervisor and only a single kernel shared between all systems, there is nearly zero overhead in most container systems making it perfect for many highly demanding tasks. Containers are especially popular in Linux where many container options exist today. The almost total lack of system overhead in container systems used to be a key feature of the approach, but as systems have moved from resource tight to having a surplus of power in many cases, the value of squeezing every last drop out of hardware has begun to wane in comparison to the greater flexibility and isolation of full virtualization. Because of this, what sounds like the greatest virtualization option ever is often overlooked or even forgotten about!

Containers do not require any special hardware support and are implemented completely in software allowing them to exist on a broader variety of platforms (it is easy to implement on old 32bit Intel hardware or a Raspberry Pi, for example) at lower cost. This made them important in the era before hardware acceleration was broadly available for full virtualization technologies.

In the Linux world, which is what we care about, we can run many disparate Linux-based operating systems on a single container host because only the kernel needs to be shared. So, running virtual machines of Ubuntu, Debian, Red Hat Enterprise Linux, SUSE Tumbleweed, and Alpine Linux all on a single container host is no problem at all.

Linux is blessed (and cursed) with a plethora of options for nearly every technology, and containerization is no exception. Multiple open source and commercial container products exist for Linux, but today the undisputed reigning champion is **LinuX Containers** (**LXC**). LXC is unique in that it is fully built into the Linux kernel so utilizing it is simply a matter of doing so, it does not require additional software or kernel modifications. If you are going to be implementing real containers on Linux, chances are it is going to be LXC. LXC is fully supported in nearly all Linux-based operating systems.

A little history of containers

Full virtualization was introduced by IBM in the 1960s but proved to be complex and did not make it into general availability for mainstream servers for decades as high end hardware support and extensive special case software was necessary to make the magic happen until the very end of the 1990s.

Containers were first introduced in System 7 UNIX in 1979 using a mechanism called *chroot jails* which is rudimentary by today's standards, but is functionally pretty close to modern containers. In the UNIX world containers, of one type or another, have almost always been available. In 1999 what we might consider truly modern containers, starting with FreeBSD's Jails, were introduced and rapidly other UNIX platforms like Solaris with Zones and Linux with OpenVZ and then LXC began to emerge. By the mid-2000s containers were everywhere and quite popular before truly effective full virtualization had taken off.

Containers saw a Renaissance of sorts in 2013 with the introduction of Docker. Docker is not exactly a container, however, even though the term container has become more associated with Docker than with actual, true containers. Prior to Docker, containers were never really seen as being very sexy but rather basic process isolation workhorses doing a rudimentary security job for the operating system with the possible short lived exception of Solaris Zones which were heavily promoted for a short time in conjunction with the release of ZFS.

Today, because of the popularity of Docker and its association with containers, the majority of people (even including system administrators!) think of Docker when someone mentions containers rather than true containers which have been around for decades.

We cannot talk about containers without mentioning Docker. Docker, today, is the name most associated with containers, and for good reason. Docker originated as a set of extensions built on top of LXC to provide for extreme process isolation with a *packaged* library environment for said applications. While Docker uses containers, first LXC and now their own container library, it itself is an application isolation environment providing a more limited range of services than something like LXC will provide. With LXC, you deploy an operating system (sans kernel) and treat it all but identically to traditional full virtualization. With Docker you are deploying an application or service, not an operating system. The scope is different and so Docker really finds itself more applicable to *Chapter 5, Patch Management Strategies*. Because Docker is an application layer containerization, it would most often be run on top of a virtual machine in either full virtualization or containerization to provide its underlying operating system component.

Containers, in general, represent a trusted, mature, and ultra-high performance virtualization option (or alternative.) While more limited in their capabilities (a Linux container host can only run Linux VEs, while FreeBSD VEs are not possible, for example) they are otherwise easier to maintain, faster, and generally more stable (there is simply less to go wrong.) They can be created faster, turned on or off faster, patched faster, are more flexible in their resource usage (they don't require the strict CPU and RAM assignments typically needed with full virtualization), need fewer skills, and have less overhead when running. The only significant caveats to containers are the inability to mix operating system workloads, or even kernel versions. If anything, that you do requires a specific kernel version (that is not uniform across the entire platform), or custom compilation of the kernel, a GUI, ISO or similar based full installs then containers simply are not flexible enough for you. But if you are dealing with a pure Linux environment where all workloads are Linux and can share the kernel, which is not that uncommon, then containers can be ideal for you.

Containers, at least thus far, are not able to leverage the graphical interface of the operating system and so are relegated to server duties where pure text based (aka TTY) interfaces are used. Thus, they are not an option for graphical terminal servers or **virtual desktop instance** (**VDI**) deployments. Those kinds of workloads still need full virtualization until someone builds a workaround for that. But as that is generally not a heavily desired workload to support, there is little chance that someone is going to invest heavily in tackling that already easy to solve problem just to be able to say that they did it with containers.

Through the use of containers, you allow the operating system itself to act as a hypervisor, but one that is still the operating system as well and can be managed using all of the normal Linux tools and techniques because the system is still Linux. There is no separate hypervisor to learn or maintain. In this way the ability to leverage existing Linux skills and tools is very good.

It has to be noted that, because KVM and containers both use a standard, bare-metal Linux installation as their base, and because both are baked directly into the stock vanilla Linux kernel, it is not only possible but actually not uncommon for systems to run both full virtualization and containerization on the same host at the same time with vanilla Linux workloads running in containers. This ensures the lower overhead and extra flexibility and non-Linux (primarily Windows) or custom-kernel Linux systems running in full virtual machines on KVM. There is no reason to have to choose only one approach or the other if a blend is better for you. Container technology has become extremely popular and important today, but the use of containers in their traditional sense has dwindled heavily. This is mostly because of deep misunderstandings of the terminology and technology, which has caused it to be ignored even when it is highly appropriate. As a Linux system administrator especially, it may be very beneficial to consider containers instead of traditional virtualization in your environments. Having containers as another tool in your proverbial toolbelt makes you more flexible and effective.

Next, we will apply what we have learned about virtualization and containers, and add management, to learn about cloud computing.

Cloud and VPS

Any discussion of virtualization today is inevitably going to lead us to cloud. Cloud has become, that hot decade-long buzz-worthy concept that everyone wants, most people use, and no one has a clue what it is, what it means, or why anyone uses it. Few technologies are more totally misunderstood, yet widely talked about, than cloud. So we have a lot to cover here, much of it clearly up misconceptions and the misuse of terms.

The bizarre confusion of Cloud

It is a rare combination of being vastly technical and non-applicable to normal business conversations while being constantly discussed as if it were a casual high level non-technical business decision at nearly all levels. Considering only a minuscule fraction of IT professionals have any serious grasp of what cloud is, and even fewer have a clear understanding of when to choose it, that the average non-technical mid-level manager will toss around the term as if they were discussing the price of postage stamps is mind-boggling. What do those people even think that they are discussing? No one truly knows. And I mean that, sincerely.

Ask a group of people who have been throwing about the term cloud. Separate them so that they are not stealing each other's answers. Now, ask them to describe what cloud means to them. Mostly you will get gibberish, obviously. When you drill down, however, you will get a variety of descriptions and answers that are nothing like one another, and yet people listening to others discussing cloud will generally state that they believe that all of those people were meaning the same thing and often that *one thing* is something none of them actually meant, let alone more than one of them. Truly cloud means something different, random, and meaningless to nearly every human being with no rhyme or reason to it.

If cloud as a term had a musical equivalent, it would be Alanis Morisette's Ironic, the song where the only thing ironic about it is the title. The term cloud is the same, everyone uses it, no one knows what it means. Just like ironic.

For quite some years, cloud was commonly used to mean *hosted*. Simply replacing a long established, well known industry term with another one for no good reason. Of course, cloud means nothing of the sort. This horrific misinterpretation led to the meme of *there is no cloud, just someone else's computer*. Of course, even a passing knowledge of cloud would make one cringe to hear someone state something so profound while getting what cloud means so incredibly wrong.

Today, you are more likely to hear cloud used to mean *built with HTML*, I kid you not. Or sometimes it means *platform independent*. Other times it means *subscription pricing*. You name it, and cloud has been used to refer to it. The only thing you can be confident in is that no one, ever, actually means cloud. It could mean almost anything else, but it never means what the term actually is meant to refer to.

The only benefit to the mass hysteria around cloud definitions is that no alternative definition is used commonly enough to rise and overtake true cloud. The problem, however, is so bad that there is no reasonable way for you, as an IT professional, to use the term *cloud* with anyone except a truly well read and trusted technical colleague that you know actually knows what it means.

What is truly amazing is that the use of *cloud* has replaced terms like *synergy* as the default joke in business – that is, as a term only used by those who are truly lost. It is such common knowledge that *cloud* is complex and completely misunderstood that you can never use it to communicate an idea. It has become a standard example of a *marker* in the language for someone who is just spouting off management-speak without having any idea what they are saying and not realizing that everyone else is silently laughing at their ineptitude, and yet you hear it repeated almost constantly! No matter how much everyone knows that they are misusing it, somehow it remains addictive and in constant use.

One of the most important takeaways from this seeming rant (and rant it is) is that you cannot use the term cloud to any but the most elite professionals and you cannot explain cloud to any but well educated IT professionals. Avoid using the term because, no matter how much you think that you can use it in the same mistaken way that everyone else is using it, you cannot. There is no way to use cloud in a way that can be understood because everyone believes that they know what it means, even though they all think that it is something unique.

When you absolutely must refer to cloud for some reason, use more complete terms like *cloud computing* or *cloud architecture* to clarify that you actually mean cloud and not just throwing out a word for the listener to interpret at will.

In many ways, it is almost easier to describe cloud by what it is not, rather than what it is, because everyone thinks that it is one thing or another. Cloud has no association with hosting, none with the web, not even any with the Internet (the thing sometimes referred to as THE cloud, as opposed to A cloud.) We cannot here go into all details explaining every possible aspect of cloud computing, nor would much of it be applicable as the majority of cloud is not related to systems, and therefore not related to Linux. But we should address, to a small degree, what it means to the Linux system administrator, when it is applicable, and so forth.

First, we must start with the **NIST** (the **National Institute of Standards and Technology** in the USA) cloud abstract definition which works from Amazon's original definition. Always keep in mind that cloud computing is a real, strict, technical term created by Amazon for a real-world architecture and therefore has a strict, non-fungible definition and no amount of misuse or misunderstanding or attempts to co-opt its use by others changes that it means an extremely specific thing. It is common for those who do not understand cloud to argue that it is a loose term that can mean what you want it to mean, but it is not. That is simply not the case, it is not a random English language word making its way into the lexicon organically, it was defined carefully within the industry before first use.

The NIST definition is as follows: *Cloud computing is a model for enabling ubiquitous, convenient, on-demand network access to a shared pool of configurable computing resources (e.g., networks, servers, storage, applications, and services) that can be rapidly provisioned and released with minimal management effort or service provider interaction. This cloud model is composed of five essential characteristics, three service models, and four deployment models.*

The most important parts of this definition to us, as system administrators, are the parts that include *pool of resources*, with *rapid provisioning and release*. So shared resources (meaning servers, CPUs, RAM, storage, networking) that we can create and destroy access to quickly. While cloud certainly means more than that, those are the basics. If you immediately thought to yourself *that sounds a lot like what virtualization already does*, you are correct, there is an extreme degree of overlap, and virtualization is the key building block of cloud computing (both full virtualization and/or containers.) If you thought to yourself *wait, pooled resources that I can build AND destroy rapidly - those do not sound like useful characteristics to me or any environment I have worked in previously*, you are also correct. Cloud computing is not logically applicable to traditional workloads or environments, it is designed around extremely specific needs of purpose-built application architectures that few businesses are prepared to leverage on any scale.

What makes the use of *cloud* more confusing is that many vendors (and quite logically at that) use cloud themselves as a component of their products. So, when you ask your vendor if a product is cloud-based, they might be answering if they are providing you cloud itself as a product, or they might be answering if they use cloud in the building of the tool somewhere under the hood. The two are both applicable to how most people ask the question, and since no one knows what cloud really is, no one is sure what you are really asking or want to know. Let me give a contrived, but reasonable, example.

If I were to purchase a legacy application, say a client-server application that uses MS SQL Server and an old Delphi (Objective Pascal) front end and then use a true cloud product to create a virtual machine and deploy the server-side components so that we can truly say that we built the solution on a cloud. Yet the resultant product is not cloud computing, in any sense. Just because one piece of an architecture is built on cloud does not imply that the final product is cloud. Cloud is a layer in the stack.

For us, as system administrators, we are concerned with the type of cloud knows as **Infrastructure as a Service (IaaS)**. That's a fancy way of saying cloud-based virtual machines. Other types of cloud, like **Platform as a Service (PaaS)** and **Software as a Service (SaaS)**, are very important in the cloud space but exist only when the system administrator is *somewhere else*. If we are the system administrator for PaaS or SaaS then, to us, cloud is the workload, and it is not cloud to us. If we are not the system administrator for a PaaS or SaaS system, then it is of no concern to us as system administrators to talk about those systems as they do not apply to our role.

From our systems perspective, cloud is almost like an advanced, flexible virtualization layer. Of course, like virtualization, we may find ourselves tasked with being the ones to implement the cloud platform. That is a completely different animal and worthy of a book (or two) on its own. But in practical terms system administrators may consume platform resources from a hypervisor, a container engine, or either one orchestrated through a cloud interface. To us it is all the same - a mechanism on which we deploy an operating system. So, from a pure system administrator perspective, think of cloud computing no differently than any other virtualization because it is exactly the same. The only difference is how it is managed and handed off to us.

In practical terms, we likely have to be heavily involved in decision making around the recommendation for the use of cloud versus other paths to acquire virtualization. Like any business decision, this simply comes down to evaluating the performance and features offered at a price point and comparing to the performance and features at the same price point with other options. It is that simple. But, with cloud, due to all the reasons that we mentioned before, we are often fighting against a mountain of misinformation and a belief in magic. So, we need to talk a little about cloud in a way that we should not have to simply because these misunderstandings are so common and deeply rooted.

First, there is a belief that cloud computing is cheap. And while in special cases cloud computing will potentially save a lot of money, this is rarely the case. Cloud computing is typically an extreme price premium product chosen because it allows a great degree of flexibility so that less is needed to be purchased overall to meet the same needs. Cloud computing is very expensive to provide and so vendors are forced to charge more than for other architectures in order to provide it to customers (keep in mind that the *vendor* might be in your internal cloud department, nothing about cloud implies an external vendor.)

Horizontally scalable elastic workloads

Attempting to describe what workloads cloud computing was built to address can be a challenge for those not already familiar with certain types of application architecture. We must take an aside and dive into some application concepts here to really understand how we related as the systems team and to see why different approaches play such a large role in our platform decision making at this level.

In a traditional application design the expectation is that we will run the entire application on just one or maybe just a few operating system instances. Typically, one instance would be used as the database server and one as the application server. More roles might exist, and you could have a redundant database server or similar, but essentially the number of instances usable was quite limited and static. Once deployed, the number of instances would not change. In many cases the entire application would exist on a single operating system instance.

Scaling a traditional application mostly focuses on increasing the power of a single operating instance. This might be done through some combination of faster CPUs, more CPUs, more CPU cores, more memory, more storage, or faster storage. Or as we would typically say, if you need your server to do more, you need a bigger server. This kind of performance improvement can go a really long way as top end servers are very powerful, and few companies need to run any workloads that exceed the performance capabilities of a single large server. This style of scaling is called vertical scaling, meaning that we improve the performance of the single thread or server *within the box*. This kind of scaling is by far the easiest to do and works for any kind of application no matter how it is designed (this is how you improve video game performance or any desktop workload).

For most people, workloads designed for vertical scaling are the only types of workloads that they know. Of course, for end users working on desktops, everything is vertical. Even system administrators almost exclusively have to oversee applications that are built for this kind of scaling only. Nearly all deploy in house applications assume that this is how you will scale and only recently do many developers know alternatives well and still many (potentially most) still do not, even though those that do are the more prominent in media.

The alternative approach is to design applications that allow for the application to scale by adding additional, isolated operating instances. For example, running multiple database server instances (likely in a cluster) not just for resilience, but also performance. Running multiple application servers with the ability to simply add more operating system instances running the application while keeping each individual instance small. In a traditional application architecture, we might require a single application server with four high performance CPUs and 1TB of RAM to handle our application workload. A horizontally scalable application might use sixteen smaller servers each with 64GB of RAM and a smaller CPU to handle the same load. Traditionally we would say that our systems *scaled up*, but in adding more instances we say that they *scale out*. Now, of course, you can always scale both *up and out* which would mean increasing the resources of each individual instance while also increasing the number of instances.

As you can imagine, few applications that we work with in the real world as end users or as system administrators are designed for or could leverage horizontal or *scale out* platforms effectively, if at all. It requires that the application architectures, analysts, and developers plan for this style of deployment from the very beginning. And no amount of planning or desire makes every workload capable of scaling in this manner.

Some applications like common web-based business processes, most websites, email systems, and so forth are very conducive to this type of design and you can easily find or make these kinds of applications to take advantage of these resources. Other applications like financial processing or inventory control systems may struggle with the design limitations and take far more work to be able to work in this way, if at all.

Just because a workload is designed by the development team to be horizontally scalable does not mean that the workload itself will, however. This is easiest with a simple example. You create a website that helps people choose a healthy breakfast. You market to the United States. From 6am Eastern until about 2PM eastern (when California is wrapping up breakfast) you are really busy, but outside of those hours your website is really slow. But another website helps people choose food for any meal and is marketed worldwide. This second site never gets quite as busy as the first but stays roughly as busy all day long. The first site can leverage scalability, the second site cannot because its resource needed never really change.

The key advantages to horizontally scalable workloads are that they can be grown rapidly. Adding an additional operating system instance (or an additional one hundred!) is easy and non-disruptive. Adding more CPU or RAM to your existing server, is hard and slow by comparison. The next step of horizontal scaling is making it elastic. To be elastic, your system does not only have to scale out quickly but allow you to also scale back in quickly: that is to spin down and destroy unneeded operating system instances when capacity has changed. This is the unique proposition of cloud computing, to provide capacity on demand for elastic, horizontally scalable workloads so that you can use resources only when needed and stop using them when you do not.

Vertically scaled resources are far less expensive to provide than horizontally scaled ones. You can test this by trying to assemble several computers with roughly the same specifications. With only rare exceptions, it is a fraction of the cost to build a single large server than several smaller ones using real-world economics. A single system requires only a single operating system and application instance, but multiple systems require the overhead of the same operating system and applications loaded into memory in each case wasted many resources there, as well. It just takes less management power to oversee something that is *faster*, rather than many slower tasks. It is not unlike managing humans. It takes less overhead to manage one really fast, efficient employee than it does to manage several slower employees trying to coordinate doing the same work that the faster one was doing.

So a horizontally scaled system, in order to make sense to choose, has to be able to leverage both scaling out, as well as in, have a workload use case that actually leverages this, and that does so to an extent large enough to overcome the lower cost, lower overhead, and great simplicity of traditional designs. Unless your workload meets all of those requirements, cloud computing should not be a consideration for you at all. It simply does not apply. And while contrary to how the media and trend-happy IT professionals want to portray cloud, it is only a small percentage of workloads that can effectively leverage cloud and only a small percentage of businesses have those rare workloads at all.

Of course, we are talking IaaS aspect of cloud. Other cloud aspects where only the application portion is exposed to the business will often be cloud based. But this is essentially unrelated and certainly a decision process totally different from anything we would be looking at in a tomb such as this.

Second, there is a belief that cloud computing is reliable. Absolutely nothing in cloud definitions or designs implies reliability in any way. In fact, this runs completely contrary to all standard cloud thinking. Cloud computing, because it is useful exclusively with scale out design, is built on the assumption that any redundancy or reliability is built into the application itself as scale out implies - as you have to have this inside the application itself in order for scale out to work properly. So including any redundancy at the system or platform level would be nonsensical and counterproductive. A basic understanding of cloud computing should, with any thought, make us surprised if anyone expected redundancy beyond the minimum at this level. In the real world, cloud computing resources tend to be far more fragile than traditional server resources for exactly this reason. Cloud computing assumes that either reliability is of trivial importance or that it is provided elsewhere in the stack. Cloud computing is just a building block of the resulting system, it is not in any way a complete solution by itself. Of course, theoretically, a high availability cloud provider could arise, but their cost and performance caveats would make it hard to compete in a marketplace driven almost entirely by price.

Third, there is a belief that cloud computing is broadly applicable, that every company should be using it, and that it is replacing all other architectures. This is not true at all. Cloud has been around now for more than fifteen years (at the time of writing) and it made its inroads quite quickly towards the beginning of that cycle. Today, cloud computing is mature and well known. Companies and workloads that are going to move to cloud (or design for it) have largely already done so, and new workloads are created on cloud at a roughly constant rate. The industry saturation rate for cloud computing has been more or less achieved. Some new workloads will go there as older ones or anti-cloud holdouts retire or give in to other pressures. Some will come back as overzealous cloud fanboys and buzz-word driven managers learn their lessons of having gone to cloud without any understanding or planning. Sending standard workloads to cloud computing without redesign is typically costly and risky. But by and large cloud computing has already settled into a known saturation rate and the computing world is as it will be until another exciting paradigm shift occurs. Basically, what we see today in advanced bespoke internal software and grand scale multi-customer software is ideal for the cloud paradigm, and traditional workloads for single customers remain the most beneficial on traditional paradigms. This is all as originally predicted when cloud computing was first announced long ago.

Using cloud does not require any specific skills or training, as many in the industry would like us to believe in order to sell certifications and training classes. In fact, just knowing what cloud truly is often enough to enable you to utilize it effectively. That said, individual cloud vendor platforms (such as **Amazon**'s **AWS** or **Microsoft**'s **Azure**) are so large and convoluted that there can be real value to getting vendor certifications and training to understand how to work with their product interfaces. But to be clear, the value in the training is learning how to work with the vendor in question, not in learning about cloud.

That does not change the fact that most organizations seeking to get significant value out of cloud computing will most likely need to do so with deep vendor integrations that will almost certainly require an investment in specific vendor product knowledge.

Cloud is an amazing set of technologies that serves an incredibly important purpose. When your workload is right for cloud computing, nothing else can come close to it. Whether you build your own private cloud or use a public shared one, whether you host your cloud in house or let a hosting firm handle the data center components for you, cloud might be the right technology for some of your workloads. With your understanding here you should be able to evaluate your needs to know if cloud is likely to play any reasonable role, and be able to look at real work vendors and costs and make solid, math-based valuations of cloud in comparison to other options.

Now that we know cloud computing, we can step back and look at the older concept of virtual private servers and see why they are so closely tied with, but not actually related to, cloud computing today.

Virtual Private Servers (VPS)

Similar to and, on the Venn Diagram of things, nearly overlapping with IaaS cloud computing is the modern concept of virtual private servers or VPS. VPS actually predates cloud and comes from simpler virtualization (or containerization) allowing a vendor (which could be an internal department, of course) to carve out single virtual machines for customers from a larger, shared environment. Instead of needing to provide an entire server of their own, customers need only buy a small slice or set of slices of the vendor's server(s) to use for their needs.

As I mentioned, this sounds very similar to IaaS cloud that we just described, and certainly it is. So much so, that most people using IaaS cloud actually use a VPS aspect of it without realizing so. The idea behind VPS is to allow companies to purchase server-class resources at a fraction of the scale typically required to a single physical server. If you think back to our discussion on virtualization and how by using a hypervisor we might be able to take a single physical server and, for example, create one hundred virtual machines that run on top of it, each with their own operating system, then we could sell those resources to one hundred separate customers, each of which could run their own small server inside its own secure space. This allows small companies, or companies with small needs, to buy enterprise level datacenter and server hardware capacity and prices within any realistic budget.

Before we go further, we need to do a quick breakdown and comparison of VPS against IaaS cloud to see why VPS is so commonly confused with cloud computing and why they often compete:

- **First, the goals**: An IaaS Cloud's goal is to provide rapid creation and destruction of resources on demand via automation - primarily used by the largest organization or workloads. The goal behind VPS is to carve up traditional server resources in such a way that they can be affordable to be used by small organizations and/or workloads. So, one goes after the biggest scale and the most complexity, the other after the smallest scale and least complexity. One expects custom engineering effort on both the application and infrastructure team's sides where the other expects traditional applications and no special knowledge or accommodation from any team.

- **Second, the interface**: In cloud computing the expectation and purpose is for systems to be self-provisioning (and self-destroying.) Cloud is not designed for humans to have to interact manually to request resources, nor to configure them, nor to decide when more (or less) are needed, nor to destroy them when done. So, cloud's focus is on APIs to allow software to handle provisioning. VPS is meant to work just like any normal virtual machine with a human initiating the build, installing the operating system, maybe configuring the operating system, and turning the VM off, and then deleting it when no longer needed. It is standard for cloud products to not offer any interaction directly with the virtualized hardware such as access to a console so any system requiring console level interaction (such as a GUI) are impossible. To qualify as a VPS console and GUI access are required to completely mimic a hardware device in a standard way. If you can use a normal server, you can use a VPS.

- **Third, the provisioning**: Cloud assumes a need for rapid provisioning. Of course, rapid is a relative term. But it is just expected that in a cloud ecosystem that systems must be able to go from first request to fully functional in minutes, and sometimes seconds. In the VPS world, while we always want everything available as quickly as possible, having to wait a few minutes before being given the access to begin a manual operating system install that could take potentially tens of minutes is common. We assume that a cloud instance will be created via software, but a VPS instance we assume will be created manually by a human.

- **Fourth, the billing**: Because the value of cloud computing is assumed to come from its ability to be created and destroyed rapidly in order to keep costs managed it follows that billing must be granular to accomplish this. To this end billing it generally handled in increments of minutes or possibly hours, or in other extremely short measurements like processor cycles. VPS will sometimes charge in these short increments but may easily use longer intervals such as daily or monthly as it is not a rapid create and destroy intended service. (We can say that cloud leans towards stateless and VPS leans towards stateful.)

What often makes VPS and IaaS Cloud harder to distinguish is that today, VPS providers almost always use IaaS Cloud as their own mechanism for provisioning the VPS under the hood, and most IaaS Cloud providers have opted to offer VPS additionally. This was bound to happen for two reasons. First, VPS providers use cloud because it is a really logical way to build a VPS (if you think about the requirements that the VPS *vendor* would have, they would sound a lot like what cloud is meant to do) and because by being built on top of cloud computing, you can advertise the VPS as being cloud in a sense and be a sort of *simple interface to cloud resources*. It makes sense for cloud providers to offer VPS because the majority of customers who look for cloud have no idea what it is and only use it for political, not business or technical reasons, so offering something simple that allows them to purchase from you and use your resources (because cloud is so hard and complex) allows you to capture the majority of revenue.

Vendors like Amazon used to offer no VPS services and using their resources if you were not truly in need of cloud was difficult, at best. To address this, Amazon added LightSail as a VPS product layered on top of their cloud product.

Other cloud providers, such as *Digital Ocean, Linode*, and *Vultr* use VPS as their primary product offering and focus on it almost entirely while keeping their cloud interface quietly to the side so that customers truly looking for cloud can find it, but those seeking cloud when they intended to use VPS will be able to get what they need right away.

VPS is one of the most popular and effective ways for real world companies to run workloads, especially smaller companies, but companies of any size can leverage them. Cloud is effective but primarily for special case workloads. The majority of companies talking about already leveraging cloud are actually using VPS and not even aware that they missed cloud computing entirely.

It is worth noting that when we talk about rates of cloud adoption we have a fundamental problem: no one knows what cloud is, including people who think that they are or are not using it currently! Vendors like Amazon can tell you how many customers that they have, but they cannot tell you if their customers are using their products as cloud or just using cloud in some other way. In a survey about cloud adoption you have zero reasonable chance that the person being asked about their adoption, the person doing the asking, and the person reading about the adoption rates all understand enough about cloud to answer or ask meaningfully and, in reality, generally none of them know at all what is being asked. So any information about cloud adoption rates border on being totally meaningless. There is no honest mechanism by which any person or organization could possibly know what the cloud ecosystem really looks like. You would get just as meaningful data if a group of squirrels surveyed a bunch of hamsters about astrophysics and then handed the results to a bunch of hyper puppies to interpret. People love reports and data and rarely care if the survey in question was real in any way.

You should, at this point, feel both overwhelmed and depressed about the state of cloud understanding within both IT and business, but you as the Linux system administrator should now be prepared to explain it, evaluate it, understand what is built upon it, and know when and how to choose to use it for your own workloads and when to look at VPS instead.

On premises, hosted, and hybrid hosting

Now that we have talked about so many aspects of the underlying components that are used to provide us with a platform on which to deploy an operating system, we can finally talk about where those systems should exist!

This is, at least, the simplest of all our topics. Physical location is easy to explain, even if many businesses get confused about it in practice. Conceptually we really only think about two locations for a workload and that is as being either on premises or off premises. This can be a little convoluted, though, as companies own multiple locations so what is off premises to one site might be see as on premises to another. But we generally consider on premises to be all of a company's owned sites and off premises being any sites that are operated by a third party. Because of this we generally refer to off premises physicality as being hosted, as physical systems are being hosted on our behalf. However, there are reasons why this can prove to be very misleading.

Most people assume that when a system is kept on premises that that also implies that it will be being operated by an internal team. This is most often true, but having on premises systems managed by third party teams is not unheard of, especially in very high performance or high security environments. For example, if you required Amazon's specific range of cloud computing products, but could not allows for any off-premises hosting, you can have Amazon operate an AWS cloud instance on your own premises. This is anything but low cost or simple and requires housing at minimum a small, self-contained data center and all of its associated staff!

Hosting gets more complicated in practice, but the core issue remains the same: the demarcation points. When we decide to start having our systems be hosted off premises the questions rapidly become about defining what portions of the hosting will be provided by the hosting provider and which by us.

In its most extreme (and impractical sense) you could rent a house, office, or storage unit and provide your own rack, servers, Internet, cooling, power, and so forth as needed, but if we did this one would rightfully argue that we had basically made the site our own premises. Touche.

Classically it was assumed that nearly all workloads should be run on premises. This was for the very simple reason that early business networks had no Internet connectivity so hosting elsewhere was effectively impossible or at least impractical. Follow that in the early days of the Internet wide area network links were slow and unreliable keeping remote servers almost unusable. And software was built around LAN networking characteristics, unlike today when enterprise software of any quality assumes that it needs to perform adequately over a long distance connection, most likely on the Internet.

Because of these old assumptions, the tribal knowledge that servers need to be local to an office where people work has been passed down by rote generation to generation without many people evaluating it. This information went from generally true to seldom true pretty quickly during the early 2000s.

Today most workloads work effectively over the Internet and so can be located almost anywhere. Using some form of off-premises hosting or centralized hosting that is not based at any specific company location is now the norm rather than the exception.

In all on-premises and off-premises evaluations we have certain factors that are universal: who will access the data and from where, how does latency and bandwidth impact application performance, which people accessing the data have priority and what is the cost of performance issues at different locations.

There is no hard and fast rules, we simply have to carefully consider as many factors as possible. On and Off Premises solutions are just locations and should be treated as such. The ability to use an enterprise datacenter off premises might be significant, especially if we do not have a real server room on premises. And disaster recovery might be better at an off premises location. But will the user experience be good enough if the server is far away? These questions are all situational and need to be answered not just about the status of the business infrastructure today but also for the near future.

Colocation

When a site provides the real estate, cooling, power, Internet, racking, networking, and so on but we provide our own physical servers it is called a colocation facility. Colocation is one of the most popular and effective ways that we, as system administrators, can acquire enterprise class datacenter services outside of our own premises while retaining the flexibility to use any hardware that makes sense for our organization and its workloads. Colocation is effective for very small businesses up to the most absolutely massive. No company outgrows the value that colocation may bring, nor does any government. It is a strategy that lacks a *top end* size.

Colocation is one of the most useful and effective forms of moving IT equipment off premises because it allows the IT department to retain full control of hardware purchasing and configuration not just for systems but for networking and appliances as well. Only non-IT functions necessary to support the technology hardware is provided. This allows the colocation facility to focus on a strong facilities management skillset and IT to retain literally all IT functions and flexibility: basically doing remotely with a third party what you would hope you would be doing internally with your own teams assuming that you had enough volume to do so. It is expected that a colocation provider will also have *remote hands* to assist with bench tasks when necessary, such as changing or verifying cabling, power cycling devices, and things of that nature.

Flexibility is key with colocation. Whether it is because you want to custom build your own hardware, maintain legacy systems, use special case hardware (for example, IBM Power hardware), or what the freedom to do a lift and shift of an entire existing environment from on premises to the colocation facility you can do it all. Most colocation facilities allow for a range of scales as well, from housing a single 1U server on your behalf to fraction racks (tenth, quarter, half, full rack sizes are common) to providing cages that can house many racks to even renting entire floors of the datacenter!

The biggest challenges that colocation faces is that there is no effective way to go extremely small because the smallest size you can reasonable host is a single server. If your needs are smaller than this, then colocation will generally struggle to be cost effective for you. But do not simply reject colocation with an assumption of it being expensive. I run these calculations often for companies who were ignoring colocation as too costly for them only to find out that it would be less than half of the cost of their alternative propositions while having more flexibility for growth without additional expenditures. Most people assume that servers are more expensive to buy than they really are and that colocation costs are higher than they really are. Colocation costs are often inappropriately associated with legacy systems, as if only twenty year old equipment can go into a datacenter today, and decades old cost models are often envisioned. Twenty years ago servers were much more expensive than they are today and had noticeably shorter operational lives and datacenter space was more costly than today as well. Like everything in IT, cost over time have come down and for workloads of any size colocation tends to be much less costly than most alternatives.

Colocation is just one approach to hosting systems off premises. Other approaches like public, hosted cloud and cloud-based VPS systems are standard alternatives.

The biggest challenges around locality are really all associated with understanding current marketing pricing for different approaches and being able to evaluate the benefits and caveats of hosting equipment on premises or off premises, and if that equipment should be dedicated or shared. You should now be ready to make that evaluation and choose appropriately for your deployments. Next, we dig into the much more complicated topic of platform level system design architectures.

System Design Architecture

One of the more challenging aspects of system administration is tackling the broad concept of system architecture. In some cases, we have it easy, our budget is so low or our needs so simplistic that we simply do not need to consider any but the most basic options. But for many systems, we have broader needs and a great number of factors to consider making system architecture potentially challenging in many ways.

We now understand platform concepts, locality, and the range of services normally associated with providing an environment onto which we can install an operating system. Now we have to begin describing how we can combine these concepts into real world, usable designs. Most of system design is just common sense and practicality. Remember nothing should feel like magic or a black box and if we get services from a vendor, they are using the same potential range of technology and options that we are.

We are going to talk about risk and availability in the next section, but before we do, we should mention here to make it more clear why system designs rely on this data, that any redundancy (whether for performance or risk reduction) that we add to our overall system can be done at different layers. Principally in the system layer (where we are looking now) or at the application layer (which we do not control.) So even in the most demanding of high availability workload situations, we may have no need for a robust underlying system design and have to consider this when thinking about design options.

I am going to break down common design approaches so that we can understand how they best apply to different scenarios. These are physical system architectures that include both the storage and the compute. It is assumed that some sort of abstraction, meaning virtualization and/or containerization, is used in every case and so will not be mentioned case by case.

Standalone server, aka the snowflake

You really cannot get more basic than this design. The simple server, the baseline against which all else must be measured. The self-contained, all in one server with storage and compute in a single chassis. No external dependencies, no clustering, no redundancy (external to the single box.) Of course, we assume standard hardware practices are followed such as minimums like RAID and dual, hot swappable power supplies.

Today, many IT people are going to frown on attempting to use a stand alone server, but they should not. The classic single server is a powerful, effective design appropriate for the majority of workloads. This should be the *go to* design, the default starting point, unless you have a specific need to do something else.

Because of its simplicity, single server designs have the best cost ratios to all other factors, are more robust than they appear, and have excellent performance. Many people think of servers as being rather fragile creatures, and years ago, they were; but that impression stems from servers of the 1980s and early 1990s. By the late 1990s server technology was becoming mature and reliable and today failure rates on well maintained servers are extremely low. The idea that a single server is a high risk is an antiquated one, but like many things in our industry old feelings often linger and get taught mentor to student without reevaluation to see if factors remain true (and in many cases without initial evaluation to know if they were ever true.)

Single servers benefit from having many fewer components and lowered complexity compared to any other approach and with fewer parts to fail and fewer configurations to get wrong it is that much easier to make really reliable: hence why we sometimes refer to this as the *brick* approach. Bricks are simple but effective, while they can fail, they rarely do. Emotionally it is common to associate complexity with robustness, but in practice simplicity is far more desirable. Complexity is its own enemy and an unnecessarily complex system takes on unnecessary risk (and cost.)

While hard to measure for many, many reasons, we generally assume that a properly maintained and supported stand alone server can delivery average availability rates close to five nines (that is around one hour of downtime per year.) It is a rare workload in any business that cannot function well with that kind of downtime. What is difficult in stand alone servers is that this is an average only (of course) and we will have isolated systems experiencing much higher downtime, and others experiencing none.

Simplicity also brings us performance. By having fewer components in the path single servers have excellent performance. Attempting to gain total performance greater than what can be achieved using a single server is difficult. Single servers give us the lowest latency and nearly the best throughput of any approach.

When it comes to single server systems, use math and logic to explain why it may or may not make sense. Many people rely on emotions when it comes to system architecture and this should never happen. Our concerns with system design are about performance and availability and these are purely mathematical components. Emotions have no role here and are, in fact, our enemy (as they are the enemy of any business process.)

Single servers can scale far larger than most people assume. I often hear arguments that they cannot look at single servers because their needs are *so large*, but then deploy systems only a tiny fraction of the standard scale, let alone the maximum scale, that a single server can achieve. Remember that vertical scaling is highly effective compared to horizontal, and generally cost effective as well. The biggest single server systems can support hundreds of the most powerful CPUs, and many terabytes (or more) of RAM. The challenge is not finding a single server that is large enough for a task, but rather finding any workload that can leverage so much power usefully in a single location!

A big advantage to standalone servers is that each physical device can be scaled and custom designed to address the needs of its workload(s). So different servers can use different CPU to RAM to storage ratios, different servers can use different CPU generations or architectures, one system might use large hard drives while another uses small but screaming fast solid-state storage. Tuning is very easy.

Simple does not necessarily mean simple

Having a standalone server does not mean that we give up all of the options and flexibility that we might believe that we need from more complex designs. It simply means that we have to think about them differently. Many of the concerns that one may have about standalone servers likely stems from a pre-virtualization world with relatively slow networks. Today we have virtualization, fast storage, and fast networks and these can move the goal line by a bit.

We refer to servers as being standalone in reference to their architecture, everything is self contained in a single piece of hardware. This does not mean that we do not have (or cannot have) more than one server. On the contrary, giant Fortune 100 firms will often have thousands of standalone servers. What makes then standalone is that they have no dependencies on each other. The complete failure (or theft) of one does not negatively impact another.

A tiny organization might choose to rely on a single standalone server for their entire business and depend completely on backups and the ability to restore to replacement hardware should disaster strike. This is a totally valid approach and quite common.

If your organization is larger, or workloads require more immediate protection against loss of availability, then it is standard to run multiple standard servers. This spreads load between physical hardware devices and, because of virtualization, provides an opportunity for there to be natural ways of mitigating hardware failure by rapidly rebuilding lost workloads on other hardware. If deployment density is too high, spare hardware is an option as well. With modern storage, networks, system management, and backup techniques restoring many workloads can be done in as little as minutes allowing even complete hardware failures to often carry only the tiniest of system impacts. In fact, keeping backups stored on other standalone nodes can allow for essentially instant recovery of lost systems while maintaining strong decoupling.

Standalone servers also do not imply that there is no form of unified management. Tools like ProxMox or VMware vSphere allow a consolidation of management while keeping system hardware independent. Modern tooling has made managing sprawling fleets of standalone servers very simple indeed.

Most every aspect of a stand-alone server can be improved by adding more to it and making it more complex, the two things that it always leads on are cost and simplicity. No other approach will reliably be able to keep our costs or our simplicity as low and in business, these are generally the factors that matter most.

Many to many servers and storage

As companies grow there can be an opportunity to consolidate different aspects of the architecture in order to save money. Separating networking and storage is the most common approach to this. Creating a layer of compute nodes, and a layer of storage nodes allows for a lot of flexibility. The primary benefit is allowing for the easy movement of resources and better system utilization.

For example, an organization maybe need fifteen physical compute nodes (traditional servers) but only half a dozen storage nodes (SAN or NAS) to support them. Each individual system can be easily custom scaled and does not need to match other systems in the pool. In this way this approach is not so different from the standalone server approach.

It should be noted that when doing this, the storage layer is the greater risk, compared to the compute layer, for two key reasons. First, it is stateful where compute is stateless, which means that here we not only have to protect the availability (uptime) of the system, but this is also where we have to protect the data as it is stored so we have the risk of data loss as well - there is simply more to lose here. Second, storage is more complex than compute and equivalent hardware and software at both layers means that the storage layer is just more likely to fail due to complexity. This is all risk that also exists in the standalone server, but when combined into a single chassis it can be more difficult to understand where the risk is occurring, even if we know what the overall resultant risk is.

In its simplest incarnation, we would have a single compute server node and a single storage node (typically a SAN array) and would connect them directly via a straight cable (Ethernet, eSATA, FC, and so on.) This is really more of a hypothetical scenario as it is so obviously bloated and illogical without any scale, but we can learn from the example to see how we take the single standalone server design and, without any benefits of scale, simply double the chassis to manage and increase the physical, as well as, logical complexity of the system design.

Typically, this type of design is leveraged most to consolidate heavily on storage, pushing as much storage into a single node as possible, while having many small to medium sized (one to two CPU) servers that allow workloads to move between them in order to best balance said workloads. This approach is flexible and generally cost-effective, and makes large scalability quite simple.

Moving past standalone servers means we start to inject dependencies that need to be discussed. At a minimum, when we move to a multi-nodal system, we have the complexities of the interconnections (which might be as simple as just a cable, or more complex like going through a switching fabric of some sort), any complexities that come from configuring the nodes to speak to each other, and the risks of the extra components that might fail.

This kind of design really does nothing to address risk, and actually is far riskier than standard standalone servers. This is why it is important to use standalone servers as a baseline and discover risk variation from that point. In a *network* system design, there is no redundancy, so each workload has a full dependency on both its associated compute node and its associated storage node(s). This risk may be uniform across a compute node or each workload located there might have unique storage configurations so that risks may vary widely between different workloads on a single server. This is where risk becomes much more complicated to measure because we have to deal with the cumulative risks of the compute node, the storage node, the connection between the two, and the configuration! Each individual piece is extremely difficult to measure on its own – putting them together, we mostly have to look in relative terms only and understand that it is much riskier than a standalone server.

Viewing the world as a workload

System architecture is a *by the workload* task and there is no specific necessity for all workloads in your organization, or even all workloads running on a single compute node, to share architectures. Mixing and matching is totally doable and somewhat common. Each workload should be being evaluated as to its own needs, and then the overall architecture evaluated.

Often overlooked is the ability to use a complex and less reliable (but potentially less expensive at scale) option like network design for workloads that are less important, while on the same compute nodes also having local storage that is extremely fast and/or reliable for more critical workloads. Mixing and matching can be a strong strategy in a large environment where storage consolidation is considered necessary without endangering an isolated number of highly critical services in order to do so.

In the same vein, each workload can have its own backup, replication, failover and other risk mitigation strategies for deal with disaster. Sharing a compute node generally dictates very little as to how reliability and availability from workload to workload must be handled. Of course typically all workloads are treated the same either out of a desire for standardization and simplicity, but also regularly out of a misunderstanding of the range of customization available for each individual workload. It is often assumed that choosing a system design is an all or nothing endeavour, but this is not the case.

The main challenges of network system design is that any efficiency gained has to offset the additional cost created by needed more overall nodes (separating compute and storage means that additional hardware chassis and operating systems are necessary for the same tasks) and at any scale additional networking equipment is needed to handle the interconnects. Networking equipment can be as simple as a single Ethernet switch or as complex as clusters of Fiber Channel or Infiniband switches. Switches represent no only additional cost to purchase, but also additional points of failure both for hardware and, to a much lesser extent, configuration. Often redundant switches are purchased reducing hardware risk but increasing cost and configuration complexity. Even in extremely large environments this represents additional cost and risk that is very hard to overcome.

The Inverted Pyramid of Doom: Clustered Compute with Risky Storage, aka the 3-2-1

Sadly the Inverted Pyramid of Doom (aka 3-2-1 or IPOD) has traditionally, for the majority of the 2000s and 2010s, been the most commonly deployed architecture in small and medium business and is also the prime example of the absolutely worst possible design decision for normal workload needs. It is also the design that maximizes profits for vendors and resellers, so it is what everyone wants you to buy.

The IPOD design is differentiated from the network system design above in that the compute layer is clustered for high availability, but the storage layer is not. As we discussed in the last design storage is both more important to protect and more likely to fail. Typically the networking layer (the layer providing connectivity between compute and storage) is also clustered for high availability. This nodal count by layer creates the naming conventions used: 3-2-1 refers to the design having three (or more) compute nodes, connected to two redundant switches, all relying on a single storage device which is most typically a SAN.

When viewed in an architectural drawing, the IPOD is a pyramid with the wide portion on top and everything balanced on the point. Hence the term *inverted pyramid*, this design is designed to be as costly and risky as possible, hence the moniker *of doom*.

Top-down redundancy

Why is a design so obviously impractical as the IPOD so traditionally popular? The answer requires us to understand several factors. First, redundancy, risk, and system design are all areas that most businesses, and even most IT departments within those businesses, have received no training and are generally completely unaware and so represent an easy target for vendors to be manipulative.

The real trick comes from two things: linguistics and the simplification enabled by top down viewing. The linguistic trick happens because the term *redundancy* does not mean what most people believe that it means and this system *has redundancy*, but in an all but meaningless way. So, when a customer says, *I need redundancy* they actually mean *I need high availability*, but this allows the vendor to state that there is redundancy and ignore actual needs. Semantics are super important in all business, and IT more than most.

The top-down aspect of the system comes from how we view the architecture. As IT professionals, we know that we should view our architecture *from the side*, that is seeing the reliability as it stands layer by layer, knowing that compute is built on top of the network, and the network on top of the storage. But a vendor wanting to steer a customer to believe that there is strong redundancy will demonstrate the system *top down* showing a view that only sees the compute layer where there is redundancy. The other layers are overly complex and tend to be happily ignored by all parties as being *black boxes that do magic*. Ignoring the hard parts and just focusing on the trivial, easy part where redundancy is least important makes it really easy to mislead a customer.

Of course, if we really stop and think about it, what matters is the overall reliability of the entire system. Getting distracted by any single layer will simply lead us astray. We need to understand all of the layers, and how they interact with each other, in order to determine overall reliability, but there is a strong emotional drive to see one layer as being extremely reliable (as the compute layer here often is) and then feeling that the overall system must therefore be extremely reliable. But this is anything but true. The overall reliability of the system is driven primarily by the most fragile layer, not the most reliable. The system risk is, if you recall from earlier, cumulative. You combine all of the risks together because each layer depends one hundred percent on every other layer, if any layer fails everything fails. You can demonstrate this easily with a thought experiment... if one layer has a 100% chance of failure, and all other layers have a 0% chance of failure, the system will still fail 100% of the time. The impossibly reliable layers do literally nothing to offset the unreliable layer.

Redundancy itself is a dangerous word to use. In general English usage, the word redundant simply means that you have multiple of something when fewer are needed. This can mean that one is a replacement or backup should the other fail, but that is not implied and often the term is used to mean something else. In RAID, for example, RAID 0 has multiple disks (redundant) but the more redundancy the higher the risk, not the lower. RAID 1 is the opposite. Redundancy is polar opposite there, even within a single context. This really shows the importance of semantics in IT (and business, or really, life in general.) People often use redundancy as a proxy work for reliability, but the two mean very different things. Use the term you mean and you will get far better information.

The biggest problem with the IPOD design is one of practicality. If we were to look at it purely from a reliability standpoint we could state that it is safer than the network system design because at least some of the layers contain high availability measures, even if not all of them do. And this is totally correct, but tends to be misleading. Network system design is meant to trade high risk for cost savings versus the simple stand alone server design, using the idea of *safer than* something that is not even designed to be safe is not exactly wrong, but talking about it in that context is done to evoke an emotional response - to make the IPOD feel safe, which is not the same as *safer*. If we compare the reliability of an IPOD to the stand alone server, it feels quite unsafe and remember, we stated at the beginning, the low cost, simple, stand alone server is our baseline for comparisons. The problem with the IPOD is that the risk is extremely high, approaching the risk of the network system design, while its costs are much higher than the network system design and generally much higher than the stand alone server design all while having more complexity and effort for the IT team. It is the consistent combination of high risk and high cost that makes it problematic and generally accepted as the worst design to encounter in the real world.

Outside of production environments, the IPOD is often ideal for large lab environments where capacity matters most and reliability does not matter at all. The ability to flexibly scale compute with a single consolidated, low cost, highly unreliable storage layer can make sense to make large scale labs more affordable.

Layered high availability

The logical system design derived from what we have already seen is to take the separate layers of the network system design, and the high availability clustering from the compute layer of the Inverted Pyramid of Doom and apply it to all layers giving us a high availability storage layer, a high availability networking layer, and a high availability compute layer. In this way we can have large scale compute, storage, and networking without any individual layer being a high level of concern.

where each layer still depends on every other layer, that three highly available layers must still be evaluated with the risk of each layer added together. So, while we can almost certainly make any individual layer more, or even far more, reliable than a single standalone server would be when we accumulate the risk of each layer, and then add in the risk of the additional complexities from incorporating the layers together, it may or may not remain more reliable than the standalone server would be.

Reliability is relative

When discussing reliability and these different architectures we have to remember to think in terms of apples to apples, not apples to oranges. When we say that a single server is a certain level of reliability, and that servers clustered with standard high availability technologies have a certain relatively higher reliability, we are assuming that all of those servers are roughly identical in their individual reliability. In most situations this is true. Whether we are talking compute nodes, networking hardware, or storage nodes, for roughly the same price range we get similar quality hardware and software with roughly similar failure characteristics. So, these different devices *of the same quality* are all about the same level of reliability with networking hardware being the most reliable (least complex) and storage nodes being the least reliable (because they are the most complex.)

However we can manipulate this dramatically. A five thousand dollar server will generally be much less reliable (and performant) than a five hundred thousand dollar server. Yet each is an individual, stand alone server. So clearly we have to think in terms both of architectural reliability (the reliability of the system design that we make) and in terms of the individual components.

A common problem found here is that *you get what you pay for* applies not at all and you can easily find extremely expensive single-chassis systems for both compute and storage nodes that are not highly available at all and may not even be as reliable as average devices! As reliability is hard to measure and even harder to prove, vendors have little incentive to tell us the truth. Vendors are highly incentivized to tell us whatever is likely to make us spend more money with them whether it is making us feel that traditional servers are more fragile than they really are, or by making wild high availability claims for devices that are essentially built from straw (and by pigs.)

So we must be careful that we consider all of the factors. And we must understand that the ability to protect a single chassis (vertical reliability) compared to multiple chassis (horizontal reliability) is different. Single chassis reliability tends to be incredibly powerful for certain components (such as redundant power supplies, high quality components, and mirrored RAID for storage) but tends to be complex and problematic for others (CPU, RAM, Motherboards.) And single chassis systems, while easier to operate, cannot address some key concerns like physical damage (water, fire, forklift) in the same way that multiple chassis can.

We must also be keenly aware that marketers and sales people often use confusion around reliability as a sales tactic and will push concepts such as *dual controller* systems as being essentially impossible to fail but without science or math to back it up. Dual controller systems are simply horizontally scaled systems inside a single chassis with all of the complexity of the former and the lack of physical protection of the later. And any product sold based on being misleading is that much more likely to be poorly made as it means that the vendor is unlikely to be being held accountable to quality design.

It has become known, especially in the early 2010s, that server vendors were regularly pushing products branded as high availability or *cannot fail* that did not even begin to approach the baseline reliability of traditional servers. Since customers could not verify this for themselves, they often just take the vendor's word for it and if the businesses loses money, finger pointing is the natural recourse.

This approach necessarily is the most expensive design we can reasonably assemble because we need multiple devices at each layer, as well as technology to create the clustering at each layer. This is best for very large systems where each layer is able to scale so large that cost benefits of scale come in at every point.

It is worth noting that essentially all cloud based systems run on this architecture due to their enormous scale. Certainly not all as cloud can run using any architecture, but this is far and away the most likely to be used in a large, public cloud implementation and is most generally what would be found even in a moderate scale private implementation. Many clouds do, however, run on stand alone servers even at massive scale.

Hyperconvergence

The last architectural type that we will look at is hyperconvergence and we have now come full circle from increasingly complex designs to one of the least. Hyperconvergence as an architecture is anything but new, but for the last few decades it has been almost completely ignored before having a sort of renaissance in the mid-2010s and is now, along with stand alone servers, the bulwark of system architectural design.

Hyperconvergence, also called HC or HCI, takes the compute and storage nodes of other, more complex architectures, and recombines them back into single servers (or you can view it as taking stand alone servers and adding high availability through engineering redundancy without adding unnecessary complexity.) Hyperconvergence gives us the best of both worlds, simplicity like stand alone servers, but options for high availability like Layered High Availability.

Hyperconvergence is both so incredibly simple, but so effective that it can be hard to explain. The key strategy is taking the existing power and cost savings of the stand alone approach and doing as little as possible while still being able to add high availability clustering. By having multiple stand alone nodes that are clustered together (are they still stand alone, then?) we make all of the pieces highly available, while also reducing how many pieces are needed in total.

When done correctly, data can also be guaranteed to be kept locally to a compute node, even though storage is replicated between nodes to create storage high availability, which not only means that we can get the high performance for our storage like we can with stand alone servers, but also that we can avoid a cross-node dependency allowing any node to keep working on its own, even if all other nodes and/or the network connecting them fails! That means that unlike all other system designs, we are only adding more resilience on top of the stand alone server design! That is huge. All other designs must put in the bulk of their efforts to overcoming their own introduced fragilities and risk failing to adequately do so potentially leaving them riskier than they would have been had we done nothing.

Hyperconvergence, therefore, acts as the logical extension of the stand alone design and represents probably the most applicable system design for any large scale system. It is common to see hyperconvergence as being limited to small systems, but it is able to scale to the limits of the clustering technologies - the same limits affecting all design options. So all standard designs can go to roughly the same size, which is generally far larger than anyone would want to go in practical terms within the confines of a single system.

> **High availability vs basic clustering**
>
> In this section we talk about clustering with the assumption that clustering (whether compute, storage, or networking) is done for the purpose of making the system highly available, at least within that one layer. High availability clustering is not the only kind of clustering, however. In all of these designs, including stand alone, we can add generic clustering as a management layer to manage many systems together. This can be confusing as the term clustering can be used to mean many things.

Best practices in System Design Architecture

The best practice for system design is to keep your architecture as simple as possible to meet your needs, but no simpler. Remember that simplicity is a benefit, not a caveat. Complexity should be avoided when possible as complexity brings cost and risk.

In any assessment, start with the brick. Just one, single, solitary stand alone server. Simple and effective. Now evaluate, does this meet your needs? How could spending more money better meet your needs?

If high availability is needed, then assess hyperconvergence. Nothing can be more reliable, architecturally speaking.

If you have special cases where you need cost savings at massive scale, other designs might be applicable. But remember, no matter how reliable it might feel emotionally or how much a sales person may push the solution, hyperconvergence is literally impossible to beat for reliability by design. Make sure that any design that is not one of these two starting points is being used with a comprehensive understanding of all of the risks and costs involved.

That is a lot of material that we covered and a lot of turning conventional thinking on its ear. It is sad that here *conventional thinking* equates to *blindly ignoring needs and risk logic*, but it is what it is. This is a very difficult topic because it is so foreign to most people in any technical or business realm and such a specialty skill to master. And in many cases, you will get a lot of pushback from others who struggle to assess or communicate risks, and fail to turn risk information into actionable business decision making.

At this point you have the tools and knowledge to design systems physically. This is a big topic, and it might be worth revisiting from time to time. This is very foundational and gives us the starting point to build reliable systems farther up the proverbial stack. And now that we know how to approach different designs for different purposes we will go on to look at risk itself and learn to ask what risk mitigation is right for us.

Risk assessment and availability needs

At the very core of what we do in designing a system architecture is taking business needs around performance and availability and applying our understanding of risk and, as with everything in business (and therefore IT) assessing against costs. In the last section we already talked about risk, a lot. We have to - risk and performance pretty much define everything (other than strict capabilities and features) for us during our design stages.

If we ask our businesses about risk, we almost always receive one of two stock answers: *we are not willing to pay anything to mitigate risks* or *we cannot afford to go down, it is worth anything to be up one hundred percent of the time.* Both answers should be obviously seen as insane and have no reason to ever come out of the mouth of any business person or IT professional, and yet they are nearly the only answers that you will ever receive providing you with no guidance whatsoever. They represent management simply *blowing off* IT and leaving IT to take on all decision-making risks without management providing any guidance.

We have some amount of basic guidance that we can almost always work with. On the *low* end of the spectrum the rule of thumb is that if data is valuable enough to have stored in the first place, then at a minimum it is worth backing up. This is the simplest aspect of data and availability protection, and if your business thinks that the data that they store is not worth even backing up, you should be asking yourself why you are there yourself. There are extremely special cases where storage data is truly ephemeral and does not need a backup, but this situation is so unique and rare that it can be safely ignored.

On the other end of the spectrum no system, anywhere, ever is so important that it is worth anything to not have downtime. First of all, totally avoiding downtime is impossible. No one can do this with any amount of resources. We can make a system ridiculously reliable and easily recoverable from nearly endless potential scenarios but no government, military, secret cabal, alien species, investment bank, or otherwise can possibly meet requirements often demanded by small businesses without any discussion whatsoever. The theoretical maximum that can be invested into making systems reliable is the entire value of the company in question and even if every penny that the largest firms had was invested into reliability and nothing else, risk still remains, no matter how small.

> **Risk and diminishing returns**
>
> Attempting to invest in risk mitigation technology is a tough thing to do because as systems become more reliable the cost of *moving the needle* significantly towards ever improved reliability becomes more and more costly. For example, getting a stand alone server that is well built may give us as much as five nines of availability without any special *high availability* features.
>
> We may find that we need much higher availability. Perhaps six or seven nines. To get those order of magnitude jumps in reliability will require, almost certainly, at least double the investment in hardware as the standalone server. This may be well justified by our needs, but the cost per workload just jumped significantly.
>
> If we want to move the needle again an order of magnitude beyond that, the price jumps yet again. We get less and less protection as we spend more and more money.

And so, because our business will rarely be willing or able to clearly define for us what our risk aversion level truly is, it often falls to IT and within IT to systems administration to carry out this all-important task on behalf of management. This will generally take some math, a lot of interviews with many different parts of the company, some common sense, and of course, some guesswork. Working with risk requires maintaining a logical view while it is tempting to become emotional, which is often the mistake made in business. Business owners or managers tend to react emotionally either seeing money spent on protection as not generating revenue directly and therefore undesirable, or seeing their business as not justifying protection and so tending to spend too much to provide an impression that the company is seen as valuable because downtime would be such a terrible thing.

Of course good management will always been heavily involved in risk assessment tasks. This should not fall to IT. While IT has great insight and is a valuable contributor to any risk discussion it is the core management, operations, and financial departments that truly have the full risk picture necessary to create a corporate infrastructure risk strategy.

When looking at workloads, we must attempt to evaluate what downtime will truly cost our business. This is not straightforward in nearly any scenario, but it is what we need to understand to have any means of logically discussing risk. Most businesses want to simplify downtime cost into the simplest possible terms. So dollars per minute or hour is generally how downtime is discussed. For example, losing the company's primary line of business application will lose the business one thousand dollars per hour of downtime.

While simple, almost no real world workload actually loses revenue evenly hour by hour. In the real world it is most common to see a complex curve. For example, in the first minutes, and possibly even hours, we might see almost zero revenue loss. But then commonly we see a spike as outages go long enough to first be perceived by customers, then to cause customer concern. Lack of confidence and lack of operations spikes tend to hit a peak relatively quickly. Then long term revenue loss tends to start to kick in after days or weeks as customers leave. But this curve is different for every business. Of course, making an entire curve graph of all downtime scenarios is difficult and probably impractical, but the business should be able to predict significant inflection points along a timeline that represent major changes in impact behavior.

It is tempting to look at outages as being all or nothing. Basically, ignoring workloads and seeing the entire company as completely down, as if the zombie apocalypse is happening and all of the staff have been infected. It is a rare workload that is going to impact any business in that manner. For example, if a company loses an email workload there will likely be an impact, but as email is not real time, it might take hours or even days before there is an actual loss of revenue (but if email is used to win real time bids, losing even a few minutes might be very impactful - it just all depends.) But assuming email could not be restored for hours, or days, a normal business would immediately begin mitigating the loss of the email workload through other channels. Maybe employees talk to each other in person, or use the company's instant messaging products. Perhaps sales teams begin to call customers on the phone rather than emailing. Working around a lost workload is often far more possible and effective that one realizes until a triage process is performed to see what a real world recovery might really look like.

Workload interplay

Something else that we need to understand is how workloads interact with each other. As systems administrators we might have excellent insight into technical dependencies such as that a key ERP or CRM system depends on another application, such as email or financial, to function and if one is down, the other is down, too. That is an important aspect of workload dependency, but one that is well known and understood just by mentioning it. What is much harder to understand is the human workflow interdependence of systems.

Some workloads may be exceptionally stand alone. Some may depend significantly on others. Others may overlap and provide risk mitigation unofficially.

Lets look at the third case first. In many an organization today there might be traditional telephones, email, a few types of video conferencing solutions, and a handful of instant messaging solutions. Even in a tiny organization it is easy to casually end up with several overlapping solutions simply because so many things come bundled with so much functionality. In a situation like this, losing email might matter little for a very long time as internal communications may move to instant messaging and customer communications to telephone or video conference. Most organizations have the ability to work around a system that is down by using other tools at their disposal.

But the converse is also true. Two technically unrelated systems, again lets say CRM and email, might not connect together but the human workflow may require that both be used at the same time and the loss of either one of them might be functionally equivalent to losing them both. So we have to consider all use cases, and all mitigation possibilities.

This interplay knowledge will help us to determine how it makes sense to deploy some workloads. For example, if email and instant messaging work to overlap during a crisis, it likely makes sense to decouple them as much as possible so that if the hardware or software for one was to fail that it would not take down the other.

If we have systems, like our email and CRM example, where one is useless without the other, then combining the two workloads to share a failure domain might make total sense. Meaning, as an example, if we had two independent servers one running the CRM and one running email, then each individual server would carry its own risk of failing with near certainty that those failures would not happen at an overlapping time. Each workload has an equal amount of expected annual downtime of X. The total downtime expected for the combined workloads is 2X. Easy math. Combine the two workloads onto a single server as each retains an equal amount of annual downtime risk, still X, and the combined is still 2X. But the effective downtime in the first case is 2X (or 1.99999X as there is some tiny chance of the outages overlapping) but in the second case is just 1X. How did we do that? Not by reducing any individual risk, but by reducing the effective risk - that is the risk impacting the business as a final result. Under the hood, we did reduce risk physically as a single server has half the risk of downtime of two equal servers simply because there is half as many devices to fail.

Even a complete company shutdown is not necessary a total loss. Sending staff home for a surprise holiday might lower insurance costs and raise morale. Given a day or two to spend at home might reinvigorate workers who may be happy to return after systems have been restored and work more efficiently or maybe put in a little extra time to attempt to recoup lost business. We have to consider mitigation strategies when looking at losses from failed workloads. Some businesses may simply lose some efficiency, while others may lose customers.

Of course we have to consider the possibility of the opposite. What if you are a business that depends heavily on customer perception of high uptime and even a tiny outage has a sprawling impact? Maybe your entire business generates only one thousand dollars per hour, but loss of customer confidence from even a two hour outage (which we might assume could only, at maximum, lose us two thousand dollars in this case) resulted in the loss of customers resulting in tens or hundreds of thousands of dollars of losses!

All of these losses are just estimates. Even if an outage actually happens, there is no guaranteed way to know what revenue would have been without an outage having occurred. So if we cannot know this number for certain after something has happened then obviously we cannot know it with any certainty before the event that only might happen, has happened. Bottom line... estimating risk is very hard.

In a large organization, consider playing *what if* games on a weekend with some staff from different departments. Run through scenarios of *X or Y has failed* and attempt to mitigate in a nearly real-world simulation on a small scale. Can you keep working with one tool or another? Which departments become dysfunctional, which keep humming along, how do your customers see the situation? This kind of *game* is best played with a combination of strong planners who are thinking about risk strategies and writing procedures as well as with a group of perceivers (for example: triage experts) who do not plan, but work tactically on the ground figuring out how to keep working with the tools at their disposal.

Defining high availability

One of my favorite quotes in all of IT comes from John Nicholson who said *High availability isn't something that you buy, it is something that you do*. It is so tempting to see high availability as an intrinsic need in IT, and then to see it as so complicated that we cannot know how to approach it, and so fall prey to vendors who slap the unverified name *high availability* onto products, or even just slip it into a product name, and act as if buying a product can deliver high availability when logically, this is impossible. Imagine buying a high availability airplane, as an example. While you can make one airplane must more reliable than another, almost all of your overall reliability and safety comes from the pilot, not the plane. The same is true in IT. The best made product does little if operated poorly. A million-dollar cluster without backups is likely not as safe as a desktop with good backups!

So first we need to establish a baseline for measurement. In our last section we said that stand alone server infrastructure serves as our baseline. This baseline has to represent what we will call *standard availability*. We now have two ways that we would hope that we can look at this availability. One is in absolute terms by giving a number such as a *nines* number and through industry evidence, it appears that well maintained, well-made stand alone servers can approach five nines of availability which roughly means six minutes, or less, of unplanned downtime per year (planned downtime for maintenance can represent a potential problem, but is not itself included in a reliability figure such as this.)

Now keep in mind, when we are talking about a server or a system design architecture, we are not including the final workload, only the platform providing an underlying system onto which a hypervisor will be installed. So essentially hardware availability. Any software running on top may have its own reliability concerns and no amount of platform stability will fix instability from bad code in the final workload, for example.

The other, and generally more useful, way to look at reliability of system architecture is not in unmeasurable absolute terms, but in relative terms comparing different designs to one another. No one really knows what system reliability really is. It is not a big secret that server vendors are keeping from us, they simply do not know. Every little system configuration difference produces very different reliability numbers and, like we said about planes, the users operating the systems create the largest impact in terms of reliability. A company with a pristine datacenter and continuous onsite support that responds to alerts immediately and spare parts on hand or nearby might be able to squeeze very different reliability numbers out of the same server stuck in a closet without air conditioning, lots of dust, and generally ignored. There are simply too many factors involved. And even if we could somehow account for all of the potential variation, in order to get meaningful statistics on systems so complex with failure rates so low, we would need to operate thousands or tens of thousands of servers for more than a decade to collect useful numbers and then all of the data would be outdated by more than a decade. So, for all intents and purposes, it cannot be measured.

So, our most important tool is not talking in terms of *nines*, that is a great marketing tool and something that managers steeped in big heavy processes like Six Sigma like to repeat, but it means nothing in this context, but rather looking at orders of magnitude of systems deviating from our baseline. A system that is significantly more available than our baseline can be classified as *high availability* and a system that is significantly less available than our baseline can be classified as *low availability* and any system that is roughly the same as baseline remains *standard availability*. Beyond these general terms it becomes all but impossible to discuss.

A system design like hyperconvergence would be generally classified as *high availability* as it is the most reliable design approach. And an IPOD would generally be classified as *low availability* as it is closer to the least reliable design approach, which is the network system design. Layered clustering is generally considered high availability, but not *as high* as hyperconvergence. Of course, in this case we are only considering the availability of the system design and ignoring individual components. If we use extremely highly available individual components at every layer of an IPOD, we can theoretically get it back up to standard availability, but likely at great cost.

It is far more valuable to think of reliability in relative terms, rather than absolute ones. It is almost trivial to look at a standalone server, an IPOD, and hyperconvergence and see how there is a clear *high*, *medium*, and *low* availability based on nothing but common sense and the location of risk, risk mitigation, and risk accumulation in the design. It requires no special training or math to see how dramatically each is separated from the next and how improving the overall quality of components moves the absolute reliability number, but the relative does not change. And at the end of the day, this is all that we can know.

By knowing what downtime impact will look like financially for our business, even if it is only a very rough estimate, we have something to work with when attempting to decide on how to invest in risk mitigation. We should never invest more in risk mitigation than what calculated risk shows as our potential losses. This sounds obvious but is a common stumbling point for assessment in many firms. For example, if a possible outage might cost us one thousand dollars and protecting effectively against that outage would cost two thousand dollars, we should not at all entertain paying to mitigate that risk.

We should think of risk mitigation as a form of outage itself, for mathematical reasons. This makes calculations easier to understand. With good risk mitigation we would incur a minimal financial penalty now (say spending one thousand dollars) to protect against a large potential outage (that might cost us one hundred thousand dollars.) The upfront cost is guaranteed, the future risk is only a possibility. So any risk mitigation must therefore be much smaller than the potential damage that it is meant to protect against.

An analogy of my own that I have been using for years to describe paying more for risk mitigation than the potential damage of the outage itself is: *That's like shooting yourself in the face today to avoid maybe getting a headache next year.*

When comparing standard availability systems and high availability systems we might be talking about a difference of only several minutes per year, on average, of downtime. High availability, therefore, has to justify its cost and complexity very quickly. A massive public website where just a minute or two of being unavailable could cost millions in purchases or worse, erode customer confidence, therefore could easily justify a large expenditure in high availability systems even if the time saved seems trivial. But an internal system servicing employees where customer confidence is not a factor, and downtime does not lead users to turn to competitors (for example: email system, financial, CRM, and others) the lose of even several minutes a day, let alone a year, is likely to have no real financial impact whatsoever and investing heavily to protect those systems would be wasteful.

So, how do we apply all of this to best practices? The hard answer is that risk assessments and resulting system design are very hard things to do. Determining risk is a long process involving a lot of math, logic, and to some degree, guessing. It requires that we understand our businesses and our technology stacks deeply. It demands that we engage the business at all levels, and from all departments, and consolidate information that is generally siloed. It forces us to evaluate other risk assessments against logic and expected emotional reactions.

Rules of thumb tell us that the majority of systems that we deploy should be standalone servers, and nearly all remaining systems should be hyperconverged. These two standard patterns represent the near totality of what proper design will look like in the real world. All other designs are realistically relegated to extremely niche use cases with the IPOD being the ultimate *anti-pattern* of what not to do except for the most extreme of special cases.

We have covered a lot of material in this chapter. But now we have an idea of how we make risk determinations, how we design our architectures based on that assessment. We understand how and why we use different kinds of virtualization, and why we always virtualize. And we know how to evaluate the use of cloud and locality for our deployments. Now to put all of this together! We use all of these tools in deciding the deployment of every workload! So many options, but that is what makes our careers challenging and fulfilling (and what makes us worth our salaries.)

Summary

In summary, system architecture is complex and requires us to really dig into business needs, how operations works, talk to key roles throughout the organization and elicit input, and take a broad view of technological building blocks to construct solutions that deliver the performance and reliability that our workloads need at the minimum cost.

We looked at fundamental components with virtual machines and containers and should now be able to defend our use of them and choose properly between them, as well as be able to use traditional containers without becoming confused with more recent application containers. And we learned about locality. You should be able to navigate the complicated linguistic minefield that is managers attempting to talk about the placement and ownership of server resources, analyze costs and risks and find the right option for your organization. Colocation, cloud, traditional virtualization, on premises are all options that you understand.

And finally, the big piece, system design and architecture. Taking the physical and logical components of our system and building a full functional platform that empowers our workloads rather than crippling them. This has been a long chapter and touches on a lot of topics that are very rarely taught individual, let alone together. These are some really hard topics, and it is probably worth covering a lot of this material again before moving on.

For many of us in systems administration we might use the material in this chapter almost never. For others it might be nearly everyday skills. These topics are often ones that allow you to completely elevate your career by demonstrating a concrete ability to take seemingly mundane technical minutia and applying background system design decisions to key organizational needs. Of all of our topics in this book, this one is probably the one that should empower you more than any other to stand out among your peers and cross organizational boundaries.

I hope that with the information presented here that you can filter through sales and marketing misinformation, apply solid logic and reasoning, and build on concepts that will remain timeless. Taking the time to really understand failure domains, additive risk, false redundancy, and more will make you better at nearly every aspect of your information technology journey whether your goals are purely technical, or you dream of sitting in the board room chairs.

In our next chapter, we are going to return to the seemingly more pedestrian topic of system patching, and move from the high level system strategies to in the trenches security and stability warfare.

5
Patch Management Strategies

In day-to-day operations of your Linux systems probably the most common task for an average system administrator is going to be patch management (aka patching.) Unlike the Windows and macOS worlds, it is standard for the system administrator to handle a broad variety of operating system and application patching tasks covering both primary and third-party ecosystems. It is also standard for there to be built in, and sometimes third party, application management ecosystems to assist with this potentially daunting task.

Patching, and of course system updates, are large parts of what we do and while it may feel mundane it is very important that it be something that we get right. And production Linux systems today have become much more complex and diverse than they were just ten years ago. And of course, patching has become more important than ever, something that we expect to only see increase over time as well.

We will start by understanding how patches and updates are provided and what we mean by different installation methodologies.

In this chapter we are going to learn about the following:

- Binary, Source, and Script Software Deployments
- Patching Theory and Strategies

- Compilation for the Administrator
- Linux Installation and Redeployment Strategies
- Rebooting Servers

Binary, source, and script software deployments

Software can come in all shapes and sizes. So, software deployments are not a one size fits all affair. The standard means of deploying software are directly as a binary package, through source code that needs to be compiled into a binary package, or as a script. We will dig into each of these as it is necessary to understand what each is and when they might be appropriate.

Compiled and interpreted software

Many system administrators have never worked as developers and often are not aware of how software exists. There are two fundamental types of programming languages: compiled and interpreted.

Compiled languages are written in one language (source code) and run through a compiler to produce binary executable code that can run directly on an operating system. This can be an oversimplification, but we are not developers and need only be concerned with the original code being compiled into a binary format. For Linux, this format is called **ELF**, which stands for **Executable and Linkable Format**. Compiled binaries run on the operating system.

Interpreted languages are different. Instead of being compiled into a binary, they remain as written, and are processed in real time by a program called an interpreter which itself is a binary executable and which treats the code as an input file to process. So interpreted software requires that an appropriate interpreter for the language in which it is written is installed on the operating system in order for the software to work. For example, if you have a Python program that you want to be able to run, the system on which you want to run it will need to have a Python interpreter installed to process that file.

Both approaches to software are completely normal and valid. As system administrators, we will work with both all of the time. Modern computers (and interpreters) are so fast that performance is of little concern between the two types and other concerns (mostly of the developers) are generally more important in deciding how a given piece of software will be written.

Software is not quite this simplistic, there are bizarre concepts like code that looks like it is interpreted but is actually compiled at the last moment and run like a binary. Some languages like those based on .NET and Java are compiled, but not to a binary, and so are essentially an amalgam taking some benefits of both approaches.

However, by and large we think of all software as either binary executable (runs directly on the operating system without assistance) or interpreted (requires a programming language environment or *platform* on which to run, on top of the operating system.) For the purposes of understanding code deployment, languages like .NET and Java, as well as **JIT (just in time)** compiled ones like Perl are lumped with interpreted languages due to behavior.

Common languages that are generally precompiled include C, C++, Rust, and Go. Common languages that are interpreted, or act as though they are, include Python, PHP, Perl, and Ruby. To make matters more confusing, any interpreted language can be compiled. A standardly compiled language could even be interpreted! It is less *what a language does* as much as *what is it doing in this specific situation*? Essentially, any given language *can* be compiled or interpreted depending on how we treat it in practice. However, in the real world, no one is interpreting C++, and no one is compiling Python. But if you wanted to, it is possible.

As system administrators, we really have no say in how software is built, we simply have to deploy software as it is given to us. What we have to understand most is how much control and influence we may have on the total platform. If we run a binary application, it is possible that our only real options are around the version of the kernel that we run. But if we are installing a PHP script, we may have to decide how to install PHP as well, what version to run, which PHP provider, and so forth. It can become rather complex in some situations.

The majority of software that we will deploy is going to be binary in nearly all business scenarios. Often we might not even know (or care) about specific software as so much of the process will typically be handled for us. It is quite common to have to install software blindly.

It is increasingly common for system administrators to install software that is built of scripts which are readily readable code. These files are simply processed by an interpreter. So, the software never exists on disk in binary form. Since modern computer systems are so powerful, the seemingly problematic lack of efficiency in this process is often no problem at all. Many popular software packages are now written and delivered in this way, so most system administrators will commonly work with script installations.

In many cases, scripts are installed using the same automation methods as binary software packages making the entire process often transparent. As a system administrator, you might not always even know what kind of package you are deploying unless you dig into its underlying components. This becomes especially true of non-critical packages, and packages deployed using a dependency resolving system that handles any platform inclusion (for example, PHP or Python) for you, or if those dependencies were already installed for other components ahead of time.

Today we expect that the installation of scripts will be a common task that may not represent the majority of all packages that are deployed to a system but that can easily form the majority of primary workload code on a system. By that I mean that the operating system, supporting utilities, and large system libraries will often be binary packages. But the final workload, for which we are running the system in the first place, will have a very good chance of being a script rather than a compiled binary.

And, finally, the last type is the source code that cannot be run as is and must first be compiled into binary packages before being run. We are going to cover this topic, in depth, in just two sections, so I am only going to touch on it briefly here. You could argue that this approach is still a binary installation because the resulting deployed package is binary and that is totally correct. However, the workflow that must be followed to get and deploy that binary is quite different and makes this a valid tertiary deployment option. Some systems implement compilation steps automatically, and so it is plausible to have a deployment package that is compiling and installing a binary, and the system administrator is not even aware that it is happening.

Misleading use of source installation

For reasons we will dig into a little later in this chapter, installing from source code generally developed a bad reputation. In some ways this was deserved, and in some ways it was not.

Because source based installation is, for all intents and purposes, unique to the free and open source software world it was heavily targeted by vendors and IT practitioners in the 1990s and 2000s in an attempt to discredit it because it was cutting heavily into more traditional closed source products (and the jobs of people who only supported that software.) This was, of course, completely fabricated but as source licensing is complicated to understand it is easy to instill fear and doubt into those that fail to grasp the nuances of the topic.

More importantly, however, source installation got a bad reputation because it was seen as being a generally unprofessional and unnecessary practice being promoted by system administrators who acted more like hobbyists and installed in this manner without real consideration for business needs. It was fun or looked impressive or their friends did it, so they did it, too. This, I am afraid to say, was broadly accurate. There was an era when lots of software was installed in unnecessarily complex and convoluted ways without regard for the business efficacy of the process. Not to say that source code compilation never has a place, it most certainly does. But even twenty years ago or more that place was in a niche, not the majority of deployment situations. So, in many ways, the bad reputation was earned honestly, but not completely.

Today, however, it is not a big deal as source code compilation is nearly forgotten and almost no one today knows the standard processes with which to do it and the installation of the necessary tools is often banned or at least the tools are not readily available making compilation quite difficult, if even possible. Only software that has a strong need to be compiled is distributed in this fashion. So, the market has all but eliminated this in practice.

But the fear and shaming of those that used to do compilation often still exists. Saying that someone is performing a source code installation remains a derogatory statement. Sadly, in an attempt to discredit even more software today, it has become common to use the term not to reference software that has to be compiled from source into binary, but to refer to script-based software which does not have a compilation step in this way. The term source code implies that the code has to be turned from source into binary. Scripts are not considered source code, at least not in this context. Technically, however, they are the original code, the source, but the implied bad step does not exist. But few people would follow up and can be easily misled by this little linguist trick. So, it is an easy way to take a manager who is just looking for an excuse to make an emotional decision, rather than more difficult rational, one, and mislead them. It sounds reasonable, and few will bother to actually think it through.

The trick really comes from semantic shorthand, something that is always dangerous, especially in IT. The concern with self-compiling software has nothing to do with the availability of the source code, but from the need to compile it before using it. If that step did not exist, the existence of the source code is purely a positive for us. People then incorrectly refer to the compilation as a source code installation. This semantic mistake opens the doors for someone to take something that technically truly is a source code install, without compilation, and because the term has been used incorrectly for so long it becomes a negative connotation applied to the wrong thing, and no one understands why any of it is wrong or backwards.

Linux offers us many standard enterprise methods for software deployments. The plethora of options, while powerful, is making it far more difficult to standardize and plan for long term support.

One thing that is ubiquitous in all production Linux systems, regardless of the vendor, is a package management system that exists by default. More than anything else over the years, these package management systems have come to define one Linux based operating system from another. Several software packaging formats exist, but two, that is, DEB and RPM, have become dominant with all others remaining very niche.

It is increasingly common for Linux distributions to either have multiple software packaging systems or to use multiple formats under a single software package management system. This variety is good as it gives us more options for how we may want to maintain specific packages on our systems, but it also means more complexity as well.

As with all software deployments, we have a few standard concerns that are universal to all operating systems. First is whether software is self-contained or requires access to other packages or libraries. Traditionally, most software, especially on Linux and other UNIX-like operating systems, has been designed to reduce their size both to install and for the operating system itself, by utilizing extended system and third-party libraries (collectively called dependencies.) This means that we can have software that is as tiny as possible and other software that utilizes the same resources can share them on disk minimizing, sometimes significantly, how much we need to store and maintain. Updating a package or library for one piece of software will update it for all. The alternative is to have each individual software package come packaged with all of its own dependencies included with the package and available only within the singular package. This makes for much larger software installations and the potential for the same data to exist on the system multiple times. Possibly a great many times. This leads to bloat, but also makes individual software packages far easier to maintain as there is reduced interaction between different software components.

For example, dozens or scores of software packages will potentially want to use the OpenSSL libraries. If each of twenty packages include OpenSSL, we have the same code stored on disk twenty times. Moreover, if an update for OpenSSL is released or, more importantly, if a bug is discovered and we need to patch OpenSSL we will need to patch it twenty times, not just once. Whereas if we used a single, shared OpenSSL library then whether we have one application that relies on it or one hundred, we would only need to patch the library to make sure that any bug or update had been addressed.

Both approaches are completely valid. It used to be that shared libraries were a necessary evil because system storage was small and shared resources allowed for not just reduced disk usage, but better memory optimization because potentially a shared library might be able to be loaded into memory and shared by multiple pieces of software there as well. Today, we generally have more storage and memory than we can practically use, and this small efficiency is no longer necessary, even if potentially nice. Non-shared approaches trade this efficiency for the stability and flexibility of each package having their own dependencies included with them so that conflicting needs or an unavailability of shared resources does not pose a problem.

The biggest advantage to shared resource approaches is probably that patching a known vulnerability can be far simpler. As an example, if we assume that OpenSSL (a broadly shared library) discovers a critical vulnerability and releases an update. If our systems have shared resources, we need only to find systems with OpenSSL installed and update that one package. All systems that depend on that package are automatically patched together. If OpenSSL were instead to be individually packaged with dozens of separate applications that all depend on it individually, we would need a way to identify that all of those packages use OpenSSL *and* patch all of them individually. A potentially daunting task. We rely on the package maintainers of every piece of software to do their due diligence, patch their dependencies quickly, and provide updated packages to us right away. Not something that happens too often.

Often systems with multiple packaging approaches will use one type of package management and software repository system, such as DEB, when there are shared system components and many dependencies to handle. They will use another package management system, such as SNAP, when they are going to keep all dependencies included in the final package. But it is far more complex than that makes it sound, for example, make a DEB package that include all dependencies or one that expects them to be provided externally and shared. It is only a convention that DEB tends to be shared libraries for software packages. In Linux we also have a completely different set of concerns that you would be used to if coming from a Windows or macOS background: a software ecosystem tied to the operating system itself. In Linux, we expect our operating system (for example, RHEL, Ubuntu, Fedora, SUSE Tumbleweed, and others.) to not only include the basic operating system functionality and a few extremely basic utilities, but also a plethora of software of nearly every possible description including core libraries, programming languages, databases, web servers, end user applications, and on and on. In many cases, you might never use any software that did not come packaged with your Linux distribution of choice, and when you do add in third party software it is often a key application that represents a strategic line of business application or, somewhat obviously, is bespoke internally developed software.

Because of this, when working with software on Linux we have to consider if we are going to use software that is built into the operating system, software that we acquire and install independently (this would include bespoke software), or a mix of the two where many components of the software come from the operating system, but some are provided otherwise.

Digging into specifics of different packaging systems would not make sense, especially as they tend to overlap heavily in their general usage but be very unique as to real world usage. Now we know the options that exist for them. Software package management systems are more important in Linux than on other operating systems, such as Windows or macOS, because there is typically much more complexity in the big server systems that Linux tends to run, and the software being installed typically uses much broader sets of dependencies pulling components or support libraries from often many different projects. Linux packaging systems that maintain online repositories of the software, libraries, components, and so forth make this all reasonably possible.

Probably the most important aspect of the large Linux software package management systems and their associated software repositories is that they allow the distribution vendors to assemble and test vast amounts and combinations of software against their exact kernel and component selection and configuration providing a large, reliable platform on which to deploy solutions.

Best practice here is difficult. Really, we are left only with rules of thumb, but very strong ones. The rule of thumb is to use the vendor repos whenever possible for as much software deployment as you can. This might seem simple and obvious, but surprisingly there are a great many people who will still go and acquire software through a manual means and install it without the benefit of vendor testing and dependency management.

The real best practice is, as you might expect, to get to know the package management ecosystem of your distribution(s) of choice so that you are well prepared to leverage their features. Features tend to include logging, version control, roll back, both patching and system update automation, check summing, automatic configuration, and more.

The more common and foundational a software component is, the more likely you should have it supplied by the vendor as part of your distribution. The more niche and close to the line of business, the more likely that it will be acceptable to install it manually or through a non-standard process as end user products are far less likely to be included in a vendor software repository and are much more likely to have a need to carefully manage versions and update schedules rather than primarily caring about stability and testing with respect to the rest of the system.

It is not uncommon for software vendors making products that are not included in distribution repositories to make and maintain their own repositories allowing you to configure their repository and still manage all installation and patching tasks via the standard tools.

Software deployments are made up of so many special cases. It is tempting to want to delivery standard, clear *always do this* style of guidance, but software just does not work that way. Learn the tools of your system, use them when you can, be prepared to do or learn something unique for every workload that you have to deploy. Some, like WordPress, may turn out to be so standard that you never need to do anything but use the distributions own packages. Others may be so unique that you simple deploy a basic operating install, download the vendor's installer and it installs every needed piece of software, and potentially even compiles it! It just all depends, and more than anything else it will depend on how the software vendor chooses to build and package the software and if your distribution decides to include the package in the operating system or will require it separately.

Now that we have a scope of the software installation processes and considerations, we can dive into the real heart of our concerns with patching and updates.

Patching theory and strategies

One might think that patching is pretty straightforward, and that there would be little to discuss. This is not the case. In fact, if you talk to several system administrators you are bound to get some pretty widely varying opinions. Some people patch daily, some weekly, some wait as long as they can, some do so only haphazardly, and some believe that you should never patch at all (hey, it if isn't broke, don't fix it!)

We should first establish why we patch our software. Patching, as opposed to updating or upgrading, implies that we are applying minor fixes to software to fix a known problem or bug but not to implement new features or functionality. Adding new features is generally considered to be an update.

Most software vendors and operating system vendors honor this system and maintain patching systems that only address security or stability issues in their software between releases. In the Linux ecosystem this is primarily tied to an operating system release. So, for example, if you use Ubuntu 22.04 and you use its own patching mechanisms to patch the software that comes with the distribution, then you will safely get nothing but security and stability fixes for the existing software versions and not new versions, features, or functionality. The logic here is that upgrading to a new version may break the software, change usability, or cause other products that depend on that software to fail.

Upgrading to new versions are assumed to only happen when a new version of the operating system itself comes out and then the operating system and all the packages included in it can be upgraded together at the same time. This allows the operating system vendor to, theoretically, test the software together as a singular package to give the customer (you) confidence that all of your software components will work together even after you move everything to new versions.

So, we assume that if a patch has been made available to us, then this indicates that a vendor, quite likely in conjunction with our distribution vendor, has identified a problem that needs to be fixed, a fix has been created, it has been tested, and it is now being distributed. However, even with testing, many eyes watching for errors, and the intent to do nothing but fix known bugs, things can still go wrong. Both the patch itself can be bad and the process of patching can run into unexpected problems. This means that we have to remain cautious about patching no matter how good the intentions are of those providing the patches.

When patching, we are left with two opposing concerns. One is that if the system is currently working for us, why introduce the risk (and effort) of the patching process when we do not have to. On the other hand, why keep running a system where known bugs are left exposed once a patch has been made available to us? We have to look at the concerns and pick a reasonable course of action.

Risk aversion is really not a key concern here, we are not looking at expense versus risk but rather two nearly equal courses of action (from an effort and cost perspective) with two very different outcomes. We need to pick the approach that reduces risk the most for our business and that is all. It is not how risk averse we are but what our risk profile is like that matters most. If our business is heavily susceptible to small downtime events, then patching may be deprioritized. If our company has highly sensitive data that is a likely target or we are very sensitive to public relations blunders in the event of a breach, then we might patch very aggressively. To make a sensible determination we must understand how each approach creates and mitigates different risks and how those risks affect our specific organization.

The risk of delayed patching

Simply pushing off patching does not eliminate certain types of risks. It may prove to have benefits but may also introduce even more risks depending on the situation. Under normal circumstances, new patches are made available with great frequency, potentially as often as multiple times per day, but at least multiple times per month.

If we patch often, such as once a week, then theoretically we will normally have to deal with extremely few patches at any given time and any break or incompatibility will be relatively easy to identify and to roll back as there are so few patches to work with.

If we save up patches over a period of time and only patch, for example, once a year then we have a few problems. First, the patching process may take quite some time as many patches may be needed. Second, if there is a break it may be very difficult to identify the offending patch as it could be lost in a sea of patches that all have to be deployed. And third, the greater volume of changes made at once, as well as the increased *drift* from any tested scenario (few, if any, vendors will test a system that is specifically as out of date as yours against a large volume of sudden changes) means that the chances of there being a break caused by the patching process is greater.

Delaying patches, therefore, becomes a self-fulfilling prophecy in many cases where neigh sayers who avoid patching because *patches break things* will often see this come to pass because they create a situation where it is more likely to occur.

There is no one size fits all approach. Every organization has to tailor the patching process to meet their needs. Some organizations will avoid patching altogether by making stateless systems and simply deploying new systems that were built already patched and destroying older, unpatched instances neatly sidestepping the problem altogether. But not every workload can be handled that way and not every organization is able to make that kind of infrastructure leap to enable that process.

Some organizations run continuous patch testing processes to see how patches will be expected to behave in their environment. Some just avoid patching completely and hope for the best. Some patch blindly. We will discuss all of these options.

Avoiding patches because of Windows

A culture of *patching avoidance* has sprung up in recent years within the ranks of system administrators. Given how critical patching is in general and how central it is to our careers this seems counter-intuitive. No one could be as strong of a cheerleader for rapid, regular patching as the system administrators.

In the Windows world, patching is very unlike what it is like in the Linux, or any other world. It is often delayed, secretive, slow, unreliable, and worst of all buggy and error prone. Patching in Windows was always problematic but during the 2010s became so bad that it is no longer deterministic, can take much longer than simply deploying new systems, and fails with great regularity.

And failures with Windows patching can mean almost anything from the patch simply failing to install and needing to devote resources to getting it to work, to causing software to fail and no longer function. Some patches can take many hours to run only to fail and then take hours to roll back!

Because of this it has become common and almost expected that system administrators in the Windows realm will range from gun-shy about patches, to practicing total avoidance. This has sprawled to not just include patches but full system version updates as well. So now finding Windows systems that are years or even a decade out of date is becoming commonplace creating more and more security vulnerabilities throughout the ecosystem.

Microsoft then has responded, not by fixing the patching problems, but by attempting to force patching without permission, and obscuring the patching process creating even more reliability problems and in many cases breaking patch management systems so dramatically that even administrators who desire to stay fully updated are often unsure how to do so or are simply unable to do so.

These problems are unique to Microsoft today and are mostly unique to Microsoft in the modern era. We must not allow an emotional reaction to a uniquely bad situation influence our practices in the Linux or other realms that are not impacted nor influenced by Microsoft. Outside of Microsoft's isolated piece of the industry no other ecosystem has experienced these types of issues. Not in Ubuntu, Red Hat, Fedora, SUSE, IBM AIX, Oracle Solaris, FreeBSD, NetBSD, DragonFly BSD, macOS, Apple iOS, Android, Chromebooks, and on and on. Patching always carries risk and we should be aware of this, but Microsoft's problems are unique and have nothing to do with our practices in the rest of the industry.

Patching can be automated or manual. Both approaches are perfectly fine. Automation requires less effort and can protect against patching being forgotten or deprioritized. In a large organization formal patching procedures that require manual intervention may be simple to ensure consistency, but in small organizations with just a few servers it can often be easy to overlook patching for months. When looking at manual versus automated patching just consider the potential reliability of the process and the cost (generally in time) of the human labor involved.

The benefit to manual patching is that you have an opportunity for a human to *inspect* each package as it is patched and to test the system in real time as it occurs. If something was to go wrong, every detail of the patching process is fresh in their memory as they just performed it and they know exactly what to test and what to roll back or address if something fails.

Automation benefits from happening automatically, potentially at very predictable times, and being able to happen even if no humans are present to do the work. Scheduling automated patching for evenings, overnight, weekends, or holidays can minimize impact to humans while speeding the patching process. Automated patching is unlikely to be missed and it is easy to send alerts when patching happens or when there are problems caused by patching.

In most cases automation is going to be preferred to manual patching simply because it is less costly and nearly all manual benefits can be automated in some form as well, such as by having a human on standby to receive alerts in case of problems.

Testing patches is rarely feasible

Everyone talks about how important it is to test patches. System administrators often demand it and attempt to refuse to apply patches until testing can be done (a convenient way to avoid having to patch systems at all.) And if patches ever go wrong, management will almost always demand to know why patch testing was not done beforehand.

Here is the harsh reality: there is no practical or realistic means of testing patches on any scale. There, I said it. Say it out loud, go tell your bosses. The cost, both in time and money, to test patches is much larger than anyone believes. In order to thoroughly test patches, we need a replicated environment that gives us a mirror of our production environment so that we can test the patches that we want to deploy against software, hardware, and configurations. Attempting to shortcut this process does not work as it is the interplay of all these parts that make testing important. If you change anything, you may totally nullify the benefits (or worse, create a false sense of security) to a very expensive process.

In a real-world environment, every system is effectively unique. There are exceptions, but generally, this is true. There are so many variables that are possible, including hardware manufacturing dates, varying parts, firmware versions, and on up the infrastructure stack (that is, code and components that sit closer to the final application at the top of the stack). Computer systems are so complex today that there are millions of variables that could result in a combination that causes a bug in a patch to be triggered.

This is one of the reasons why virtualization is so important, it creates a middle layer of standardization that allows some of the more complex parts of the system to be standardized. So at least one portion, a very complex portion involving many drivers, can be reduced in complexity.

In very rare organizations real patch testing is done. Doing so is costly. Generally, this involves carefully replicating the entire production environment right down to matching hardware versions, firmware revisions, patch history, and so forth. Every possible aspect should be identical. Patches must then be run through a battery of tests quite quickly in order to test any given patch in the company's range of scenarios.

In practical terms this means duplicating all hardware and software and having a team dedicated to patch testing. Very few companies can afford that and even fewer still could justify the small amount of potential protection that it might provide.

Timeliness of patching

Patching is an activity that generally has to happen extremely quickly, and as the industry matures the importance of rapid patching continues to increase. Those of us trained in the 1990s and earlier will tend to have memories of a time when patching a system had almost no time sensitivity because most computers were offline, and patches were almost exclusively for stability issues that if you had not experienced already that you were unlikely to experience. So, waiting months or years to patch a system, if you ever did, tended to not be a very big deal.

Oh how times have changed. Software is so much bigger and more dynamic today, there are so many more interconnected layers, and all but the rarest of computer systems are now on the Internet and potentially exposed to active security threats and a dynamically changing environment all of the time. Everything, as pertains to patching, has been flipped on its ear over the past two decades, although most of the significant changes had happened by around 2001.

Today patching tends to focus heavily on shoring up security gaps that have been discovered recently and every step of the process is one of rushing the patch to market before the bad guys are able to discover the vulnerability and exploit it. Bad actors know that patches will come for any vulnerability that they find quite quickly and so exploitation is all about speed. In some cases, it is the release of a patch itself that alerts the greater community to the existence of a vulnerability and so the action of releasing a patch triggers a sudden need for everyone to apply said patch in a way that was not necessary just hours before.

Because of this, patching quickly is incredibly important. Attacks based on a patch are most likely to either already be occurring and will increase in a last-ditch effort to leverage a soon-to-be-dwindling vulnerability or will soon start as a previously unknown vulnerability becomes public knowledge. For many, this means that we want to consider patching in terms of hours rather than days or longer. We still have to consider potential stability risks or impacts that might occur during production hours, so patching immediately is rarely an option, but it is certainly possible when it makes sense.

In Linux, because patching is generally quick and easy and almost always reliable, it is reasonable to consider patching throughout the production day in some cases, and daily patching in most other cases. Potentially smart approaches might include using a built-in randomizer to patch, somewhat randomly (for reasons of system load reduction) every four to eight hours, or having a scheduled patch time every day at ten in the evening or other appropriate time.

In extreme environments, patching on a weekly schedule is possible and this was popular in the enterprise just one to two decades ago. Today, waiting up to six days to patch a known vulnerability borders on the reckless and should be done only with caution. Six days is a very long time in the world of exposed system vulnerabilities.

The choice of time frames is generally based on workload patterns. Some workloads, like email or an inventory control system, might have little susceptibility to momentary disruption and can be rolled back or failed over quickly. So, patching these in the middle of the day might make sense. Most companies have their workloads have a cyclical use pattern throughout the day and can predict that an application becomes very lightly used from one to two in the morning, or perhaps that by seven in the evening every single user has signed out and even a ten-hour outage would be noticed by no one.

Whether or not intra-day patterns exist or not we almost always see inter-day patterns on a weekly basis. A workload might be light on the weekends or heavy during the week or vice versa. Once in a while, especially with financial applications, the pattern is more monthly based with the beginning of the month probably seeing a heavy load and the middle of the month being light.

Any given workload will typically need to be assessed to understand when patching is reasonable and practical. Sometimes we have to be creative with our timing to be able to get patching in when the workloads allow. If workloads cannot be interrupted for patching, then they cannot experience other, less planned, downtime events either and we should have a continuity plan that allows us to patch, repair, or failover in case anything happens. In most cases we can, when necessary, do zero impact patching through these methods.

The old idea that patching can be saved as a special monthly activity or done only when a specific need is identified for the organization is no longer realistic. Waiting that long leaves systems dangerously exposed and any workload that claims to have only one time a month when it can be patched should be questioned as to how any workload can be both important, and unable to be patched. Conceptually the two things cannot go together. The more important a workload is, the more important that it be patched in a timely fashion.

A common excuse for slow patching processes is that testing is required before rolling out a patch. On the surface this makes sense and sounds reasonable. You could probably sell this idea to non-technical management. There is nothing wrong with wanting to test a patch. But we have to consider that we are already, presumably, paying (or getting for free) a distribution vendor to test patches before we receive them. Those enterprise operating system vendors (Canonical, IBM, SUSE, and others.) have far more skills, experience, and resources to test patches than any but the largest organizations. If our testing does not add something significant to the extensive testing that we are already relying on them for, then our own internal testing process should be avoided, in order to avoid wasting resources and putting the organization at unnecessary risk by not getting potentially critical patches deployed promptly.

Rapid, light testing can be reasonable if it is kept light enough and never used as an excuse to avoid timely patching. A common approach to useful patch testing is to have just a few systems that represent typical workload environments in your organization that run demonstration workloads where you can test each patch before it is rolled out to allow you to observe a successful install and test the patches at the highest level. This could be as little as a single virtual machine in a smaller organization. Consider if any testing at all is valuable, and if it is, keep the testing as light as you can to ensure that production patching happens as quickly as it can within the confines of the needs of your workloads.

Standard patching strategies will also generally suggest that you start either with highly vulnerable workloads or with low priority workloads to focus on shoring up exposures or using low priority workloads as tests of Guinea pigs for more critical workloads. If your environment has one hundred virtual machines to patch, you can probably arrange a schedule that allows you to patch systems that are not critical first and slowly build up confidence in the patching process as you approach the patching of more critical or fragile workloads.

Consider patching to be unquestionably one of the most important, and truly simple, tasks that you will be doing as a system administrator. Do not allow emotions or irrational advice from businesses or system administrators that do not understand what patching is (or influence from Windows admins) to lead you astray. Patching is hyper-critical and any organizational management that is not supportive of patching processes does not understand the risk and reward valuation of the process and we need to be prepared to explain it to them.

Find a patching process that meets the needs of your organization. You do not need to patch on a rigid schedule, you do not need to patch the way that other organizations do. Find the testing that is adequate for you and the manual or automated patching process that keeps your systems updated without overly impacting the organization.

We should have a pretty solid handle on patching concepts at this point and you probably even feel a certain sense of urgency to go examine your current servers to see how recently they have been patched. This is understandable and if you want to look at them now before continuing on, I am happy to wait. Better safe than sorry. When you return, we will talk about software compilation and relate that process to patching.

Compilations for the administrator

It was not all that long ago when major system administrator guidelines included a requirement that any mid-level or senior administrator had to be well acquainted with the details of standard software compilation processes. Of course, all knowledge is good, and we would never say that it should not be learned at all. However, even at the time, this seemed like an odd amount of *under-the-hood* development knowledge and knowledge about the packaging of individual software solutions expected to be known by a person in a non-development role.

It would not be unlike if when ordering a new car from your favorite car company that it was expected to be delivered as parts and that every potential new car owner would be expected to assemble the car before driving it. It is important to note that this process was only ever possible in the open-source world and that the majority of software outside of the Linux space and even a significant portion within it cannot be compiled by the system administrator at all. So, the entire concept of this requirement was for an almost niche scenario and was not broadly applicable to system administration in the general sense, which alone should have been a serious red flag to organizations promoting this as an education and certification standard.

As a system administrator, the idea that we would need to take source code from developers and custom compile it, using a compiler that we provide, in our own environment feels almost absurd. In the Windows world this would be all but impossible, who would have access to an appropriate compiler or any of the necessary tools? In Linux and BSD systems it is often plausible because a compiler or multiple compilers may be included with the base operating system.

There was a time when compilation on target systems had some value. Often, generically compiled software was inefficient by necessity in order to support nearly any potential hardware, and with on-target compiling, we could leverage every last bit of performance from our very specific hardware combination. However, this was also in an era of long compile times due to limited CPU power and would result in systems taking potentially days to install rather than minutes. Software deployment was a lengthy, complex, and error-prone task. In the modern world, it would be almost unthinkable to waste the time and resources necessary to compile most software. When we can deploy entire fleets of servers in just a few minutes while using extremely little system resources, tying up many CPU cycles and spending hours to do the same task does not make sense from a resource perspective alone.

However, compilation carries far more risks than just wasting time and system resources. It also means that we, potentially, get slightly different resulting software on different systems that we run, between times that we deploy, from systems that we have tested, or, more importantly, from systems that the vendor has tested. As discussed earlier, with the challenges that we face with testing patching scenarios, compilation makes testing vastly harder. Likewise, patching becomes much harder.

Some software requires compilation today for deployment, generally software that leverages specific hardware drivers. This kind of software tends to be desktop software and not something that we would see on a server. That does not mean that we will never encounter it, but it should be quite uncommon. It is not technically wrong to self-compile software, but you should only be doing so if it is absolutely required and serves a very necessary purpose. It should not be done casually and there is no reason for a system administrator today to have any knowledge of the compilation process.

Several additional new factors have arisen that make compilation as a standard process less possible. Twenty years ago, we had one major compiler in the Linux ecosystem, and it was assumed that all software being deployed as code would be written in C and targeted (and tested) by the development team against that one compiler. There was a very standard toolchain that convention said would be what was used, and it was almost always correct. Today, not only are there multiple standard compilers now being used with different projects using different ones for testing, but also several common compiled languages that are now in common use with their own compilers and tool chains.

Compilation was never a single, standard process as was implied. However, the convention meant that it was very nearly so. Today. it is a scattered process for all of the reasons mentioned earlier. In addition to this, many projects that do require compilations by end users now build in a compilation process so that it acts much like a standalone deployer rather than requiring the system administrator to have any knowledge of the compilation or even be aware that the software being deployed is compiling! The world has changed in nearly every way.

In the real world I have worked on tens of thousands of servers across thousands of clients and have not seen the assumed *standard* compilation process used in any organization for more than seventeen years and in the years before that it was still so rare as to be able to be generally ignored and always used under questionable circumstances when a system administrator was acting as if the business was a personal hobby rather than a serious business.

The compilation era

In the early days of Linux, being able to compile your own software commonly was a major feature compared to closed source systems, such as Windows and Netware, where compilers were not free, and the operating system code was not made available. It meant being able to move between different architectures without too much effort and at the time system resources were at a premium so the small performance advantage possible from unique compilation sometimes meant a real difference in software performance. System administrators used to be passionate about compiler flags and versions.

This trend was so dramatic that even entire operating systems were released around the concept. Most notably was Gentoo Linux where the entire operating system was custom compiled every time that it was deployed. This often led to people discussing how many days it would take to install a full operating system. The investment in initial installation was significant.

In the 1990s, it was not uncommon for operating system installations to take many days. We rarely virtualized and so installations were often from physical media onto unique hardware which was time consuming even when things went well and there were often installation hiccups that would cause you to have to attempt an install multiple times. In that environment, also taking the time to compile software, or even the entire operating system, was not as crazy as it is today. But rest assured, it was not completely sane, either.

The same era that saw compilation make sense because of resource constraints coincided with the pre-internet office world and the nascent Internet world where environmental threats looking to exploit unpatched computers were rare. Of course, computer viruses existed and were well known, but the ability to avoid them by not sharing physical media between systems allowed for high confidence in avoiding infection when well managed. So, the difficulties of patching custom compiled software did not present a problem most of the time. This was the era of *set and forget it* software. Compiled software was more likely to be forgotten than to be maintained. The effort of installation made the potential effort of patching monumental and very risky. When you custom compile code yourself there is a lot that can go wrong that could result in you being left without a working system.

CPU and RAM resources used to be so tight on the majority of systems that, in most cases, having to wait days longer to be able to put a system into production was considered worth it, even if only to gain one percent in additional performance. What was often ignored was the fact that if it took all of that time and all of those resources to compile the initial installation, that those resources or more might be needed over and over again in the future to compile future patches or updates. This carried a truly enormous risk and would commonly result in a complete lack of patching as, once deployed, there was little means to bring the system down for hours or days to attempt a compilation step in the hopes of being able to update the system.

Like so much in IT, there is a strong desire to build a house of cards and hope that, when it falls down, it is long enough in the future to qualify as someone else's problem. And sadly, that became a viable strategy as companies rarely associate failures with those that caused them and often blame, at random, whoever is at hand. Because of this, creating risky situations is often beneficial because any benefits will be attributed to the person who implemented it, and any disasters it causes will be attributed randomly at a later date.

A system administrator today would still do well to learn traditional software compilation, if only to understand historical perspective and be prepared for any unusual situation that might arise. But today (and for many years previously) compiling our own software as a standard process should be avoided as a rule of thumb, it is simply not a good use of resources and introduces too much risk for no real benefit.

When compilation is done today you cannot expect to be able to follow a generic process blindly as was often assumed (incorrectly) in the past. That convention, that protected so many administrators who just got lucky, is not a convention any longer. To compile software today we need instructions and guidance from the developers to have any hope of realistically knowing what tools, configurations, libraries, languages, and external packages may be required and how to make compilation work. There are so many potential moving parts and all of the knowledge here is developer knowledge, not IT knowledge at all, let alone system admin knowledge. System administrators have so much information that they need to know and deeply understand, it is one of the largest and most challenging technical arenas. The idea that system administrators should additionally learn deeply the knowledge and skills of a totally different field never made any sense. Why would developers even exist if all system administrators could perform these jobs in addition to their own duties? Software engineering is a huge field that requires immense knowledge to do well; it is both utterly absurd and insulting to developers to imply that a different discipline can perform their role casually without needing the same years of training and full-time dedication.

Compilation by engineering department

There is a middle ground that we should mention that can make sense. That is having an engineering group that takes software in code form from developers (internal bespoke teams or open-source projects generally) and performs internal compilation either as an additional security step or for an extreme level of performance tuning or simply because the software in question requires special compilation such as when tied to kernel versions or drivers.

By having a central group that is doing internal compilation and packaging source code into resultant binaries for the administration team they allow system administration processes to remain focused on standard, repeatable, and fast to deploy binary mechanisms while the organization gets the advantages of custom compilation.

This approach is typically reserved for only the largest of organizations capable of maintaining rapid software packaging workflows so that patching, testing, and customization happen almost as fast as it would, should it be done by the software vendors. At large scale it can be beneficial.

This process works because it does not put software compilation and packaging into the hands of administrators where it is awkward and highly problematic.

Best practice is to compile software only when it is a requirement and not to do so otherwise, unless you have a department that can handle packaging compiled software rapidly and reliably fast enough to account for proper patch management needs and the overhead of doing so is justified by a savings at scale.

Compilation is useful knowledge, but just because you *can* compile software does not mean that you *should*. Keep this skill in your back pocket for when you really need it or are directed to do so by your software vendor. Next, we will talk about how to deploy our Linux distribution itself and, even more importantly, how to *redeploy* a distribution after a disaster has struck.

Linux deployment and redeployment

Today we have many potential methods for deploying, and just as importantly, redeploying, our servers (or our workstations, for that matter.) In my opinion, this topic was relatively unimportant in the past because most companies depended on *slow deployment* methods and their disaster recovery models depended on restoring, rather than redeploying, their systems. But there are so many more modern disaster recovery methods today that depend on the ability to rapidly deploy servers that we have to look at this topic with a new eye.

With modern deployment technologies and techniques, it is not uncommon to be able to deploy a new base operating system in a matter of seconds when in the past, even heavily automated systems would often take many minutes if not hours (not even considering the possibilities that would come with custom compiled systems!). Of course, computers are just faster today, and this plays a role in speeding deployments. Vendors have improved installation procedures as well. This is not unique to Linux, but nearly any operating system.

Even doing the most traditional or *ISO-based install* where we take installation media in the form of a DVD image and install from USB media or virtual media, we can generally do a full operating system install from scratch, manually, in a matter of minutes. Perhaps ten to fifteen minutes on normal hardware. This is pretty fast compared to how installs were done nearly twenty years ago.

It was traditional in larger environments to use something akin to a response file to make these installation processes faster and more repeatable. This process would generally mean storing your installation ISO files somewhere on your network and storing a set of installation instruction files somewhere and defining systems in a list somewhere else often listing MAC addresses to assign appropriate configuration files to use. Essentially just automating the human responses used when performing a traditional install. Effective, but clunky.

Today any form of ISO, or similar media-based installation is typically reserved for truly manual installations that are generally done by very small companies (that only ever build a few servers, and each is likely completely unique anyway) or special situations. There is nothing wrong with the older response automation methodology, but so many newer options exist that it has simply fallen by the wayside and continues to lose traction as a popular installation method. It was truly most effective in the pre-virtualization days when few installation automation options existed, and installations were necessarily one operating system per physical device making MAC address-based management highly effective. Today this would work effectively for platform (hypervisor) installation, but not so much for operating system installation.

The advent of virtualization meant two big things changed. First, hypervisors installed to the bare metal of a physical server are rarely customized outside of the most basic options making the need for automation less (and the entire installation process is much smaller). And second, that operating system installation is now done in a non-physical space opening the field to more esoteric installation techniques than were previously available.

In the virtual space, we can continue to install systems manually, and many people do. We can continue to use response files, and while I know of no one that continues to do this, I am confident that it is still widely practiced, especially as so many large-scale semi-automated deployment systems for this were in place already. However, now, we have readily available options to use pre-built system images that can be called from a library to install even faster. With a method such as this, an already built system is simply copied or cloned to make a new system. The initial image can be preconfigured with the latest patches and custom packages and often installed in under a minute; sometimes, in just a few seconds. If it is using certain kinds of storage, it can perform so quickly as to appear instantaneous. With containers, we can, sometimes, observe new systems initialized so quickly that we can barely detect that there is a build process at all (because, for all intents and purposes, there is not).

What method you choose to use for deploying your servers is not the most important factor. All these deployment methods, and more, have a place. What we do want to consider is how quickly and reliably new systems can be built using the processes that you choose. When a new workload is needed, it is good to know that we can build a new server in a certain amount of time and feel confident in the final configuration of it. If a well-documented manual process achieves an acceptable result, then that is fine.

For many businesses, the ability to deploy rapidly is not very important. What becomes extremely important is the ability to redeploy. Of course, for certain types of applications, like those well suited to cloud computing, rapid initial deployments are absolutely critical, but this remains a niche and not the norm and will for the foreseeable future. But redeploying implies typically that some level of disaster has struck and that a system has to be returned to functionality and in that case, it is exceedingly rare that we are not under pressure to put the system back into production as quickly as possible.

So, it tends to be that redeployment speed, rather than deployment speed, is what matters more in our environments. However, because disaster recovery is rarely thought about in this way, the importance of this is often ignored during the only time that it can be realistically affected – meaning, we can only really address this during our initial design and implementation phases but typically ignore it until it is too late to effectively change. Additionally, redeployments need to be done with confidence so that what we have built quickly is built as we expect and behaves as we expect. Under these rushed and pressured conditions, it is easy to miss steps, ignore processes, take shortcuts, and make mistakes.

The faster and more automated our systems are, the better chance we have to being able to turn out the same identical system time after time even under highly pressured circumstances. We should be planning for this situation when we make our initial deployment plans. Being able to recreate systems, without needing to resort to backup and restore mechanisms, can be a game changer for many companies who often feel forced to rely on a single approach to bring systems back online, when in many cases, superior alternatives may exist. We will dig much deeper into this when we talk about backups in a future chapter.

Processes that allow us to recovery quickly also give us greater flexibility throughout our professional workflows. The ability to test patches, deployments, configurations, build temporary systems, and so forth. Flexibility is all about protecting against the unknown.

Best practices in deployment processes are all about evaluating the time to engineer reliability deployment methods that work best for your environment and to streamline this as is sensible to allow for being able to restore a baseline operating system in a time period that makes the most sense for your environment.

Some environments are able to build new servers, and configure them for their necessary workloads, so quickly that they actually choose to do this over performing activities such as patching, updates, or potentially even reboots! Instead, they will rapidly build and deploy a completely new virtual machine or container and destroy the old one. A really effective practice if you can make the process work for you.

In our next section, we will discuss the importance of rebooting and testing your environment under regular, frequent, and planned conditions.

Rebooting servers

Ask your average system administrator, or even a non-technical but interested third party, and they will tell you the importance of long uptimes on servers and how they want to see those ultra-high *time since reboots* on them. It feels natural, and nearly everyone brags about it. *My servers have not needed a reboot in three years!*

There are two key problems with this, however.

The first problem is that *time since reboot* carries no business value, and business value determines IT value. So why should we care, let alone brag, about something that has no value? It might be interesting to know how long a system has managed to stay online, but an investor is not going to reap a reward from the fact that a computer system has gone an extended period of time without a reboot. We work for the good of the business, if we start to care about something other than resultant business value, we have lost our way. This happens when we focus on means instead of ends, server uptime easily carries an emotional value that *feels like it* might lead to good things and so we, as humans, often like to put proxies in place in our minds to simplify evaluating results and uptime is easily seen as a proxy for stability which is seen as a proxy for business value. All of this is false, none of those proxies are correct and, even worse, might be inverted.

The second problem is the big one - high uptime itself represents a risk. The risk of the unknown. Our system changes over time from wear and tear on the hardware to patches, updates, and general changes to software. There can be data corruption or unrecoverable read errors. There might be configuration changes that do not work as expected. There are so many things that can happen regardless of if anyone has made intentional system changes or not. And the longer that we go from our last reboot, the less confidence that we can have that we know how the system will react.

We always have to be confident, within reason, that a reboot will work. We cannot always control when a reboot will happen. We can attempt to have redundant power and redundant components, but every system has a chance of restarting and when they do, we want to have a high degree of confidence that it will reboot smoothly.

The process of rebooting triggers many risks for a server. It reloads a lot of data from disk or other storage locations that likely has not been read in its entirety since the last reboot so the potential for corruption or other storage problems is heightened. It puts stress on the physical system, especially if a physical restart is coupled with the reboot. This is the time that it will be discovered that a file is missing, a file has corrupted, or a memory stick has finally started to fail.

At first glance we might think that intentionally rebooting a system seems crazy, why would we want to encourage a failure to happen? But this is exactly what we want. We intentionally induce the potential for failure in order to hopefully avoid it at other times.

The logic here is that we reboot at a time that is convenient or safe, *a green zone*. If at the time that we reboot we have a hardware failure or discover an unknown software problem, it is at a time where we know what triggered the problem (planned reboot), how long it has been since the last reboot (hopefully not very long), what changed in between (in the case of a software or configuration problem), or that a reboot was the exposure event for hardware failure. This should happen at a time that we have designated for fixing a potential problem.

Finding your green zone

Your maintenance window or green zone is a designated time at which a workload is accepted to be unavailable in order to allow for regular administration tasks. Typical tasks might include a reboot, software installation, patching, database re-indexing, disk defragmentation, or other intensive tasks. Every industry, company, and even individual workload will be expected to have different time(s) when it is appropriate to assign a green zone. Do not expect that the correct green zone from one company will apply to a different company or even from one business unit to another within a single company.

A common green zone is weekends. For many companies some, if not all, of their workloads can safely be unavailable from Friday evening until Monday morning without any business impact. Often, no one outside of IT would even be aware. A good strategy if this is the case is to perform any patching or similar tasks almost immediately upon the commencement of the green zone on Friday evening and then follow that work immediately with a reboot. If the reboot causes anything to fail, you have more than two and a half days to get it back up and running before anyone comes in and complains that systems are down. Two days of impact-free time allows for far better repairs than attempting to get a system back up and running when pressure is high, money is being lost, and the business is pushing for answers from the same team they want focused on fixing the problem.

In my own experience, I once managed an application in which we measured database records that were known to have a consistent period of zero use for at least one minute every week, across all customers of the system. This gave us an effective zone of just sixty seconds, but if we planned carefully, we were able to do system reboots, software updates, patching, and more in that window. It was hardly convenient, but it was very cost effective compared to asking customers to give us a universal maintenance window or to run extra systems to cover for that one minute.

Green zones can be creative and they might not be when you expect. They might be easy like long weekends, or maybe it happens during a recurring Tuesday lunch meeting. Work with your workload customers to learn when a workload is unused or not at risk.

This is really all about planning. Only trigger problems when we are actually available to recover the system. This is so that we are far less likely to encounter have the same thing happen at a time when perhaps we are not available, or the workload is heavily in use. Never let a business say that a workload is too important to have planned downtime, that is oxymoronic. We have planned downtime specifically because workloads are critical. If a workload does not matter, then saving maintenance until things fail is not a problem. The more critical a workload is, the more planned downtime is crucial. In fact, any workload given no downtime (at least at a system level) could be designated as non-critical or non-production level simply from being given no planned maintenance.

Compared to a car, the more important a vehicle is to you the more likely that you will take it out of service to have regular maintenance like oil changes and brake checks. You do this because planned maintenance is trivial, and an unplanned seized engine is not. We know naturally that it is better to maintain our cars than to let them fail. It is no different with our servers.

Avoiding planned downtime is planning for unplanned downtime

If a resultant workload has no reasonable allowance for downtime, then additional strategies are necessary. If you are Amazon running a global online store, for example, even a minute of downtime might cost you a great many sales. If you are an investment bank, a minute of downtime could mean that orders are not completed properly, and millions of dollars could be lost. If you are a hospital, a minute might mean critical life support fails and deaths occur. And if you are a military, a minute could cost you a war, so that there are types of workload outages that we truly want to avoid is not in dispute. Clearly there are times when we need to go to extreme measures to make sure that downtime does not happen.

In these cases, we need high availability at a level that allows us to take any arbitrary component of the infrastructure offline. This could be storage, networking, or any number of platform and compute resources. That means some amount of high availability at the application level so that we are able to properly patch everything from hard drive firmware to application libraries at the highest level and everything in between.

Critical workloads need smooth running well secured infrastructure to keep them running. A good maintenance plan is at the absolute heart of making workloads reliable.

There is no hard and fast rule about the frequency of reboots. But a good starting point is weekly and gauge from there what is appropriate for your environment. There is a tendency to opt for weekly or monthly because we often know that reboots are necessary, but we still think that they should be avoided. But, with rare exception, this is not true. We truly want to reboot as often as it is deemed safe and prudent to do so.

Rebooting monthly, under current patching regimes, is about the longest that you would want to consider waiting for a standard schedule. Remember that any schedule needs to have some accommodations for a system being missed and having to wait for an additional cycle. So, if you plan for monthly, you need to be accepting of some systems going two months, from time to time, without maintenance due to technical or logistical problems.

Weekly tends to be the most practical of schedules. Most workloads have a weekly usage pattern that makes it easy, or at least plausible, to allot a maintenance window. Weekly schedules are also good for users as they are easy to remember. For example, if a system reboots every Saturday morning at nine, users will get used to that and not try to use the system then even if they felt like working. It just becomes a habit. Weekly is frequent enough that the increased risk of the reboot process is very likely to evoke a pending hardware or software failure that would have otherwise occurred in the subsequent six days. This tends to be the best balance between convenience and reliability.

We should always evaluate the opportunity to reboot more often. If our workload schedules allow for it, a daily reboot can be perfect, for example. This is how we generally treat end user workloads, encouraging systems to restart at the end of the day so that they are fresh and ready for the next day when staff arrive to work (whether virtual or physical does not matter.) Doing exactly the same with servers might make sense.

Your reboot schedule should take into account your application update schedule. If the software that you run updates only rarely then a monthly reboot might make more sense. If you have workloads receiving nearly daily updates, then combining system reboots with application updates might make sense.

System reboots are especially important after software updates and primary workloads (generally assumed not to be managed by the operating system vendor) as there are so many possibilities for services to not start up as expected, to need additional configuration, or simply run into bugs when rebooting. If you do not reboot in conjunction with software updates you lack the full confidence of knowing that when it was installed that a reboot worked successfully. If an issue arises later, knowing that reboots were working when the last system changes were made can go a long way to hastening the recovery process.

If forgetting to reboot systems regularly is a common problem, then forgetting to have reboot monitoring is nearly ubiquitous. Even those IT departments that take reboot schedules seriously often never think to add *uptime monitoring* to their list of sensors to monitor in their environment. Reboot monitoring is generally pretty simple and can be generally done quite loosely. For example, in many of my environments where we desire the servers to reboot every week, we add a sensor for *uptime exceeding nine days*. If our monitoring system determines that a server has been up longer than nine days, it will email an alert. Missing one reboot event is not a major problem, and this gives plenty of time to avoid false positives and plenty of time to plan for manual intervention to find what caused the planned reboot to fail and to get it fixed before the next planned reboot should happen.

Best practice here is to seek to reboot as often as is practical and to not avoid reboots for anything but a solid business need which cannot include the false need of *the system cannot go down*. Shoot for weekly or even daily and accept monthly if it is the best that can be mustered and be sure to add monitoring as a system not being rebooted is difficult to catch casually.

Summary

System patching, updates, and reboots may feel very pedantic. And in some ways, I suppose that they are. But sometimes really important things can also be kind of boring. And really, patching and basic system maintenance should be boring. It should be predictable, reliable, and scheduled. And if at all possible, it should be automated.

Patching should not become a challenge or a scary proposition. With proper planning, backups, testing and so forth, it is generally easy to have a reliable patching and even update processes that very rarely experience major issues of any kind. If we fail to make our patching and updates regular and reliable, we will begin to fear the process which will almost certainly lead us to avoid it more which will just exacerbate the problem.

In the modern world of computing, there is always someone looking to exploit our systems and while nothing can protect against every possible attack, we can heavily mitigate our exposure through rapid, regular, and reliable patching.

You should now be confident to evaluate your workloads, get each up to date, and begin implementing a formal patching, update, and reboot schedule across your fleet.

In our next chapter we are going to look at databases and how they should be managed from the perspective of system administration.

6
Databases

Technically database systems are applications that run on top of the operating system and should not be of direct concern to us as system administrators. That is an excellent theory. In the real-world, databases are tightly tied to the operating system, tend to be general knowledge items, require deep technical knowledge, and relatively little overall time. Because of this it almost universally makes sense that database management duties fall to system administrators; this is especially true on Linux distributions today because most databases that we are likely to use are bundled with the operating system itself.

In this chapter we are going to learn about the following:

- Separating a Database from a DBMS
- Comparing Relational and NoSQL Databases
- Discovering Common Databases on Linux
- Understanding Database Replication and Data Protection Concepts

In the good ol' days system administration and database administration were almost always two discrete tasks. The SA and DBA roles would work closely with each other, but the specialized skills and large time requirements of each role meant that they were typically discrete, dedicated staff. Typically, only one DBA was needed for every five to twenty system administrators, so teams were smaller, and the career field was always much smaller as the system administrator role exists even in companies that have no databases to administer.

As both operating systems and database management systems have become simpler and more robust, the need for an extensive amount of database platform tweaking has been reduced to the point that the DBA career path is all but gone. The need to build a career around a single database platform just does not exist today. Those that manage databases generally do so for a wide variety of different databases, possibly on multiple operating systems.

Over time most critical workloads become commoditized. By that we mean that they move from being specialist knowledge to being generalist knowledge. Good examples of these would be email servers (MTA) and web servers. Twenty years ago (around 2001) running either one of these was highly specialized knowledge and only a very rare system administrator would understand the workings of either one. You would typically have a system administrator just for the operating system and then a specialized email administrator or web administrator dedicated to managing just the single application that they specialized in. It would not be *all email servers* or anything like that. It was often a specialized skill to just a single email server such as Postfix or Exchange. Your knowledge would be completely centered on this one product and its unique quirks and needs. This was so extreme that products like Postfix, Exchange, Apache, MS SQL Server, Oracle Databases, IBM DB2, and so forth would have entire certification and professional development paths for each of them. Today, we just assume that any experienced system administrator will be more than capable of managing any email, web, or database server that is thrown at them.

Because databases are highly complex and have evolved to be far more than they were in past decades, there are many concepts that we, as system administrators, should universally understand about them. They are highly stateful and represent a secondary storage layer to the file system and general storage subsystem. They do not represent a special case, but the broad application of a general case that we need to understand deeply. Web servers, email servers, and other applications are still important, but they represent common examples of the standard, expected situation in application design and so do not require special consideration in this manner. To be blunt, databases store things and are extremely fragile. Other applications use databases or the filesystem to store things and so are simple from an administration perspective. It is because a database represents a second-tier data storage layer that it is seen as special.

To tackle this in a system administration way, we are going to start by delving into what a database really is and how it works and why it becomes a form of storage. Then we can look at database types and specifics common to Linux distributions. And then finally we will wrap up by looking at the most important aspect of databases: how to protect them.

Let's get started and find out exactly what a database is and why we care so much about them.

Separating a Database from a DBMS

As with most things in life, the terminology used casually around databases is often inaccurate with highly technical and specific terms being used to primarily refer to something different than what the term is meant to describe. But by digging into what a database truly is and how they almost universally work, and by building up correct semantics around the topic, we are going to build a nearly intrinsic understanding of database needs from a system administration perspective. This is often true, simply finding an accurate way to describe a thing allows us to understand it. Databases are not magic, but too often are treated as such.

The Database

We have to begin by asking what a database is. A database is defined by the Oxford Dictionary as *a structured set of data held in a computer, especially one that is accessible in various ways*. Well that does not tell us very much. But, in a way, it kind of does. Databases cannot be unstructured; this is the most important piece and computers store things such as files. So, databases are structured data on a computer either stored on disk as a file or set of files, or they live in the computer's memory and never get written to disk. In the latter case the database is ephemeral and of little importance to a system administrator. But in the former case, which represents far more than ninety-nine percent of use cases, this matters a lot. Let's make a definition that we can use effectively as system administrators: a database is a file or a set of files that contain structured data.

You might immediately jump to statements like, *but I have run a database and it is a service, not a file - I can interact with it!* We will address that shortly. Stick with me. Databases are just file(s).

We should start with some examples because this helps to quickly understand what we mean. Let's take a standard text file, perhaps one in which we just write down some notes. This is a file, but is it a database? No, it is not. Sure, the file is encoded in some ANSII character standard, but the data is ad hoc, not structured. We *could* add structure manually to the file and that is okay to do, but the file and the tools that read that file are unaware of the structure. Any structure is purely accessible or visible only to the human using it, not the system itself. This is unstructured.

Okay then, text files are not databases. What about a Word file? Same thing. It might have a strong format that a text file does not, but what is stored in the Word file is completely unstructured. We could put anything in the file, anywhere. It is unstructured.

Now what about a spreadsheet file? Like we would get with Excel or a CSV? This is a tough one. This gets us into a grey area. Under normal conditions these files are not considered databases, but nominally they are. I would refer to them as *semi-structured*. They contain some structure (which you visibly see as cells) and they have a way to designate what is contained within that structure, and certain groups of data (namely what is represented as columns or rows) that are expected to relate to each other. But none of the structure within the cells or even between them is suggested or enforced. So, there is structure, but we generally assume that it is not as much structure as a database would require. We will call it a maybe. Do not break out a `CSV` file and try to defend it as the new database format at the next IT cocktail party, it is not defensible. It is only useful as a thought experiment. But we can see that a spreadsheet format has very easily the potential for acting like a database should.

With a little more structure, we start to see real database-lite files emerge. XML makes a great example. XML is a little more structured than a spreadsheet format, like CSV, and can be used as a full-fledged database. There are some database formats that are based on XML. XML is still just ANSII text, but with enough specific structure to take us all the way to *very simple database*. XML maybe does not make for the most robust of databases, nor the fastest to use, but it is very simple and can be effective for the right purpose.

Purpose built databases tend to avoid the unnecessary bloat of ANSII or similar text formats as this slows the system down when used by software, and databases really are not generally intended to be human readable on their own. Generally, some form of binary format or even compression is used to reduce read and write time from storage and minimize storage space needed today. A great example is *SQLite*, a free, open-source database that uses a highly optimized, but openly publicized, format that you can study if interested. It remains simply a file, like any other, but with even more structure than XML provides.

All these examples use just one file per database. But there is no reason that a database might not use many files. A common purpose for this might be to simply break up data so that it is not all stored in a single file that is too large to easily manipulate or data might be broken up by purpose. Imagine a database that stores a list of user addresses and also user telephone numbers. One file that holds only the addresses and one that holds only the phone numbers would be a sensible *on disk* storage design.

The Database engine

We now know what a database is. It is just a file (or files, of course) that we store that contain our structured data. We *assume* that in production circumstances, we will be using a database format that is highly efficient and not meant for a human to open it and read it directly, that would just be silly. So, the computer is going to read, write, and manipulate this file or these files.

But how will the computer know how to work with the database file that is in such a specific structured format? Before we answer that, we should make a comparison to storage.

In storage we have a standard raw format that we refer to as block storage. On top of that we implement filesystems. These filesystems interface with their operating system by providing a standard file access format that the operating system can use to store and retrieve data. Each filesystem can do this in quite different ways from each other. For the operating system to work with a filesystem it has to have a filesystem driver.

Databases work the same way, but they generally do so one layer higher up the stack. To a database, the filesystem is the generic storage. The highly structured data goes on top of the filesystem, contained in a file, or files. But in the same way as we need a driver for the operating system to know how to talk to the filesystem, we need a driver for applications to know how to talk to databases! It is legitimate to refer to this as a database driver, but it is generally known as a database engine or a database library.

Filesystems are actually databases

Here is one of those weird things that those of us working in IT should understand and that you can never whip out at that proverbial cocktail party because you will never be able to convince anyone that this is true, and yet it is. Filesystems themselves are actually a form of database. They are an on-disk format of highly structured data that can contain other specifically formatted data, in the form of files, contain structured data about those files, and generally contains complex mechanisms for searching through the data structure. A filesystem is a database in every sense. Even down to how it is commonly used. They are simply databases with a unique, but extremely common, purpose that is so important that we have forgotten what it really is under the hood.

This is similar to how web servers are actually file servers but are such a special case that no one ever talks about how that is what they really are. It is useful to understand that that is how they work because it helps us to understand them better and keeps our brains from getting confused when it realizes that the two overlap and you cannot figure out how they are separate - because they are not.

This is so true that it used to be common for some high-performance databases to forego using files to store their data and would store directly on the block devices themselves so that they were in the same position that a filesystem would normally be. But because there were no files, they were not actually filesystems. But for all intents and purposes, they were.

This filesystem replacement process was popular as a means of improving system performance because filesystems used to represent noticeable overhead to the servers. Today, filesystem overhead is completely trivial and attempting to replace the filesystem today presents so much complexity and so many problems that it is all but avoided.

Once you think of a database as a sort of filesystem for your data, it all starts to make a lot more sense.

The database engine is where all the real magic happens. The database itself is just the data sitting on the disk (or in memory.) Sure, it has a structure, but the structure is already there. The database does not create or enforce the structure or maintain it in anyway - all of that comes from the database engine. The database engine is the workhorse of the database stack. This is where the processing power goes, this is what we have to install, this is where the magic sauce gets applied.

In some cases, we talk about database engines heavily. In some cases, we use them directly, like **SQLite** or **Berkeley DB**, where we just install the library (database engine) and access a file using it and voila, a working database system that we can use. Or in other cases, like if you use **MongoDB** or **MariaDB** products, you will often talk about the database engines that are being chosen under the hood (**WiredTiger**, **InnoDB**, **MyISAM**, and others) as key factors in database features and performance.

Database engines get the raw deal in many cases with most people totally ignoring this all-important layer of the database services stack as it is hard to understand and generally hidden from view. Out of sight, out of mind applies heavily here.

In some cases, like BDB and SQlite that I mentioned previously, we, as the system administrator, would be responsible for installing these libraries onto our servers to make them available for software to access. But, of course, these are often listed simply as dependencies and get installed and maintained by the package management systems on our servers that come from our Linux distribution making this simple and potentially even transparent to us. Because of these, system administrators are often nearly oblivious to what database engines are deployed or in use on the systems that they maintain.

It may seem obvious, but it is worth noting, that database engines are code libraries (aka drivers) and are not programs that run or services that we find running. So, detecting a specific database engine on a system might be rather difficult. Sure, if SQlite is installed by APT or DNF we can query the system and find it easily enough. It is a library that sits on disk, has a directory, and is in the package configuration logs. We do not get to see it running in a process list or find it in the services directory, but we can find it one way or another.

But there is every possibility that the code of a database engine library will get included into and compiled into a software project making it all but impossible to detect as it is simply binary data on disk somewhere inside another application. Sure, a block level tool could scan through every piece of data on disk to look for on disk patterns but that is getting into an extreme level that is more for forensics or law enforcement and not very useful in system administration. From a practical perspective, a compiled-in database engine is completely invisible to anyone using the system.

When we are discussing the performance or features of one database or another, nearly always these features or performance characteristics are coming from the database engine in use. In most cases the database engine will also implement either a query language or an application programming interface API for querying the data on disk.

For those following along you have probably noticed that since a database engine does not *run* that you cannot access it remotely. In order to interact with a database engine, you have to use the driver in a program to interact with the database.

The Database management system

The most visible component of the database stack, and the only piece that is totally optional, is the database management system, often shortened to DBMS. For most people, even very technical people with a lot of database experience, the only part of the database system of which they are aware is the database management system.

Before we dig into what exactly the database management system is, we should give some examples of real-world DBMS: **MySQL**, **MariaDB**, **PostgreSQL**, **Oracle DB**, **Microsoft SQL Server**, **IBM DB2**, **MongoDB**, **Redis**, **Firebase**, and so many more. These names should be much better known to most of you compared to database engine names.

A DBMS does surprisingly little on its own. A DBMS is a program that uses one or more database engines to interact with the physical database on the disk and then provides access controls to that database. A DBMS might have a single database engine that is always associated with it, or like MySQL might provide access to a range of database engines while providing them a common interface to make it easy for developers to work with them. Those familiar with MySQL will be aware that when creating a new database inside of MySQL you must tell it which engine you want to use. Sure, there is a default in case you do not choose, but not choosing is very much a choice in that case.

When working with a database engine you must specify the file or files that you are working with. So, any interaction with a database engine is only to a single database, whatever database is in that file. A DBMS does not typically work in this way. It is standard for a DBMS to have many databases connected to many *instances* of database engines which may be the same library instantiated multiple times or possibly different engines for different databases, all accessible at the same time. So much of what we picture as a database comes from the DBMS.

It is the DBMS that provides, optionally of course, a way to connect to a database over a network or even over the Internet. Any networking capability comes from the DBMS. Moreover, it is the DBMS that *runs*, generally as a service, on your computer. This is where you see memory consumption, CPU usage, and other database usage details. The DBMS often provides extended features that cannot exist in the engine itself, such as in-memory caching.

It should be obvious that in a pure database engine scenario that access control to the data in the database comes entirely from file system permissions. No different than opening a Word file with Microsoft Word. If you have read and write permissions on the file, and you can run the application that reads it, then you can read the file and write changes back to it. The same is tried with a database engine like SQLite. To allow a user to use the database you simply give them the filesystem permissions to do so. Very simple, and very limiting.

With a DBMS we have more options. Most DBMS add networking and with this we can use the database's own tools to control access on a granular level, plus we can use operating system and networking tools to control access additionally. This becomes much more complex, but from this complexity we get power and flexibility. It is common to the point of assumption that the DBMS will offer user level controls and machine level controls and often very granular control within the sets of data that are managed under it such as row, table, or document level controls depending on the type of database engine in use.

A DBMS adds power and flexibility, along with some ease of use in most cases, to the database ecosystem. In production environments there is no reason that you cannot use a database engine directly and many people do. But by and large it is DBMS that rules enterprise data storage. When people use the term *database server* it is the host containing the DBMS that they mean, by definition. It is the networking capability of a DBMS that allows for databases to be stored on and served out of a dedicated server (or set of servers) rather than requiring the database to always be local to the application that is using it. This is a very important flexibility as applications grow and need more resources. For small applications where there are plenty of resources to host an entire application and its entire database on a single server you will normally get the best performance out of using a database engine directly. But once you scale beyond that a DBMS is what allows databases to get so much bigger.

A database engine on its own is not strictly a single connection limited system, but it is effectively so. It is possible for multiple people to connect to the same file on the filesystem at the same time, but this presents obvious problems. What if two people attempt to write changes at roughly the same time, do they overwrite each other, how do you update others to changes being made, and so forth. It presents the same problems that are seen when you have a SAN and connect multiple servers to the same filesystem. It solves them using the same clustered filesystem mechanics inside the database file. (See, I told you that databases acted like filesystems!)

File level locking and access control can work for light usage up to about five users before it starts to exhibit real performance problems. For heavy use it will be problematic for even just two connections. Microsoft Access is famous for encouraging the use of the JetDB database engine (often just called the Access Database) and having horribly performing file locking that makes it untenable to try to use the system with more than a handful of users. That same system can switch to using **MS SQL Server DBMS** and handle thousands of users without a problem.

So, for any real serious multiple connection situation (which can be caused by individual users or by many instances of a running service that needs to pull data from the database) a DBMS is required. There are so many more caching, locking, and permission control mechanics possible with a DBMS.

Identifying a shared database engine versus DBMS

If you are new to databases, you may not be aware of common ways to identify the use of one approach or another. Some situations are extremely easy, like if you must configure networking connection details for the database with hostnames and ports then you know that a DBMS has to be involved. But not all situations are so obvious.

Many applications handle a lot of their own connection details, and you may have little way to know how they are working under the hood. So other than looking for open ports or something similar you might be left in the dark.

One common mechanism, however, is requiring shared mapped drive access to a file or files. This is not needed by a DBMS in all but the rarest of cases and if it is, you have some serious performance issues. Sharing a database file directly to users or applications to access it is a sign of direct database engine use. This comes up with many legacy or poorly built applications and so you are likely to have come across it or to come across it in the future and know how database engines have to work helps to explain access, locking, performance, or even corruption problems with these deployments.

As system administrators, and potentially as database administrators, the assumption is that other than installing a database engine library, that our interaction with running databases will all be in the form of the DBMS. This is where we will have to manage and monitor services, resource utilization, security, patching, access controls, and the like.

We now have the knowledge of what databases are and how they work and what their components are so we can think critically about security and performance implications. Lacking this understanding of databases makes it very difficult to be able to deal with more complicated issues like performance tuning or effective backup measures. In the next section, we are going to talk about types of databases at a very high level to give us some insight into how system administrators may work with these different forms of data storage.

Comparing relational and NoSQL databases

Databases come in two major categories: *Relational* and *NoSQL*. These are terrible categories, but sadly it is how the world sees databases. These terms are truly awful for several reasons. First because NoSQL is a reference to being *not-relational*. Which means that databases are either relational or not relational. That's pretty bad taxonomy right there. But it gets worse. SQL is the *structured query language* commonly associated with relational databases; it was a language written for querying relational databases. So, the term NoSQL refers to non-relational databases, but that's like trying to refer to people who aren't from England by calling them non-English speakers. The two can overlap, but often do not.

SQL is not some intrinsic language of relations; it is just a common convention used to query them. You can make a relational database that cannot use SQL language queries and just as easily you can make a non-relational database that does! Not only can you, but this is very common. So just to be clear, you can use SQL with a NoSQL database, and people do it all the time! What madness is this?

In the NoSQL world there are generally one or more query languages used by any given database. These tend to be unique to the individual database, but without some form of query language it is all but impossible to get data into or out of a data. It is these languages that form the basics of database communications with the applications that they support. As an example, MongoDB implements their own MongoDB Query Language.

So we have to accept that these terms are ridiculous from the beginning and just realize that we are talking about relational databases on one hand and all other non-relational databases on the other which is a collection of many different database technologies. This weird situation simply exists because most popular, well-known databases are relational and for a great many years it was assumed that only relational databases were *good enough* for production use. Something that has proven to not be in any way true but it has lingering effects in the industry.

NoSQL is therefore daunting on its own because it encompasses so many things, so many types of data structures. That said, we really do not have to understand them all. What we really need to understand is just that a NoSQL database can use any kind of on disk data structure (except, of course, to be relational) and may or may not use SQL or any other query language to be queried. This leaves more questions than answers. As a system administrator we will often simply be tasked with learning whatever database the applications that we support need. This may end up being something very common about which there is a broad amount of knowledge, or it might be something very obscure.

The first databases were, by today's standards, NoSQL. This makes it odd that the earliest databases are thought of today in relation to not being something that came later. Early databases were extremely limited in capability. In the 1970s relational theory came along and the first truly modern databases came with it. Relational databases were born and proved to be so safe and effective that other database forms all but fell by the wayside in short order. The relational database was the king and was here to stay.

For nearly four decades, relational databases represented nearly all business class databases with all other database types existing as little more than historical footnotes. When used in applications they were often implemented uniquely by an application rather than being provided by a large database vendor. Because of this, even when used they were relatively unknown. A database engine implemented inside of an application is essentially invisible to a systems administrator so even if these were being deployed with some regularity only the original developers would have known. In fact, I got my start in my career writing a NoSQL database engine and a GUI data retrieval system for it at the end of the 1980s. I saw this artefact of developers using NoSQL and the information technology department having no visibility firsthand from the other side at the height of the relational database dominance time period.

The strength of a relational database comes from how efficiently it can store data, something that was extraordinarily important in the early decades of databases when finding storage systems large enough to hold them was continuously a challenge; and from how well relational databases can handle things like transactions and data integrity. This makes relational databases extremely good for any kind of system that deals with financial or other critical transactions where we need a high level of assurance that a transaction is completed entirely or not at all. Early databases were naturally very strongly focused on financial transactions because these were the data storage operations that were so critical that they could justify the use of expensive computers to make sure that they were done accurately. When a database costs a few millions dollars to implement it is easy to see why it would be useful to a bank, but very hard to justify running a blog from it.

Relational databases get their name because they are built to specify actual relationships between pieces of data. For example, if you create a piece of user data, you might create a piece of telephone number data, and create a relationship between them. Then you might make more data, an address, for example. The address will also relate to the user. The basic idea behind a relational database is that the database engine will actively control these relationships. Maybe it automatically deletes the phone and address if the user is deleted, maybe it guarantees that a phone number can only belong to one person, or it can show when multiple people live at the same address. It can even stop you from creating a phone number that does not match a certain format or might block you from filling in complete fields.

Relations might sound simple and superficial, but when put into well designed use they offer a lot of power to protect data integrity. They take a large load off of software developers and put data integrity into a position where the constraints used to protect the data remain even when the application that uses them is bypassed. Relational databases are a powerful mechanism. But that power comes at a cost to complexity.

NoSQL databases, being free from the rules that govern relational databases and the assumed necessity of speaking a SQL dialect, can explore any number of data storage and integrity approaches. Some approach data with a *joie de vivre* carelessness that is downright shocking to traditional database administrators. Data can be just stored anywhere, without any controls. Sure, there might be structure, but the structure feels more like a suggestion than anything else. Coming from the hard and fast rules of relational theory, a document database, for example, feels like the wild west: throwing data willy nilly all over the place. One document's structure might not even match the next document. It is datalogical mayhem.

The real power of NoSQL comes from this flexibility. Instead of having to work in a single, strongly predefined way or with heavy constraints, we are free to use data as it makes the most sense for our specific needs. We can use blogging as a great example of something that is generally the polar opposite of financial data in terms of our concerns.

With financial data we are generally concerned heavily with accuracy and consistency and transactional completeness. With a blog we tend to care about speed and little else. If we have a popular website and need to make our blog available all over the world with lightning speed, we will likely want to have nodes serving out that content from locations all over the world so that most people can pull a relatively local copy of the blog rather than using a distant, centralized blog that might be hosted on another continent.

A database that serves out the content via replication to many regions and does so with speed being the top priority can do things like miss the latest updates or get them out of order temporarily while data is updated, something that a relational database would be designed to avoid. Being able to replicate data as time allows while still serving out whatever data is available locally can make for very noticeable performance improvements. For many workloads, this performance tradeoff is ideal.

As we move to a world where more and more systems become computerized, more and more of those workloads use databases and this means that the ways in which databases need to be used are getting broader. Databases are often *free* or incredibly low cost to implement today and that means that nearly everything is utilizing them at some point. It is no longer a world where they are exclusively for highly demanding workloads.

Databases that provide only very simple lookups, session caches, and similar are not common. Or databases that replace text files for logging allowing for faster access to log data and, more importantly, robust searches of those logs can now be found almost everywhere. NoSQL is making the world of databases more powerful and flexible.

As system administrators it will be very rare, if ever, that we are able to choose which database type will be used for a task. Even knowing which database types are most useful in different situations is likely to be unnecessary, while potentially very interesting. Far more importantly, we need to understand that databases now come with wonderous variety, share certain common factors as they relate to systems administration, no longer have the built-in assumptions such as SQL being a universal language, and that our applications will determine what database types and products that we need to learn and support.

Understanding that relational and NoSQL represents our *two camps* of database products we will next take a brief survey of actual products most likely to be found in your Linux ecosystem today.

Discovering common databases on Linux

Most operating systems have one or two really popular, key database products associated with them. On Windows this is Microsoft SQL Server, for example. Linux is very different in this regard. Not only is one singular database product not closely ideologically associated with the operating system, but there is typically a plethora of database options available already included in nearly every Linux distribution. This makes it so much more challenging to be prepared to be a Linux system administrator because the expectation that you are knowledgeable of and ready to manage any number of various database products exists. Your theoretical Windows system administrator counterpart would, culturally, need only have knowledge of one very predictable product to claim base knowledge of the entire field. Many databases *can run* on Windows, but anything other than MS SQL Server is considered an oddity and specialized knowledge. There would never be an expectation that you had any experience or knowledge of them.

On Linux there may not be an expectation in most cases that you have truly deep knowledge of every possible database option, but that you know many of them and are prepared to administer nearly any is quite common. It is common for single servers to deploy multiple database management systems because they are built in and tend to be more purpose-built compared to say MS SQL Server and so using one database management system for one specific set of tasks and another for something with a very different data storage need is common and can be quite effective.

We will maintain the natural assumed division in database categories and look at common relational database products as well as non-relational or NoSQL database products.

Common relational databases on Linux

Probably the four best known databases on Linux, and some might be the best-known databases anywhere, are the relational databases. The big four are **MySQL**, **MariaDB**, **PostgreSQL**, and **SQLite**. Easily in fifth and sixth place, in no particular order, are **Firebird** and **MS SQL Server**. Yes, you read that right, the key Windows ecosystem database product is available on Linux.

MySQL

More than any other database, MySQL is synonymous with Linux systems. MySQL might be best known on Linux, but it is officially available on Windows as well and gets a reasonable amount of use there. MySQL is powerful and very fast and extremely well known. Nearly every Linux system administrator has worked with it at some point. MySQL is a full database management system that includes multiple database engines within it.

MySQL gained its first real popularity by being the database used to power early dynamic websites. It was free, fast, and its lack of more advanced features often required for financial transactions did not matter to content management engines for blogging and similar dynamically generated site content. MySQL became known as the *go to* product for website needs but was often eschewed for other needs because of that stigma.

Today MySQL is mature, advanced, and loaded with nearly any feature that you are likely to need for any type of workload. If you are going to learn only one database management, MySQL is certainly going to be it (or MariaDB, which I will explain shortly.) MySQL enjoys the broadest deployed user base, and by far the best overall industry knowledge penetration with almost any system administrator with Linux experience and a good number without being able to administer it with confidence. A large number of standard tools also exist for it, such as phpMyAdmin, which can make working with it even easier when you want to avoid, or move beyond, the command line. MySQL is used by nearly every major application project, at least optionally, that is made for Linux. It may not be the database most often deployed, but it is the database (when combined with MariaDB) most often deployed *intentionally* on not just Linux, but all operating systems combined.

MariaDB

There is really no way to talk about MySQL without mentioning MariaDB. MySQL's community and direction was divided a number of years ago and many of the original MySQL team left the product and took the open-source base with them and created MariaDB to be what they felt would be the spiritual successor to MySQL. To many, MariaDB is the *real* MySQL, given that it is ideologically aligned with the original product, is equally built from the same code base, and is built by the original team. Many, and probably the majority, of Linux distributions dropped MySQL and switched to using MariaDB instead. So much so, that most people who say that they use MySQL today actually use MariaDB - often without even knowing it.

Drop In replacements

MariaDB's real claim to fame is that it is a full *drop-in replacement* for MySQL. That means that it is designed to be able to be used, completely transparently, anywhere that MySQL would be used. It uses the exact same protocols, interfaces, tools, names, ports, conventions, and others. Everything, in theory, is the same.

MariaDB does this so well that many people say that they are using MySQL, when in fact they are using MariaDB. Some say it is kind of like a code. Others say it because management expects one thing, and it is not worth trying to explain why MariaDB is used instead of MySQL. Still others just have no idea that that is not what they installed. The most common tool for managing MariaDB is the `mysql` command line tool and that is quite often the closest thing to an actual view of the system that many people get. A database designer or a developer working on a system would only need to know that the system is MySQL compatible. There is no real artefact to lead them to suspect that it is one product or the other. They truly do look and act the same.

It is common for people to now refer to the famous LAMP stack, which used to be Linux + Apache + MySQL + PHP, as being Linux + Apache + MariaDB + PHP. It seems that MariaDB has truly taken over MySQL's former position in the market. But it is extremely difficult to gauge this accurately as an indeterminately large percentage of the MariaDB market reports itself to be using MySQL either because they do not realize that they are different, they casually report inaccurately, or they truly have no idea what they are using.

Drop in compatibility gives the additional benefit that learning MySQL means learning MariaDB and vice versa. You do not need to learn one or the other, since everything you do with one is identical on the other.

PostgreSQL

Pronounced *post gress* no matter how unlike that it looks, PostgreSQL is arguably the most mature and advanced database available on Linux systems today. PostgreSQL (originally written POSTGRES) was started in the 1980s as a successor to the successful Ingres database product (**Postgres** meaning **POST inGRESs**.)

Today, PostgreSQL is often considered the fastest, most stable, and most feature rich database product available on Linux and possibly at all. Database distinctions of this nature are typically more opinion than anything else as performance measurements are rarely directly comparable and variances in features often outweigh straight query performance, but PostgreSQL's reputation is one of unmitigated excellence, but at a cost of being more complex and less well known compared to its competition which are generally perceived as being simpler systems.

In recent years, PostgreSQL has seen a major resurgence in popularity. More and more today you will find that software that you deploy supports and even recommends it as the database of choice, a major change in the database winds since the heyday of MySQL a decade or more ago.

As a Linux system administrator, the PostgreSQL database ecosystem is a solid second, if not first choice, alongside MySQL (and MariaDB which are identical) to learn as most Linux system administrators will need to manage PostgreSQL at some point.

SQLite

Far more commonly used than most system administrators realize is **SQLite.** So much so that SQLite claims to be the world's most deployed database! SQLite, as we have mentioned earlier, is a database engine, but not a database management system. Because of this, SQLite installs simply as a driver or *library* to the disk and is called by an application that uses it. There is no running SQLite service or SQLite program that you need to run or can even run if you want to. The closest thing is the `sqlite` client utility that you can use to read and alter a SQLite database file, but this is nothing like a database server.

SQLite's power is in how simple and subtle it is. Outside of having its access library installed, a system administrator typically needs no knowledge of SQLite existing on the system. It does not get tuned or configured. It just exists. And because of the magic of software repositories and automated dependency management, SQLite tends to get deployed simply as a dependency to another piece of software and we might not even be aware that we have installed it or that it is available. It just appears automagically and does its job. It is possible that it is even built into an application that uses it and it may not even appear in a form that we can search for on the system!

Because of how it exists and gets deployed most of the time, the average deployment of SQLite is unknown to everyone that uses the system. A normal installation does not hide the driver: a system administrator looking to discover it or patch it or to just find out where it exists would have little problem in doing so. But unless you are looking for it specifically, a typical server will have hundreds or even thousands of packages of this nature and we cannot generally invest the time into knowing which libraries exist on every system, let alone know which software depends on which packages. It just is not practical.

Firebird

Dramatically less well known than MySQL, PostgreSQL, SQLite, and MariaDB is **Firebird**. Firebird is a complete and mature database management system based on Interbase 6 but split off from that project in 2000. While considered relatively minor software in the Linux world, Firebird is nevertheless a capable and mature database management system whose primary claim to fame is around the overall small amount of system resources necessary to run it.

Firebird is far more likely to come up in conversation than to actually end up being deployed on one of your servers. It simply is not very common. Linux has no shortage of database options and that makes it difficult to get any traction, even when you have a mature and serious database option like Firebird.

Microsoft SQL Server

This entrant on the list is always very shocking to those not aware of its availability. Long considered one of the best database products on the market, MS SQL Server was always limited only to Microsoft Windows operating systems. But in recent years Microsoft, in an attempt to gain traction in the more lucrative database licensing space (when compared to operating system licensing) has been releasing MS SQL Server to Linux as well as to Windows. Today, MS SQL Server is a completely rational and viable option for Linux system administrators to see.

Like most of the more traditional options on this list, MS SQL Server is a complete database management system with multiple database engines under the hood. It is a big product with a lot of bells and whistles.

MS SQL Server remains uncommon on Linux, and likely will for quite some time. But it is already far more common than Firebird, for example. Database departments are often focused very heavily on Linux, or at least UNIX in general, and so it has been problematic if applications require MS SQL Server and that a team that otherwise typically does not use Windows Server is required to manage a wholly different platform just for one database product.

Common NoSQL Database Products on Linux

If the list of relational database products seems long, that is nothing compared to the potential list of NoSQL or non-relational database products that we can potentially discuss. Unlike the relational list where all of the products work in similar ways and would make sense for use by similar software tasks, the NoSQL list varies widely. What is an amazing product for one task might be completely useless for another.

We cannot possibly go into all of the details of each of these database products, or even just all of the different types of databases that are lumped into this catchall of a category. We could easily fill a book this size with NoSQL database examples alone. We will do our best to race through and give enough highlights to make you appreciate what products are out there, why you care, and where to start your own investigations into products that you may want to learn about on your own.

MongoDB

Definitely the mind-space leader of the NoSQL field is MongoDB. MongoDB is what is known as a document database management system. Like many of the database management systems that we have mentioned, MongoDB can use multiple database engines, but nearly everyone who uses it today uses its **WiredTiger** database engine. Document databases are one of the closest NoSQL database types to a relational database with many situations where both would be capable of doing the job. A document database is almost as if we took a relational database and a traditional filesystem, and they met in the middle.

Document databases

Document databases store structured documents. While every document database can use its own format and approach, the concept is universal. A good example could be XML. XML could be the format of an individual document inside of a document database. The database may dictate the structure of the XML, or it might be freeform. The database might contain millions of these XML documents.

Since the database knows the intended use of the documents that it contains, it is able to use common fields throughout the documents to generate indexes and other artefacts to empower the database to do so much more that you could do if you were to simply save many XML documents to your hard drive.

But it only takes a simple example like this before it begins to be obvious why a file system is truly not just a database, but a document database specifically!

Document databases tend to be easy to use and very straightforward making life easy for developers and system administrators alike. They have proven to be a highly desired and effective alternative to relational databases for many common workloads.

MongoDB is broadly deployed and used in a large variety of situations. Like most of the relational database examples that we mentioned initially, MongoDB can be found commonly with third party software packages, with Linux distribution vendor packages, while also remaining popular with internal software teams for bespoke development. If you are going to start experimenting with a NoSQL database on Linux, I would generally start with MongoDB as this is almost certainly going to be the most useful experience to have, even if MongoDB itself never ends up being a database management system that you support in production.

MongoDB is a great way to explore many alternative database approaches, many of which have become somewhat standard in the *post relational* world.

While document databases have proven to be the most common alternative to relational databases for common software usage, not many document databases have risen to prominence. MongoDB is the only really large, well-known example. Other examples are **Apache CouchDB** and **OrientDB** as a database management system and **NeDB**, a database engine with nearly identical data structures to MongoDB.

Redis

A completely different approach to storing data is Redis, which is what is known as a key-value store database management system. This type of database is much more popular as an assistant datastore rather than a primary one. The idea behind a key-value store is that the application using the database supplies a key and the database returns the data associated with that key. It is an extremely simple mechanism compared to the types of databases that we have already encountered, but it is a very useful one.

Key-value stores (sometimes called dictionary lookups) are commonly used for high-speed online caches and can be a great way to manage such data as we might store when managing sessions. It would be rare, but not impossible, for an application to use a key-value database as the only data storage mechanism. It is almost always just one piece of a multi-part database strategy.

Redis brings many of the advanced features needed for giant online applications to the key-value space, such as the ability to cluster across nodes and convenient access methods that make it very popular in large applications. It's performance and simplicity make it popular in web hosting realms.

Other key-value stores remain popular such as **memcached** which is extremely popular on Linux for web hosting, **LevelDB** and even a key-value database engine under **MS SQL Server**.

Cassandra

A key (pun intended) competitor with relational and document databases for the *general purpose application* space is the wide column database and the big name here is Cassandra. Wide column databases deserve more attention and description than we can afford here, but needless to say they primarily tackle the same workloads as relational databases with an eye towards greater flexibility and scalability in most cases.

Along with Cassandra, **Apache HBase** and **ScyllaDB** are major wide column databases common to Linux. You will not find this kind of database as frequently as you will see key-value and document, but it has some traction and is easy to acquire and experiment with if you are looking to expand your knowledge of database specifics or types.

And beyond: NoSQL is not constrained by definitions and special purpose databases and database types keep arising. I would recommend also investigating **Amazon OpenSearch**, a search database, and **InfluxDB** and **Prometheus**, time series databases. All three of these databases, and both types of databases, are typically used in storing log or log-like data at high speed and great volume.

Do not be afraid to search for new or interesting database approaches and products on your own. This is a fast-moving area of the industry and one where a book will become outdated quickly. Linux is the leader in database platforms in every sense, from market share to stability to performance to variety. You should have some familiarity with the major products, what is included in your distribution, and what is likely to be used by your applications. And a general sense of what involvement will be necessary from you. Remember that some organizations will continue to use dedicated database administrators to handle database tasks separate from the system administration team, but nearly all companies will combine these roles leaving the need to understand many database platforms often on your shoulders.

We now have a great understanding of what products we are likely to see in the real world of Linux system administration. All these database concepts are fun and interesting, and it is always exciting to be able to get our hands on many different products, but what really matters is how we protect these systems and that is what we will address in our next section.

Understanding database replication and data protection concepts

From the perspective of the Linux system administrator, nothing is going to be as important as database protection which includes, just as it does with systems in general, both disaster avoidance and disaster recovery. Because databases are so critical, and because they are so common, and additionally because their needs are so different from what we typically encounter otherwise in our systems, we are breaking them out here so that we can tackle the nearly unique needs of the database world with respect to data protection.

Because databases store structured data, they come with all of the challenges to protect what we face with heavily used storage systems, which they effectively are. Because databases are highly stateful we must be very careful that we do not break state when looking at data protection.

What does all of this mean in the simplest of terms? Basically, databases are all about storing data and to do what they do with any effectiveness at all they need to both hold open the files that represent their data on disk as well as keeping generally a large amount of that data in memory at any given time. This presents several challenges to us for when it comes to data protection.

Open files are always a problem for backups. There is no good way to take a backup of a file that is currently being held open by an application because we have no way to know what the state of writing to the file is. Perhaps the file is perfectly fine as it is, or perhaps it has been half modified and the data is gibberish until more data that is current in memory or possibly has not been calculated yet has been added to the file, or perhaps the format of the file is not corrupt but the data that is in the file is no longer accurate. From outside of the application using the file we cannot know what the actual state of an open file is other that it cannot be trusted.

Because of this, most file-based backups will simply ignore the file as they cannot lock the file to take a backup. For most files this is no problem because files only get locked occasionally and if you take backups on a regular basis you are expected to eventually get a good backup of any given file. It may not be deterministically safe, but it is statistically safe. And if you need determinism, you can always use a log to see if files that are critical have been safely backed up or not.

Block based backups, that is backups that work from the block device layer instead of the file layer and are not aware of individual files or filesystem mechanics, can easily take a backup of an open file, but they cannot know if the file is in a safe or accurate state or not. So, in the first case, we assume that an open file will simply be skipped. In the second case, we assume that a backup of it will be taken but that the accuracy of that backup can only be determined at the time of restore. Neither option is ideal, of course. Both are better than nothing.

The traditional method to move from *maybe we got a good backup* or *we almost always get a good backup* to fully knowing that a backup is good is to use an agent that informs applications to complete transactions and puts all data onto the disk and to close the file for the duration of a backup operation. Some applications have mechanisms for doing this, API calls to be made that tell them to prepare for a backup, and for others you do this via brute force by closing the application completely prior to taking a backup and restarting it when the backup is complete. The problem here being that only very few applications support this kind of communications, and any backup software agents have to support each unique piece of software individually to be able to do this. So, it requires both parties to work together, something that does not happen frequently given the large number of applications and backup tools on the market. It is impractical.

What many applications resort to doing is taking their own *backups* of a sort by executing an internal data dump process and saving the data in a safe way to another storage location. This file, a complete copy of all of the data that is in the application, is kept closed and is only used for the backup software to read and use instead of the live data. This provides a universal mechanism for working around data corruption problems caused by open files. It is easy for the application writers to implement and works universally with all backup software.

This method is totally effective but has a major caveat: we have to shut down the application in order to have it work. That means we must have a way to consistently bring the application down at the right time, and a way to bring it up at the right time. That part is generally feasible without too much work, although it is generally quite manual outside of a few common applications. But the real problem is getting organizations to agree to regular downtime for applications to allow for the backup process. Unlike a system reboot which we generally only look to do weekly or possibly monthly, we typically want to run backups daily at a minimum and sometimes hourly or even nearly continuously. This is not always something that we can do.

Beyond scheduled backup frequency, it is generally desired that we be able to take *ad hoc* backups at a moment's notice, as well. If an *ad hoc* backup is going to trigger application downtime that is rarely going to be acceptable. Something else is needed.

This all comes together to make database backups a dramatic problem. Databases are almost always the most critical *under the hood* workload components in your IT infrastructure, the least able to withstand any extended period of or unplanned downtime, the only ones that typically hold their files open indefinitely, and the ones for which backups are most critical.

This challenge extends far beyond just backups. If we were dealing with a stateless application, rather than a database, such as a typical website, we have replication options that can be as simple as just copying the application directory between servers. Load balancing options can be as simple as directing some traffic to one application server or to another. In most cases, non-database application replication, backup, and even load balancing is easy.

With a database we do not have the option of simply copying files between systems. For the data to be consistent and updated, the database management systems running on each host would need to coordinate with each other and ensure concepts such as locks, caches, and flushes to disks were done consistently and communicated around the cluster. Replication, if it is even an option for the database management system that you are using, is generally quite complex and comes with many caveats. There is no simple way to make databases have replication or clustering without introducing significant performance challenges.

So, both protecting databases from failure, and making them easy and reliable to recover should they fail, is challenging. And while the techniques to do so, at the highest level, may be somewhat common, it is really configured and applied at the specific tool level every time. So, learning how it is handled in one ecosystem may not reflect in another.

In all cases, backups or replication and clustering, is going to be handled by the database management system itself or a dedicated tool to that database. The capabilities of the system will be unique to that database. Some database management systems, for example, are limited to very simple clustering, perhaps on to a pair of mirrored servers or they might be limited to only a single node that is able to make changes to disk, but other nodes in the cluster maintain a cached copy of part or all of the data and serve out requests to read the data. Others have massive scalability and may allow hundreds of independent nodes, each of which is allowed to fully read and write data!

So, because of this complexity we must learn each product completely individually. There is a strong *tendency* for NoSQL databases to have been replication and redundancy options. This generally comes from having fewer controls and constraints. As well as from the fact that most NoSQL databases are built in recent years and most relational databases are decades old so the considerations at the time of design were vastly different. Most relational databases that you know today had to adapt first to networking and then to the Internet over a period of many years. Nearly all NoSQL that you will encounter were made decades after the Internet was a part of everyday life.

In many cases, databases being put into a cluster will work by locking some combination of file, record, document, row or other discrete portion of the database and signaling other cluster members of the lock. Then the original system waits for all members to communicate back with an acknowledgement of the lock. Once locked the system cannot write new data to at least a portion of the system and must unlock it before it can continue fulfilling data storage requests. This locking can be quite fast in some cases or can have storage impacts large enough to impact application usability in others. Doing this on a single server database can have noticeable impact. Doing this on a cluster where cluster nodes need to wait for each other to complete their locked tasks and report back to each other can magnify that effect by many times. If we then need to do the same thing while waiting for Internet latencies and possible outages happening during a locked operation the scale of potential impact can magnify many times again. Locks ensure consistency but always come at the cost of performance and complexity. The bigger our database system grows, the larger the potential impact of locking even a portion of that system.

Traditionally, and mostly with relational databases, we have assumed that increasing performance was done by *scaling up* - that is adding more CPUs or faster CPUs, and more memory, and more or faster storage. This is really effective as long as we need to access the data more or less centrally and only until we are able to keep squeezing more and faster CPUs into a single box. After that, we hit a performance wall and we are stuck.

Today more and more we see *scale out* designs where more, smaller nodes are added to a cluster like we have discussed in earlier chapters. Relational designs are not completely unable to use this model, but they tend to struggle to do so efficiently, especially if we expect write, as well as read, operations to scale with the system. It is NoSQL that has taken this need for data access and really run with it. New databases designed from the ground up to do this with amazing efficiency have emerged and are tackling data problems in whole new ways. Some are even using NoSQL *under the hood* while presenting common relational interfaces on top (via SQL language queries, for example) to add new performance options to otherwise old designs.

Clusters of databases introduce new potential complications to our data protection plans. Depending on the database and the data integrity design a backup or replication operation may require something dramatic such as temporarily *freezing* the entire database cluster while the backup is taking place. Or it may require collecting data from individual nodes, when each contains unique data, to assemble a complete backup of data that does not exist in its entirely in any one location until the time of the backup operation. This approach, generally called a sharded database, can be logistically challenging as it is possible to have a scale of data that no node is well prepared to handle and restoration can be complicated as the data has to be fed back into the cluster and distributed out to the nodes. Reassembling the data may be a large task. Other databases might simply *take their chances* and provide backup of what they have without checking in with other cluster nodes. It all depends on the database and the setup of it.

When working with database clusters there are many considerations and while we can talk about high level approaches here, in the real world each database and sometimes even the database deployments are unique. We will need to investigate the documentation for the unique setup and be mindful that we need to have a way to ensure consistency end to end through the process of collecting and storing the data.

We also have to consider the case of straight database engines that do not include a database management system. In this case, there is no database management system to create consistent backup files to disk or to handle replication. Any data protection features will need to be handled by the application that uses the database rather than having a database product do the work. This situation is extra difficult for system administrators because every application is potentially very unique and challenging. Of course, we can *always* resort to shutting down the application or even the entire server in order to make a backup, but it is not a desirable process.

With a database engine if we want any functionality around data protection without shutting down the system ourselves, we need to rely on the application in question to provide it. This *can* provide for the best possible options (along with applications doing this same thing on top of a database management system, of course) for data protection because the application itself will generally have exclusive access to the database as well as total knowledge of the current use case of the entire application stack. In theory the application layer, with its additional knowledge of the state of the system and intended use cases, can take backups that are more meaningful, at times that are more useful, and store them in more dynamic ways.

For example, the application layer can replicate partial data, only data that is deemed truly critical, in near real time to an arbitrary data storage location, perhaps offsite. Maybe it records data in a log-like structure so that it can be recreated. Or maybe it knows when there is going to be downtime on writes and can lock the database and replicate it with greater intelligence than the database management system itself could do or the application layer could replicate a transaction across application nodes to ensure consistency before ever sending data down to the database for storage. There are many ways that the application layer, with its greater insight and flexibility can make the data storage layer better.

An application layer backup has the potential to do interesting things such as take backups based on slow times of day and to be able to automate both backups and restores. Automating the setup of a new node can be an incredible benefit if the application allows for it.

As was said about the database cluster situation, each scenario may be unique and will require knowing the application, knowing what accommodations it provides for data integrity, and adding our own knowledge to ensure that we are able to accomplish a consistent and integral data set. An application that is clustered may present the same challenges and opportunities that a clustered database does on its own.

Summary

Databases are pretty much the most important thing that we will need to work with as system administrators. Whether we run the database ourselves, or only administer the operating system on which they run, they will require more of our attention and will cause us more stress than pretty much anything else that we will do. Our skills and expertise will matter most when working with databases and it is here that the greatest range of our skills will likely be tested.

Wherever we have databases running or in use we need to evaluate how the data is stored on disk, how we can ensure consistency, and how that consistent data can be moved to a backup location whether tape, online, or other. This is probably the single most important task that we will do in system administration, Linux or otherwise.

Our best practices around databases really focus on data protection. We would love to talk about how we should choose the right database type for the job that we will be performing, but in all but the rarest cases these decisions are made long before anything gets to the system administration team.

Database best practices for backups are to ensure that a full consistent, fully safe, closed set of data is used as the source for a backup to ensure data protection is predictable. Whether this is handled by the database, by the application using the database, or manually, there must be a mechanism that ensures that data is not in flight at the time of the data acquisition.

In clustering scenarios, the same logic applies. But now we must ensure that the data accessible to our node is accurate and complete in the context of the entire cluster.

In our next chapter, we are going to start digging into the less glamorous, but super important world of documentation, monitoring, and logging.

Section 3: Approaches to Effective System Administration

The objective of Section 3 is to step back and take the business context and apply it to our view of Linux administration. With this context, we will consider many process and design decisions in a better way and show how conventional thinking so often fails to accommodate real-world business needs.

This part of the book comprises the following chapters:

- *Chapter 7, Documentation, Monitoring, and Logging Techniques*
- *Chapter 8, Improving Administration Maturation with Automation through Scripting and DevOps*
- *Chapter 9, Backup and Disaster Recovery Approaches*
- *Chapter 10, User and Access Management Strategies*
- *Chapter 11, Troubleshooting*

7
Documentation, Monitoring, and Logging Techniques

Now we are getting into the real *meat* of system administration work, although I think that most people would feel like this is the *potatoes*. In this chapter we are going to be dealing with all of the parts of system administration that no one can see on our servers. These are those nearly invisible components of our jobs that are so critical and can do so much to separate the juniors from the seniors; the extra steps that make all of the difference when things start to go wrong.

In this chapter we are going to learn about the following:

- Modern Documentation: Wiki, Live Docs, Repos
- Tooling and Impact
- Capacity Planning
- Log Management and Security
- Alerts and Troubleshooting

Modern documentation: Wiki, live docs, repos

The thing about documentation is that everyone admits that it is important, everyone talks about it, and almost no one does it or if they do, they don't keep it up to date. Documentation is boring, often harder than it seems to do well, and because almost no management will ever follow up and verify it, extremely easy to ignore. No one ever gets promoted because of excellent documentation, no one throws documentation parties, and no one talks about it on their curriculum vitae. Documentation just is not cool enough for people to want to spend time talking about.

Documentation is, however uncool it might feel, amazingly important for so many reasons. It can go far for moving someone from being an acceptable system administrator to being a great one.

Documentation does some interesting things. Of course, it allows us to recall how systems work and what tasks need to be done to them. It allows us to hand off tasks to others. It protects the business should we go on vacation, get sick, or move on to greener pastures or even retire. But beyond these obvious points, documentation allows us, forces us in fact, to think differently about our systems that we maintain.

In the software engineering world, a new technique of writing tests before writing the functions that they test has become popular and has shown that it can force people to think differently about how they approach problem solving and can lead to greater efficiency. We have very similar benefits in system administration. Approaching documentation more aggressively can lead to faster processes, better planning, less wasted time, and fewer mistakes. Taking a documentation-first approach, that is writing documentation before systems are built or configured, can help us think differently about our system designs and to document thoroughly: An intentional process of documenting what should be, rather than attempting to document what we did. This provides a wholly different way of thinking, and a way to double verify veracity, and an actual process to encourage completeness of documentation.

If we force ourselves to document everything before we enter it into a system, we can improve our chances of having accurate and complete data. Avoiding the need to go back and attempt to remember everything that was needed or done on a system is important. Using a document first process we have an opportunity to catch missed documentation at the time that we go to use it, which may only be a few minutes later. If we work first and then document it, it is very easy to forget small details and there is no triggering event to remind us to verify that something is written. Alternatively, if we document first, we have the moment when we need to put data into a system or a configuration to make. We have a triggering event, the actual moment of entering configuration data to remind us that we should have pulled that out of a document. It is not foolproof, only more reliable than typical processes.

No one really disputes, at least not in polite company, that documentation is needed or that it is one of the most important things that we can do working as system administrators, or really working in information technology at all. It is practically a mantra that we repeat, yet few of us really internalize this decision. Instead, we pay lip service to the ideology of documenting everything and still push off documentation as a secondary concern that we might do tomorrow if, and only if, we get bored during some mythical free time. This is where things break down. We cannot simply claim to believe that documentation is all important, we have to truly believe it and act accordingly.

We can talk all day about the importance of documentation, but all that really matters is taking that knowledge and putting it into action, however that works for you and your organization. To make that more likely to be successful we need to use good documentation tools. Having a high barrier to documentation encourages us to avoid it or to see it as too time consuming to do at the correct time. If we make documentation fast and easy, we are much more likely to find ourselves just doing it, perhaps even enjoying it to some degree. I know that I am personally very satisfied finding my documentation to be complete and up to date - having that satisfaction of knowing that I could show it to someone, at any time, and feel good about what is there.

In choosing a platform for documentation we have many considerations. How will the documentation be stored, backed up, protected, secured, and accessed? What kind of data will be stored: text, audio, images, video, or code? How many people will use it? Do you need to make it accessible inside an office? In multiple offices? Globally? Will third parties need to access it?

Wikis

Over the last two decades, the wiki has arisen to be the de facto tool for documentation of all sorts. Wikis are designed around being fast and easy to edit and at this they excel. Wikis also traditionally use simple markup languages, like the MarkDown language, that make it easy to store exact text and technical data without it being manipulated by a formatting system. This creates a minor learning curve but rewards a small amount of very standard learning with the ability to make very accurate, well formatted documents quickly.

The wiki format is all about simplicity - the simplest possible system that still allows for enough formatting to be able to be used for nearly any documentation task. This simple format makes it easy to have a variety of wiki products on the market that satisfy nearly any specific need. From small, light, and free open-source products that you can run yourself to large, hosted commercial offerings that you simply sign up for and use. It covers nearly all bases. Organizations of any size can use it effectively and there is nearly always a wiki option that integrates with other systems that you have.

A wiki will generally suffer from a need for some degree of organization which is not native to the platform. The strength of a wiki, that it is fast and flexible, is also a great weakness: it is just far too easy to start to throw data someplace that it does not belong and to leave no trail as to how to find the information again. Some wikis will go above and beyond the basics and include meta data tagging options or structured data organization options. These are the exception, not the norm.

Wikis have, for a number of years, been used as a component of or even the basis for larger products. A great example of this is Microsoft's SharePoint which uses a wiki engine as its core rendering engine and all of its interface details are simply advanced components being rendered on top of a wiki.

An issue typical to wikis is that they are rarely able to have the same data modified by multiple people at the same time. Their simplistic design often assumes that they will be treated quite simply - a single author, as the only reader, during the time of edits. This makes a wiki more useful in single user environments, or environments where users rarely use the documentation platform at the same time, or in organizations where different users tend to be segmented off from one another so that they will use different documentation pages at different times. If your team needs to have multiple people making active edits, or viewing updating information, in real time of the same data then other documentation options are likely going to be better.

Live docs

The newest way to approach documentation feels much like a step backward: word processor documents. Yes, you read that correctly. Hear me out.

Traditionally, that is in the late 1990s and early 2000s, the idea that you would use a word processor document as documentation seemed ridiculous unless you were such a tiny company that you only had one person who would ever need to use and access these documents and then it was reasonable as it made storing everything as a file relatively easy. This has changed heavily in the last several years as new technologies have turned nearly all mainstream word processors into online, web-based, multi-user tools that only resemble early word processors in their superficial capacity, but not in usability or technology.

These next generation document systems provide a surprisingly powerful and robust mechanism to use for documentation purposes. While there is no single universal standard for how these systems should behave, a set of conventions have arisen that are sensible and are followed by all major systems and are available both commercially and in free, open-source packages as well as in hosted or self-hosting modes. Of the greatest importance to us are the ideas that these live documents are able to be edited by multiple users at the same time, show changes as they are made in real time, have secure access controls, track changes, and use web interfaces that are easily published online or anywhere that they are needed, as well as, being able to output documentation to an easily portable or transferable set of formats.

Modern document handling systems like these will often times use a database behind the scenes, rather than resorting to sets of individual documents, and only expose individual documents as views into a single, large data set rather than truly individual sets of data. These systems are becoming increasingly powerful and can fit easily into other document management or replacement workflows. Nearly all organizations today are already tackling the need for modern document systems in other parts of the business, and these will easily be systems into which system documentation can be added without incurring any additional cost or effort. Doing more with the systems you have to maintain already is a great way to get high value at low cost.

Because these modern document systems allow for multiple users on the same document at the same time, they are especially useful for times when you have a multi-person team working on a single customer or system at the same time and the documentation needs to be shared. That way one person making a change keeps the data on everyone's screens constantly updated. The documentation system itself becomes a mechanism for team collaboration instead of being a risk of using outdated data because someone did not know to refresh their view.

Alternative interfaces to similar data

As these kinds of tools have become more and more popular, alternative interfaces to similar database drive document data have started to arise. Popular alternative formats like notepad applications are beginning to become more popular. These formats are less well known than traditional word processing and spreadsheet tools but can be very good for system documentation.

Because of the multi-media and often changing ad hoc nature of documentation, journal-style applications can be ideal. Over time I expect to see more and more applications designed around flexible documentation to become more mainstream.

Personally, I have become a large fan of these systems. They utilize standard tools that nearly all staff know already and tools that are likely to be already being used for many other purposes and repurposes them in a way that is surprisingly well suited for them. Less retraining, fewer special case tools to manage, and easy access and usability by teams that may use the documentation less often.

Repos

A new standard for documentation is to use online code repository systems. These systems generally work from a collection of text files that are loosely formatted but they get updated and version controlled centrally. This is a very different approach than what is taken with the examples given previously. This system does not address live collaboration between team members, but does allow for offline usage quite easily using standard tools used in the development space.

The real reason that the use of version-controlled code repositories has become an area of interest for documentation is that it is already being used heavily in the development and DevOps spaces and so is a natural system to adapt for use in IT documentation. The ease of using documentation offline using local reading and writing tools, and the ability to also have online copies makes it very flexible.

There are beginning to be ways to even use this type of documentation in a live, shared manner with some of the newer editing tools. Likely we will see this advance significantly in the near future as more focus is put on expanding the robustness of this process.

Ticketing systems

Traditional documentation as we have been discussing is really state documentation: documenting the way that systems *are* or, at least, *should be*. There is much more to document. The other system that we should use is a ticketing system or, to think of it another way, *change documentation*.

Tickets are a form of documentation just like your wiki might be. Unlike a wiki, tickets are focused on recording events in time. They track errors, problems, issues, requests, observations, reactions, changes, decisions, and so forth. Unlike traditional documentation that is a final document showing the results of all decisions and changes made until the current time, your ticket system should reflect the history of your systems and workloads to allow you to, theoretically, *play back* events as they happened to not only know what changes were made to a system but also who made them, who requested them, who approved them, and why.

Tickets, when used properly, play a huge role in the lives of a system administrator. While possible to function without a good ticketing system, this will add so much unnecessary work. Using tickets to track tasks as they get assigned, the process of completing the work, and the final disposition provides the missing half of the documentation puzzle.

If your business does not have or is unsupportive of getting a ticketing system, consider implementing a private one just for yourself. Ticket software comes in many shapes and forms and free software and services are available that work quite well if spending money on upgraded products or more extensive features is not an option. You can think of your ticket system as a personal work journaling mechanism if that makes more sense.

You do not have to go overboard attempting to integrate tickets into your company's greater workflow if you cannot get top level buy in or it does not make sense for how the organization should work, but it is hard to imagine any IT department that would not benefit dramatically for being able to track IT change events, including denied events, to be able to demonstrate the history and activity of the department and to be able to trace potential issues caused by changes in our systems.

Like traditional documentation, ticketing systems come in all shapes and sizes. Play with a few, give them a try, do not be afraid to change to something else. Find something that works for you and allows you to effectively document the changes that you make, when you made them, how long it took you, why you did it in the first place; and all with a minimum of effort.

Approaching documentation

Chances are you have some amount of documentation in your job today. Chances are even better that what you have is incomplete, out of date, and essentially useless. It is okay, nearly all companies suffer from bad documentation. But this is a tremendous opportunity for improvement.

If you have documentation like most businesses, the best thing is often to literally start over. Look over your options, think about how your company will need to approach documentation and collaboration and pick an approach. It does not even have to be the right one. Any documentation process is good, even if you risk having to do it again. Pick a system and try it out. See if it fits the style of the data that you need to store, if it is comfortable for you to use, and if it allows for the style of collaboration (if any) that your business needs.

Do not try to document absolutely everything right away. Take a single system or workload and try documenting that one item in a very good way. Format it to look really good. Organize the data to make the data that you need quickly be clearly visible and available near the top so that someone trying to address a problem does not have to search far to find what they need. Remove redundancy and ensure that data exists only one time, in a single, predictable place. Think of your documentation like a relational database that needs some normalization, and the first major step is organization and the second is removing redundancy. Documentation is always hard to maintain, and redundancy of data makes it all but impossible. Attempting to change unknown occurrences of the same information gives no clue to how to find it all and what needs to be updated.

When you find a system and process that works for you, stick with that. Start documenting everything. Make it a huge priority, do nothing without documentation. Add tickets and start making everything get tracked.

Maybe you work for a company where documentation is already good. Chances are, though, you do not. And if you do, chances are you will never work in a place like that again. Good documentation is a rarity even before we consider the importance of tickets in the overall documentation equation.

There are no best practices as to what tools to use or in what format to put your documentation, but there are some high-level best practices to consider:

- Use both state and change documentation systems to track all aspects of your systems.

- Avoid data redundancy in state documentation systems (it is fine in change systems.)

- Keep all documentation up to date and secure.

- Do not document data that can be recreated reliably from other data.

It is easy to say that we need to be religious about our documentation, and everyone agrees that it is of the utmost importance. Yet actually moving from saying it, to doing it, is understandably hard. Management rarely verifies documentation as we are working, but they do reward getting other work done and frown upon delays. It is unfortunate that often the most important aspects of our careers are not seen as important enough or interesting enough for those outside of our field and they get deprioritized by people with no knowledge of how they play into what we do.

In our next section, we will move on from purely manual system tracking to beginning to use tools on our systems to measure and track them in a more automated fashion.

Tooling and impact

One of the fundamental natures of physics, as well as a rule that you learn straight away in industrial engineering, is that you cannot observe or measure events without in some way impacting them. In computing, we face the same problem. If anything, we face it far more than in most other places.

The more that we measure, log, or put metrics on our systems the more of the system resources needed for our workloads is taken up by the measurement processes. As computers have gotten faster over the years the ability to measure without completely crippling our workloads has become more common and now, we often even track checkpoints inside of applications in addition to operating system metrics. But we always have to maintain an awareness of what this impact is.

At some point there is more value to just letting the systems that we have run as fast as they can rather than trying to measure them to see how fast they are going. A sprinter running flat out is faster than a sprinter running while carrying measurement devices to determine their speed. The measurement process works against them. However the sprinter getting more feedback might be able to improve with the additional knowledge over time. But you will never see someone attempting to outrun a charging hippopotamus (they are one of the fastest and most dangerous land mammals, you know) first stop to turn on measuring devices. They will just run as fast as they can. Knowing how to run faster is only useful if it gives you both potentially useful data that by using you can enact improvements and you get the chance to implement those improvements. If the hungry hippopotamus catches you, all those measurements will be for naught.

Different tools will have very different levels of impact. Some simple everyday tools that we use on our systems may have almost no impact at all, but will generally give us only an extremely high level view of what the computer is doing. Other tools, like log collection, can require a great many resources and can even put noticeable strains on networking and storage resources.

Collecting data from a system is not the only activity that uses system resources. Collating that data and presenting it in a form useful for humans also requires resources, as would shipping that data off to an external system. Each step of the process requires that we use more and more resources. All of this is before we even consider how much human time may be involved in examining the data, as well. It is always tempting to simply opt for the most possible insight and monitoring into a system, but unless we can derive true value from that process it is actually a negative to do so.

In general, we will use a variety of tooling on our systems to provide some degree of regular measurement. There are both data collection tools, like the sysstat SAR utility, and immediate, *on the spot* observation tools like `top`, `htop`, and `glances` that allow us to watch a system's behavior in real time. Both kinds of tools deliver a lot of value.

Of course, there are a large variety of both free and paid, software and service tools that can take your monitoring to another level. Researching these will be very beneficial as even the open-source offerings have become amazingly powerful and robust. Performance tooling is typically handled locally as it is rarely used for alerting or security considerations and using it to perform postmortem investigations is often fruitless so incurring the cost of central data collection for performance data is not commonly worth it. Centralized tools do exist and can be quite useful. When used, these tend to be chosen for ease of use to humans rather than to serve a specific technical need. Decentralized tools that can optionally leverage a single pane of glass style interface to display data from many locations are quite popular for this specific need.

Netdata

I typically do not want to delve into specific products, but I feel that **Netdata** makes for an exceptional use case as a way to demonstrate the variety and power of available tools on the market today for Linux system administrators.

First, Netdata is free and open source. So as a system administrator who may have to justify any software purchases, this is one that can be downloaded and implemented to enhance our monitoring abilities without needing any approvals or notifications. Installation is also very quick, and very easy.

Second, Netdata provides crazy gorgeous dashboards right out of the box. These make it just more fun to do our jobs, for one thing. If we need to show data to management or present it in a meeting, few things are going to look more impressive and polished than Netdata dashboards. This is a tool that makes it easier to sell the business on what we do.

Third, Netdata uses surprisingly few resources. For the amazing graphical output that it generates you would never expect such a light utility to be able to pull it off.

Fourth, Netdata is decentralized. It runs locally on each server and does not send its data off to a central location to be collected. You can make a combined view of many systems, but doing so is all handled in the web browser of the viewer actively pulling the individual dashboards from each system directly and simply displaying disparate systems on one screen. There is no central server used to aggregate before display.

I love Netdata as an example of truly useful, free, open source, groundbreaking software that makes our everyday experience in system administration better. And it shows a pattern that is potentially useable for a great many other products and product types to make decentralization more viable than it may first appear.

One of the more important things that you will do as a system administrator is learning what tools to use, when, and how to read them. One of the most valuable things that I have found over the years is becoming comfortable with what a healthy system will look like, both historically and in real time, and being able to look at a variety of tools and to get an innate sense of how the system is behaving. There is little way to teach this other than talking about the value of observing systems at idle, and at standard load and observing what they look like; and of course, the better you understand how system components, and software works, the more you are able to interpret what you are seeing in a meaningful way.

With enough practice and understanding it can be possible to essentially *sense* the behavior of a system and gain a confidence into why a system is behaving as it is. This is not something that can be learned from a book and requires putting in a lot of time working with systems and paying close attention to what you observe from monitoring tools and combining that with what you observe from the system's performance and a solid understanding of the interaction of the physical components.

I find that momentary tools, such as top which is included in nearly all systems by default, presents a perfect way to stare at running systems as they perform their duties and become accustomed to how CPU utilization will fluctuate under appropriate load, how processes will shift around, and how load will vary. Some of the most complicated system troubleshooting will sometimes be done with little more than staring at changing process lists over time (and performing really well-timed screenshots.)

This is an area that can do quite a lot to separate junior from senior system administrators. It is far less about knowing the basics as much as truly internalizing them and being able to intuitively apply that knowledge on the fly when a system is behaving badly or possibly being able to do so based solely on someone describing the problem! How drives perform under different conditions, how the CPU is behaving under different loads, how caches are hit, how memory is tuned, when is paging good or bad, and so forth.

Understanding all of these factors is an ever-changing target. Each system is unique and new CPU schedulers, NUMA technologies, drive technologies, and so forth regularly change how systems behave and what our expectations of them should be. There is really no substitute for experience, and only one way to get experience.

Choosing tooling can be hard. I tend to stay very light unless I have a workload with a very specific need. You should play with many different measurement tools to have a good feel for what is available and be ready to choose the right tool for you and for the task at hand whenever needed. In many cases for me, the simplest tools like **free**, **top**, and the **sysstat suite** are more than adequate for almost everything that I do, and they are available on essentially every system that I have encountered for over a decade. But on my own systems in my own environment, you will often catch me using something a bit more graphical and fun like Netdata as well.

Best Practices:

- Learn measurement tooling before you need it and learn to use it quickly and efficiently.

- Limit your usage of measurement tools to only that which is truly useful. Do not impact performance without a good reason.

Now that we have talked about how we measure what our systems are doing, it our next section we start using these tools to plan for the future.

Capacity planning

When we take our knowledge of system resource usage away from the being in the moment and begin to apply it over the long-term aspects of a system, we start to think about capacity planning. Capacity planning should be, at least in theory, a rather important aspect of system administration. Many organizations will treat capacity planning as a non-technical exercise, however, and take it out of system administration hands. It is amazing how often I am told by a system administrator that they have received hardware that they did not specify and now have to *make it work* even though it was designed by someone with no knowledge of how it would be used! So much training and knowledge of system design in system administration being ignored and critical purchasing being down with no rhyme or reason.

It Is already designed when purchased

One of the strangest problems that I run into with great regularity is system administrators asking me how they should set up hardware which they have already specified, ordered, and received. Most critically, they ask how they should configure the RAID and division of logical disks or splitting physical arrays. I am always amazed by this.

Obviously, different configurations of the server change how it would need to be configured at the hardware level. The software to be run on the server, the amount of software, the needed performance of that software, the amount of storage that will be needed, how backups will work, and nearly everything about how a server will be used over its anticipated lifespan is needed to be well understood to be able to even begin the process of specifying hardware to be purchased. How did they know how much RAM to buy, or how many cores, how fast the CPUs should be, which CPU models to start with, even which brands would work? In many cases people overspend and overbuy by such a degree that things work out and no one notices because the mistakes are made in the form of lost money that no one investigates. A server budget was given, no one follows up to determine if the server that was purchased was a good value, only if it was in budget. So overbuying is often a way to cover for failing to do capacity planning, and one that can be costing companies a significant amount of money.

Most noticeable, though, is RAID configuration. When someone asks me what RAID level and configuration that I would recommend for hardware that was purposely purchased new for this project I have no idea how to respond. Any and all decisions about the RAID configuration surely had to have been made before the server was purchased. It is only by knowing the performance, reliability, and capacity artefacts of not only each RAID level but of different configuration options and applying that knowledge in combination with available physical drive and controller options that you could have even approached purchasing the storage portion of the server in the first place.

In order to decide on storage needs you have to know what you need, first. Then you have to know how the hardware that you will specify will meet those needs. Some questions that come to my mind when someone says that they have hardware and bought it without having specified any design yet include:

How did you know which hardware RAID card to buy? Or even that you needed a hardware controller at all?

How did you determine how much cache to purchase?

How did you know which types of drives would most useful?

How did you know what speed of drives you would need in throughput and/or IOPS?

How did you know what size of drives to get?

How did you know the quantity of drives to get?

How did you determine caching capacity?

How did you determine tiering capacity?

How did you determine hot spare needs?

The decisions necessary to make any of these decisions require having made all of the decisions as a whole. The final outcome is a product of the whole and any change in RAID level, for example, would drastically change the usable capacity, the system performance, and the overall reliability. Every small change makes everything else change with it. No piece can be decided upon individually, let alone changed. The most innocuous change could result in a system that is not large enough in capacity, or fast enough for the workload to function; and more dangerously the reliability of the storage system could swing wildly between extremely safe and extremely dangerous.

It is really hard to describe just how crazy this process is; and even crazier to realize that this might even be normal for how people buy servers! The best analogy that I can muster is to say that it is like buying a transmission for a 1978 Ferrari and expecting it to just work when you do not even know if you are getting a car, boat, or small plane yet, let alone what year or model of Ferrari!

Capacity planning is about far more than saving money, at least indirectly. It is about ensuring that systems that we purchase can meet all of the *projected* needs of our business for the duration of time that makes sense to do so. This is obviously a difficult number to really nail down as what feels appropriate as a projection, what are the likely changes coming in the near future, and what is a reasonable time frame for your systems, are all rather fungible concepts.

It is a common trend in businesses to want to project astronomic growth using *pie in the sky* numbers as hardware investment bases as well as using the maximum reasonable lifespan of the hardware to calculate over. While we cannot control the political processes that drive our businesses from our positions within IT, except in the rarest of cases, what we can control is the quality of our own numbers being provided to those making the decisions.

Buy late

Good business logic says that, with rare exceptions, the cost of everything in IT goes down over time. It goes down a lot. The cost of memory, compute cycles, or storage is fractional today compared to just a few years ago and this trend has really never stopped nor reversed, nor is it likely to. Momentary issues due to scarcity during times of manufacturing or logistical crisis can happen, but these are extremely rare and short lived events. Given any amount of time to make a purchase, the cost of systems in a few months will be better than they are today. Either the money that we spend is less, or the amount that we get for that money is greater. In either case, we benefit by investing later.

A common example that we can use is what if we bought a server today with plans to use it for eight years and we have expected growth, so we buy a server that meets our eight-year projections. To acquire a server with that much power, maybe we will spend $20,000 today. Or to get a server that we project will last us for four years, we might spend $8,000 today. A big difference. Of course, this is a contrived example, but in the real world, these kinds of costs are typical in a lot of common scenarios.

In this example, we then assume that in four years we can buy another new server for an additional $8000 that meets our needs for four more years. Again, contrived, but often true. The cost tends to work out similarly to this.

The number of advantages to buying less to last an expected shorter amount of time is hard to overstate. First there is often hard cost savings because of the nature of server pricing means that buying less, more often simply costs less because of the price benefits that happen within the operational lifespan of a modern server. And then there is the time-value of money that says that spending the same amount of money, but delaying spending it, means that you have more money to make you money in the interim and that the same money that you spend in the future is worth less than that money today. Then there is newer technology - if we wait for years to buy a new server we potentially get a lot of newer technology in that server that can contribute not only to capacity advantages, but also lower power consumption, great durability, and so forth.

Then the advantage of having two servers. We assume that the second server is going to replace the first, and maybe it will. But we might also use the purchase to simply expand capacity, the first might remain in production service. If we are replacing the first server, it may be redeployed in another role within the organization or could be used as a backup server for the new one. Almost always you will be able to find a highly effective use for the original investment.

Most likely the biggest advantage is in delayed decision making. By holding off spending much of our initial budget by several years we get the flexibility to invest that money at any time, or never. Instead of doing an eight-year projection, which is wildly inaccurate to the point of being totally useless, we do two four-year projections, which are still pretty inaccurate, but the degree to which they are more accurate is pretty crazy. At our first four-year checkpoint we get to evaluate how good our last projection was and make a new one based on this new data and new starting point. We not only get to do a fresh evaluation of our own organization with four more years of insight, but also four more years of insight on the industry, and four more years of new technology. Very few businesses would make the same decisions in four years that they would make today. In business, delayed decision making of this nature can be astronomically beneficial.

Sadly, for most businesses, projecting becomes an emotional exercise because there are political benefits to making people feel good and showing faith in the business or its leadership; and it just makes us feel good to think about all of the success that we are surely going to experience. And better projections normally means more clout, bigger budgets, more to work with for many years to come. Almost no business ever goes back and evaluates past projections to see if people did a good job, so there are rewards for being overly optimistic and generally zero risk of retribution if they are falsified for personal gain (emotional, financial, or political.) This system makes projections very dangerous and anything that we can do to reduce our dependency on them is important.

At the end of the day, delaying purchasing of server resources until they are actually needed is one of the best practice strategies that we can have in technology purchasing. It is the best approach from a purely financial viewpoint, the best approach from a decision making and planning perspective, and the best way to allow technological and manufacturing advancements to work in our favor.

There are three key numbers that will come from system administration during this process. The first is simply answering the question of *how many resources are we using currently?* The second question is *how many resources did we use in the past?* And third, *what resources do we think that we will use in the future?*

In reality, answering any of these questions is surprisingly hard. Just counting up the resources that we have purchased and own today tells us nothing. We need to really understand how our CPUs, RAM, and different aspects of storage are being used and how they affect workloads.

For example, if we are running a large database instance the database might happily cache outrageous amounts of storage into memory to improve database performance or reduce storage wear and tear. But reducing available memory may not have an impact on database performance. Because of this, just measuring memory utilization can be very tricky. How much *useful* memory are we using? Storage is easier, but similar challenges can exist. CPU is the easiest, but nothing is ever completely straightforward. Just because we use system resources at a certain level does not tell us how well they are needed.

With CPU we might have a system that is averaging fifty percent CPU utilization. For one company, or set of workloads, this might mean that we can put twice as many workloads on this system to get its utilization close to up to one hundred percent. For another company, or set of workloads, it might mean that we have overloaded the system and there is enough context switching and wait times that some applications are noticing latency. It is not uncommon for companies to target ninety percent utilization as a loose average, but for others targeting just ten percent can be required.

The tradeoff, in this case, is about waiting for throughput or latency. Available CPU cycles means that the CPU could be doing more tasks, but if a CPU is tied up doing tasks all of the time then it is not necessarily available if a new task is suddenly presented to it. If you are working with low latency systems, having available system resources at the ready to process that task at the time that it is first presented can be a requirement. In order to assess capacity use and needs for systems requires us to deeply understand not just how much of the system is being used, but what that ultimately means for our workloads.

As with so many aspects of system administration, the key is to understand our workloads inside and out. We have to know how they work, how they consume resources, how they will respond to more or fewer resources. Everything that we do in capacity planning depends on this.

Of course we have tools that we mentioned previously to help us with determining how well a system is performing, and with application measurement tools and/or human user observation we can reasonably determine what kind of resources are necessary for where our workloads are today.

Many of these tools that we use can also be used to collect historical records of system performance. Chances are that if we were to do this all of the time that we would produce a volume of data that we will never be prepared to utilize. Some organizations do collect this forever, but this is the exception, not the rule. More practical, in most circumstances, is to develop and track baselines over time. This generally means doing some sort of activity where you record measurements, as well as recording what you can about end user application performance, of the system so that you can look back and see what system utilization has been. This data should be collected over long periods of time. Weeks or months, at least sometimes, to find hot spots and cold spots. Common cold spots might be Sunday overnight when many applications are not used at all. A common hot spot is month-end financial processing times. Every organization has different utilization patterns. You need to learn yours.

With this collected data we can then analyze to see how utilization has changed over time. Changes will normally come from increases, or decreases, in application level utilization. But this is far from the only aspect that might change. It is important to be cognizant that application updates, operating system patches, changes in system configuration and so forth should be recorded and noted against data recording to make the data more meaningful in evaluation.

Long term data collection has to be considered against effort. Collecting data and collating that data on any scale can be extremely time consuming and potentially resource intensive. There is the possibility that after collecting all of that data that it will provide no useful insight or that the skills to read it back will be lacking. It is not unreasonable for a system administrator to track system performance data mentally if working with systems that are used constantly. In some cases, this will be more practical.

Risk of too much data overhead

In pursuing capacity planning we risk creating a situation where we generate more overhead for ourselves, which will generally equate to more cost, than if we had not collected the data in the first place. We have to find an appropriate balance.

Large organizations will tend to find large value in using lots of data collection to save money on a large scale. At scale automating the data collection and analysis is often relatively simple. Small businesses will often find this impractical. To collect any reasonably thorough amount of system data for a single server could result in expenses as large as the cost of the systems themselves. Clearly that is unworkable. Common sense has to prevail.

Many small organizations will have just a single primary workload and will potentially never fully utilize the smallest of servers and will, from a capacity perspective, always experience overkill until moving away from running their own hardware, if that ever becomes practical. Large organizations are operating farms of servers and have many avenues to improve overall cost from playing with different software options, using many smaller servers or fewer large ones, using different processor models or even architectures, and so forth.

The effort to save money just has to be kept in check against the potential value in the data collection. This is true with all decision-making processes. Beware that data collection is part of the total that makes up the high cost of decision making and you always have to remember to keep that cost far below the expected value level of that decision.

When evaluating the cost of the data collection, we have to consider the time to collect the data, the cost of storage, and the cost of analyzing that data. We cannot forget, we have to consider the cost of considering all of this!

There are vendors that make tools specifically for tackling these difficult questions and they can be very good. Dell, famously, provides tools to customers that can be run over long periods of time and produce very detailed reports as to how systems are being used and, of course, also provide *recommendations*, which are actually sales pitches, to sell you more products. If used properly, these tools can be quite valuable.

Of course the natural question will also be *but what about cloud computing, does that not change all of this?* And yes, considering cloud computing is important and plays into this process.

As cloud computing enters into an organization's planning we have even more complexity to consider, in some ways. In other ways, cloud computing can make the process of capacity planning far simpler, if not moot.

In cloud computing, or at least in nearly all of it, we buy our capacity as needed or very nearly as needed. This is the beauty of cloud computing. Use only what you need and let the system decide what it needs in real time. This is great, in theory. But just allowing the system to do this still leaves us with a need to predict what this approach will cost in order to compare financially against alternatives, and to predict what this will cost in order to budget properly for it.

If your organization is using cloud computing currently, this can make our processes far easier. Generally your cloud platform itself will be able to tell you an awful lot about system utilization rates. Even if traditional reporting is not available, the billing for cloud computing can often tell you as much as you may need to know for many types of planning.

Our capacity planning best practices are purely mathematical. Use reasonable measurements and understanding of our systems and workloads and our best understanding of business expectations and input from other teams to plan for capacity needs for tomorrow and into the future. Study and understand the *cone of uncertainty* and use that sense of increasingly unforeseeable future combined with a good understanding of financial concepts such as the time value of money and the increased value of technological purchasing over time to provide best effort evaluations of future capacity investment needs for your organization.

Capacity planning is often boring and more political than technical, but it is a role that can rarely be handled by anyone except system administration so it is our lot in life (professional life, at least) to be integrally involved in system hardware purchasing projections. It is essentially the computing futures market on a tiny scale inside of our own business. In our next section we move from capacity and performance needs and look at tools used for security and troubleshooting starting with system and application logging.

Log management and security

If you ask system administrators in casual conversation at the bar, you might believe that it is a major task for system administrators to collect all of their system logs and to spend hours each day manually and skillfully going through them line by line looking for system errors and malicious actors. Reality is very different. No one is doing this, no one was ever doing this, and no company is interested in paying for people to do this. Log reading is a serious skill and an activity that is excessively boring. It is also a type of task at which humans are extremely poor.

If you were to attempt to have humans doing your log management by actually reading logs when there is nothing known to be wrong with a system you would run into a few problems. First, realistically no human can read logs fast enough to be truly effective. Systems log a lot of data and attempting to keep up with that kind of flow of truly mindless information would make humans extremely error prone. And then there is the cost. Anyone skilled enough to be able to handle log reading like that would be at the top end of the pay scale, and that job being so painful would have to be a premium pay position, and since typically servers run around the clock you would likely need four or five full time people *per server* to even make the attempt. Costly beyond anyone's wildest imagination and completely impractical to the point of useless. And hence, no one does it.

That does not mean that logs are not valuable. The opposite is true. Logs are very valuable and we need to collect, protect, and know how to use them.

When it comes to logs we need to know how to read them. This might seem trivial, but when you have an emergency and need to read your logs is not the time to find out that you do not understand what is being logged or how to interpret it. The system logs on Linux are rather consistent. The challenges really begin when different applications are logging as well. These logs might be independent or consolidated into other logs. Each application is responsible for its own logging and so we can face quite a potential for log variety when we start running a number of disparate applications. Add to this mix any logging done by bespoke in-house applications and things can possibly get quite complicated.

There is no need to teach log reading here. This is an activity that anyone reading this book should be well acquainted with. The exercise that you should perform now, though, is to go through your logs and determine which logs are of a format that is unfamiliar to you and make sure that you are ready to read through any of your system logs at any time without needing to do additional research before doing so. Fast and efficient log reading will do much to make you a better system administrator.

Knowing how to read your logs is not enough. You should also be familiar with your logs enough to recognize what normal activity will look like. You will be far better at recognizing when something is wrong if you first know what it looks like when it is right. So many companies bring in specialists to deal with problems after they have arisen and in doing so, there are so many more challenges created because there is no *logging baseline* to use as a basis for comparison against what things are doing now.

Without a solid baseline, errors that occur regularly can cause a lot of wasted time as diagnostic time is spent researching them to determine if they are normal log noise, a real problem, or an actual problem that is simply a component of the problem being diagnosed. When things are going wrong, we want to be extra efficient. That is the very last time when we want to be figuring out what good looks like.

It is always good to be prepared to hop directly onto a server and use traditional tools like vi, view, cat, head, tail, and grep to look at logs. You never know what situation you are going to be thrown into in the future.

With many modern systems today we expect to see extensive tooling around logs as there is much more that we can do with our logs than simply storing them on our local servers and poking around at them after something bad has happened.

Today, logging is one of the areas that has seen massive changes and advancements in server systems. We are leaps and bounds beyond where logging typically was just twenty years ago. There are some very simple advancements, such as high performance, graphical log viewers, that can be used to make the observation of logs faster and easier. There is advanced central logging to move logs away from the servers themselves and there is automated log processing.

Regardless of if logging is local or remote, modern log viewers have made a tremendous difference in how efficiently we can use our logs. Whether using local desktop GUIs, character-based sessions on the terminal, or a modern web interface, log viewing tools have been improving and for the last decade or more have made the act of reading logs pleasant and easy. Amazingly, very few organizations provide these kinds of tools or accommodate their usage and so it is far less common than it should be to find system administrators using them. If you talk to system administrators do not be surprised to find out that very few have actually had the pleasure of working in an environment with logging tools beyond the basic, included system log text files and the standard text file manipulation tools that are included in the operating system.

Good log viewers are an important starting point in a log management journey. Make logs accessible to view quickly and make their viewing a pleasurable and as simple as possible experience. When things are going wrong you do not want to be spending any more time than is absolutely necessary getting to your logs or to digging through them. You certainly do not want to be working to install a log viewing tool at a time when something is already broken.

Log viewing applications are only a starting point, we hope, for most organizations. The real leap in log management happens when we introduce central log collection. This is really where the logging revolution has taken place. Of course, even going back decades, the potential and tools for basic log aggregation existed. This can be as simple as using network copy commands or network mapped drives to store the same old text files on a central file server. So the fundamental idea of central logging is not new.

Originally the central logging constraint was that servers were not networked. Later it was performance and storage issues. Centralizing logs used a lot of network bandwidth and required a lot of storage capacity and, in some cases, was a performance nightmare for the log server as well. Those days are long since in the past. Today all of those things are trivialized simply by the natural leaps in capacity and performance of all aspects of our systems with far smaller growth in log size and complexity. Logs today are not all that much larger than they were decades ago.

Early log centralization systems were very basic and unable to scale gracefully. Large amounts of aggregated log data presents big challenges for most systems as they need to be able to continue to ingest large amounts of real time data from many sources while simultaneously being able to recall and display that data.

Modern central logging applications all use new, modern databases designed around this time of data flow and storage. No one single type of database is used for this, but many newer databases excel at handling these needs allowing data traditionally stored as large, unwieldy text files to be reduced to much smaller and more efficient database items with metadata, caching, collation, and other features that allow for the ingestion of massively larger amounts of data than ever before while being able to continue to display data effectively. This change, along with the general improvements in system power, has made for effective centralized logging not only on the LAN, but in many cases, even for and to servers running hosted, in the cloud, or otherwise not sitting on a traditional LAN.

By using this kind of system we have a few benefits. One is speed and efficiency in log reading. One (or at least fewer) places to go to read logs means that system administrators are looking at logs much faster than before and faster log reading means faster solutions. By having logs from many servers, systems, applications, and more all in a single place also means that we can correlate data between these systems without requiring humans to look at logs from multiple sources and make these connections manually.

New tools, like volume graphs, also allow us to see patterns that we may have been unable to detect before. If multiple computers suddenly show a spike in log traffic maybe applications have suddenly become busy, or maybe there is a failure or attack underway. Centralized logging tools make it easier not only for us to understand what a baseline looks like for a single system, but what a baseline will look like for all of our systems combined! More layers of system understanding.

Once we have these modern tools centralized the next logical step is using automation to read logs for us. **Security Information and Event Management** (**SIEM**) is the term generally applied for automatic log monitoring tools. Automation for logs is not new and even the United States government was putting rules in place for it by 2005. But for many businesses, log automation is far beyond their current plans or capacity.

Of course, like any automation, the degree to which we use it can vary greatly. Light automation might simply send alerts in case of certain triggering events occurring in the logs or alerting on uncharacteristic activity patterns. Complex automation might use artificial intelligence or threat pattern databases to scour logs across many systems at once to look for malicious activity.

Why central logging?

Given that centralized logging carries so much network overhead and because it tends to be so costly to implement, it is easy to question the value to centralizing logging to a single server, or group of servers, rather than simply leaving logs on individual hosts and finding better methods of examining logs from them. This is a very valid question. Of course, central logging is not going to be right for every organization. It is right for a large number of them, though.

While many advancements have been made to make central logging more capable than ever before, there have also been many advancements in making decentralized logging better as well including on-device log viewers with reporting, alerting, and attractive user interfaces and aggregation tools that display the data from multiple sources in a single dashboard even though the data itself is disparately located.

Central logging offers unique advantages, though. The biggest advantage is simply that the data is not tied to the device. If a device dies, goes offline, or that device is compromised we have isolation for our logs. It is not uncommon to be stuck trying to get a server back online and running just so that we can look at the logs from that server.

If the server dies completely we may never want to bother bringing it back online. Or if the logs tells us that there is a catastrophic failure we might know from that, that we do not want to attempt to recover a failed device. Or perhaps we want one person to be working on getting a failed server back up and running simultaneously while another scours logs to determine what led up to the failure or possibly determine what is needed to restore services.

If a server or application is compromised there is a risk that the logging mechanisms or storage systems will be compromised along with it. In fact this is generally quite likely. Modifying logs to cover up a compromise is very common with sophisticated attacks and simply deleting logs common in simpler ones. In most cases logs are attacked because they are the most likely place to easily identify that a compromise has happened or is happening and what to do to mitigate it. If the logs never show any signs of an attack, you may easily never discover that one has happened.

If we instead send our logs, either in their entirety or at least a copy, directly to a remote logging server in real time then we have the logs stored separately from both the application and the server storage with an air gap so that, in order to modify those logs, a completely unrelated system has to also be compromised, without being detected. This is orders of magnitude more difficult to do, especially as the existence of and information about any external logging system will generally not be known until a compromise is already under way at which point it may already be too late to avoid detection.

Because of the fact that logging is far more than just a way to look for bugs and primarily the key security recording system for auditing and tracking breaches, malicious activity, infiltration attempts, and others. it is critical that this data be kept safe from both accidental loss and intentional destruction.

Possibly the most important purpose of log separation, at least in larger organizations big enough to have multiple IT teams, is the separation of duties. If the logs are completely controlled by the same system administrator (presumably you) that controls the server then it is trivial to make major modifications and hide the evidence of those changes. If the logs are sent to an external system to which we are not the administrator then it is much harder for us to hide those changes as we must prevent them from being logged in the first place while not causing the logging itself to fail.

Having a strict separation of duties for this level of security may sound like something limited to large organizations, but even quite small companies, even those not large enough to have a single fill time system administrator, can take advantage of this aspect of a system like this by using an externally hosted logging platform rather than running their own. In this way all of the necessary system administration and security for the logging platform is encapsulated not only away from the individual system administrator but also away from the IT team and the entire corpus of the company itself!

Because every business is different, there is a place for different levels of logging and log management depending on your needs. Our constant IT mantra has to be that one size does not fit all.

So our best practice with logging is a difficult one. We need to evaluate logging needs. How can we use our logs efficiently to troubleshoot faster and better. How can we increase our security. Do we have factors that make local log storage outweigh the benefits of remote? Should we host our own log systems or use a third-party SaaS application that is managed for us? Will the security benefits of a SIEM or similar solution justify their cost and complexity?

Our only true best practices are to ensure that you are prepared to read logs before you need to, and to look at logs from time to time to understand what a healthy system looks like for you.

With logging we are forced to really look much more heavily at rules of thumb rather than best practices. In general, central log collection is a worthwhile endeavor for almost any environment with more than a single critical workload. This can mean a single company that has multiple workloads, or smaller firms should generally be using some form of external support vendor and that vendor would, in theory, have multiple customers and would generally benefit from a similar approach allowing them to centralize logs on behalf of their customers.

The smallest IT department

This is a topic beyond the scope of the role of system administration, but one that everyone in IT should really understand because nearly every IT role will, at some point, be put in a position of being part of an organization that is simply, too small.

In most fields we can talk about a minimum size to any professional team. Doctors, lawyers, auto mechanics, veterinarians, software engineering, you name it. In all these examples we can talk about the staff and team necessary to make a position make sense. A doctor working with no nurses or assistants of any kind is going to be really inefficient and lack some vectors for healthcare training. Same with a veterinarian, if you are a lone vet and have no receptionist, cashier, vet tech, and so on, then you are forced to do roles at a fraction of your value. In software engineering it is more about the wide range of discrete tasks and roles that go into software design that cannot reasonably be done by a single person, even on a small project.

IT is one of the more dramatic fields for this because IT is so broad and covers so many totally different knowledge areas. And every company has the need for a large scope of that IT skill set. Some skills are unique to certain types of environments, but the large base of foundational skills apply to essentially any and every company.

One of the most challenging aspects of this is that there is little room for underdeveloped skills. Because IT is the decision making and guidance around core business functions, infrastructure, support, efficiency, and security there is really no time that you do not want expert and mature guidance. A seemingly simple mistake, made nearly anywhere in the entire infrastructure, carries the risk of being a point of breach, an over expense, a decision that starts small but leads to a domino line of other decisions that will all be based on that one.

Most businesses do not need most, if indeed any, individual skill more than part time with some skills, like that of CIO, being needed potentially for just a few hours per year. Obviously if skills are not needed more than a few hours per year, or even if only a few per day, paying for the skill full time would not make sense.

Then there is the issue of coverage. Many businesses need to only have coverage for forty hours per week with strict office hours and all systems capable of being taken offline when the office is close. This is, however, not at all normal. Most businesses need to operate six or seven days per week, and long hours per day and running twenty four by seven is totally reasonable. To have full coverage just for someone to answer support tickets, let alone make decisions or solve real problems, would require at least five people just to have shift coverage, ignoring any skills needed.

When you consider all of the discrete roles that exist in even the most minimal IT environment: systems, networking, CIO, helpdesk, desktop support, end user support, application support and then any specialty roles like cloud applications, backup, disaster recovery, project management, and on and on. What many companies attempt to do is to find a single person who can fill all of these roles, a generalist. This is a great theory, there is one person with tons of skills who can do a little bit of each one adding up to one whole person.

In the real world, this does not work, at all. First because the number of people who truly possess all those skills, keep them up to date, and are so good at all of them in the universe can probably be counted on one hand. Second, anyone with good CIO level skills or system administration skills has a huge per hour billable value from those skills alone. A worker is always worth the value of their maximum skill full time, not their minimum skill. And having additional skills raises your maximum. So a CIO with all these other skills would be, in theory, worth even more than if they only had the CIO skill. So even if you found such a person, either you would have to pay them an absorbent amount of money to do the job, or they would have to be willing to do the job for a tiny fraction of their value which from an employment standpoint makes no sense.

Second because of coverage. A single person can only work so many hours leaving a business without support most of the time. And even if you have a business that only exists eight hours a day and is happy to do all support and even proactive maintenance during that time you still have the issue that generally many of the roles that one person is called on to perform will need to happen simultaneously.

Amazingly, tons of companies of all sizes attempt this approach and universally end up with bad results, although many never measure their results or even understand what good performance from an IT department should look like or even what that department should be accomplishing so often ignore or even praise the failures in this area.

Finding a theoretically useful lower limit to the size of an IT department is hard. As a rule of thumb, if you do not need at a minimum three full time IT staff who never do anything outside of IT, then you should not try to staff an IT department.

Entire ranges of IT businesses like Management Service Providers and IT Service Providers provide IT skills, management, oversight, and tooling in small portions for businesses who only need a little of many different resources. Smaller companies should never feel badly turning to these kinds of companies, they are necessary to provide the level of scale and division of that scale necessary to do IT well. That said, just like employees, the average firm is not going to be very good. So just as you want to make sure that you are hiring employees who are good, you want to hire a service provider who is good. Service providers are very much like employees, but employees that are more likely aligned with your business needs and generally with far more potential longevity - a good service provider relationship could easily outlast the career length of an individual employee.

Rethinking inappropriate IT departments from both sides can be a boost to the industry. So many employers are unhappy with IT results that are predictably bad based on the IT structures that they enforce. And so many IT practitioners are unhappy with their careers or at least their immediate jobs, because they feel that they have to, or are encouraged to, work in environments that simply do not make any sense.

Considering service providers as part of the in house IT team can make it possible to get the IT team that you need, at a price that is actually plausible. IT is not really outside of the budget of any company. If it seems like IT is going to be too expensive, something is wrong. The job of IT is to make the business money.

The most common mistakes that I see when companies engage service providers is either assuming many incorrect rules of engagement such as assuming that local resources are better or that the service provider has to match the technology that you plan to use - if you were doing this, how would you ever determine the service provider to hire as only they would have the expertise to determine the technology to be used! A Catch-22 for sure. And the other key mistake is confusing service providers (companies that provide IT services) with value added resellers (vendor sales representatives). The latter will often market themselves as the former, but it is easy to tell them apart. The first one's business is to provide IT as a service. The second one's business includes selling hardware, software, and third party services, potentially in addition to layering on some IT.

Best practices:

Avoid running IT departments that are too small to support the necessary roles and division of labour.

Never hire a reseller to do the job of an IT service provider.

Never allow your IT staff (internal or external) to have a conflict of interest and also sell the hardware, software, and services that it is their job to recommend and choose.

Now we have a good idea as to why logging is so critical to our organization. Good use of and understanding of logs and putting in place a proper, well thought out infrastructure for logs is one of the areas in which we truly see a separation between struggling and truly excelling IT departments.

Not a heavily technical discussion at this point. It is really all about sitting down and putting in the effort to develop and roll out a logging plan. Making logging happen for your organization. Centralized, decentralized, automated, whatever works for you. Getting started with something, turn your logs into a robust tool that makes your life easier.

In the next section we will continue on from logging to look at more general monitoring. Two highly related concepts that together really take our administration to another level.

Alerts and troubleshooting

Having just discussed logs we now have to consider the highly related concept of system alerting. I have to mention that of course logging systems themselves are also a potential source of alerts. If we use automation in our logging systems, that automation will generally be expected to either send alerts directly, or add alerts to an alerting system.

Alerts are, fundamentally, a way for our monitoring systems to reach out and tell us humans that they are in trouble and it is time for us to step in and work our human-intelligence magic. While we hope that our systems will have automation and can repair many problems themselves, the reality is that for the foreseeable future nearly all companies will have to keep working in a reality where human intervention is needed on a regular basis in systems administration. Whether it is to log in and clear a full disk or stop a broken process or identify a corrupt file or even to trigger a failover to a different application or notify the business of expected impact humans have a large role to play in systems still.

Having good mechanisms for discovering serious issues and alerting humans is critical to quality support. In order to understand good alerting, we have to talk about both how we discover that something is wrong, and how we are notified of it.

On-device and centralized alerting systems

We can start by looking at on device and centralized alerts. Traditionally, going back more than a few years, it was common for systems to handle their own alerts individually. Systems were already set to log issues into a central log, it was a natural extension to have them also send out emails or similar notifications should something bad be detected. Alerting was very simple, and each system would handle its own detection and its own alerting individually. While centralized monitoring and alerting has long existed, the popularity of external monitoring really did not become highly mainstreamed until it was necessary to monitor hosted Internet resources, such as websites, where an outage would often be seen by customers first, rather than by employees. When outages are first noticed by internal staff, the decision to delay discovery can be more flexible.

The simplicity of on-device alerting is enticing, and for smaller organizations or those that can risk slower error detection it can serve well. The key issues with on-device alerting are that many types of serious outages or attacks may disable alerting completely, or at least delay it. A simple example is the server that loses power or whose CPU melts, the system going offline is what we want to receive an alert about, but the system going offline suddenly precludes the possibility of the system telling anyone that something bad has happened. The same happens when there is a sudden loss of network connectivity. In the event of a system compromise, a hacker may take alerting capabilities offline before being detected leaving a system running, but unable to call out for help.

On-device alerting is generally inexpensive and simple. It is often built in and only needs a small amount of configuration. If using simple mechanisms such as email to send alerts, even simple scripts can add a lot of alerting functionality. This approach uses very few system resources and for small businesses or those that simply do not have to worry about potential downtime without employees reporting a loss of functionality, it can be adequate.

For the majority of businesses or workloads, the caveats of on-device alerting are too great. Whether a system is customer facing and you want to maximize customer confidence, or a system is internal and you want to move discovery of issues from employees to IT to improve performance, or a system has few, if any, end users that may every discover that it is not working and you want to make sure that work is continuing to be done (such as with a scanning security system or filter) then external monitoring is necessary.

External monitoring allows us to disseminate alerts even when the system in question has completely failed. Because total failure is quite common in alerting events, this can be pretty important. Complete failure might mean that hardware has failed, power has been lost, software has crashed, or networking has been lost, as examples. These are all common failure cases, and all either certainly or likely will cause on-device alerting to fail. External alerts give us a level of confidence that we will be alerted when something fails that is otherwise lacking.

Of course, external systems can fail as well, leading to a lack of alerts, but there are ways that we can effectively hedge against this. One option, of course, is external alerting on the external alerting system. Essentially backup alerting. And of course, by having only a single alert source, it is easy for us to simply check that source as humans to verify that it is working. An external alert mechanism, if decoupled from the systems that it monitors, is extremely unlikely to fail at the same time that another system fails and while not perfect, this will easily eliminate 99.99% or more of missed alerts which, for most organizations, is plenty.

Out of band on-device alerts

There can be a little bit of a middle ground in alert management. If we think of our systems as a stack, each device lower in the stack is able to monitor, in a very minor way, the services running above it. For example, an application can tell us if a database connection is working, the operating system can tell us if an application is still running, a hypervisor can tell us if the operating system is still running, and, at the bottom of the stack, an out of band management hardware device can tell us if the core system components have failed.

This system is not foolproof and tends to be quite basic. An operating system knows very little about the workings of an application process and mostly can only tell us if the application has crashed completely causing an error code be returned to the operating system, or it can tell us if the application is using an inordinate amount of resources such as suddenly spiking in memory requests or using a large number of CPU cycles, but the operating system will have little idea if the application keeps running but is throwing errors or gibberish to end users.

For those unfamiliar, out of band management is actually an external *computer* that is housed inside of the chassis with the server hardware but has its own tiny CPU, RAM, and networking. Because it is a nearly completely separate computer from the server itself, the OOB (out of band) management system can report, either directly or to some monitoring system, if there are critical hardware failures on the server itself such as a failed motherboard, CPU, memory, storage, or other component that would normally make the server itself unable to send out its own alerts.

An OOB management system does share the chassis, location, and power with the server, though. This means that it still has limited ability to monitor a system for certain types of common failures. As with many things, for many businesses this might be adequate, for others it will not be enough.

Pushed and pulled alerts

Alerting systems also have two basic ways of interacting with us, as the actors being alerted by the system. What we tend to think of is pushed alerts. That is, alerts that are sent to us with the intention of grabbing our attention when we are not thinking about alerts.

Typically pushed alerts go out via email, text messages (SMS), telephone calls, WhatsApp, Telegram, Signal, RocketChat, Slack, Microsoft Teams, Zoho Cliq, MatterMost, or other, similar, real-time communications channel. Some alerting systems have their own applications that you install to desktops or smartphones so that they can push out alerts rapidly and reliably without having to integrate or depend on any additional infrastructure. You can easily imagine an organization running their own email and messaging platforms only to have those platforms be monitored by our alert system and also be the path by which we receive the alerts. Even if the alerting system itself does not fail, it is possible that it will be unable to tell us that something has failed because the systems that it monitors are also the systems that get the alerts to us. This is why having one fewer path to fail and one fewer dependency in alert delivery is sometimes approached.

Alerting is a surprisingly complex animal. If alerting was being handled by humans, rather than computers, we would quickly find that call centers have complex, multi-branch decision trees to follow for what to do when we cannot tell someone that something is wrong. With humans, though, we know that in an extreme emergency someone will start pulling out their personal cell phone and texting someone to call someone and knock on a door and wake someone up or whatever. Computers can do all of this, too, but they need access to those tools, algorithms to make those decisions, and knowledge of how to reach people. Easier said that done.

To solve this particular problem, many companies opt to include humans in the communications path. An expensive, but effective, tool. Sometimes human decision making, and flexibility wins out. Human call centers that are always staffed and receive alerts on behalf of technical teams and managers and then manage the contact path to whomever needs to receive the alert can be a great option. And, obviously, hybrid options where computer systems alert end recipients directly but humans are always involved in verifying that alerts go out, acknowledgements are received, or whatever is possible.

The alternative to pushed alerts is pulled alerts. Pulled alerting refers to systems that display the status of any open or logged alerts when the end user logs in to look at them. These systems are vastly more reliable because the end user is looking at the system and knows if there is an inability to view the alert status or not. If the system has failed, then they can start working on the issue right away. If it has not failed, they see the alerts and know if action is needed or not.

Silence as success

The basic problem that arises with using purely pushed alerts is that we rely on silence to tell us that everything is okay. Rather than getting a confirmation that everything has been checked and that nothing is currently wrong, we depend on not having been reached to create an assumption of nothing being wrong. This is a dangerous approach.

We all know the feeling of waking up or having been on a long car ride or maybe being at a party and not actively watching our phones and mostly feeling good that the office has not reached out to us, no one has called, so everything must be fine. Then you look at your phone some hours later and realize that the battery has died, there was no service, or you had your phone on silence. Panic sets in. You plug in your phone, get service back, and turn on the ringer and find that you have been missing call after call, voicemail after voicemail, text after text telling you that there is a huge emergency, you left the office having changed a critical password and not telling anyone, the system is down, no one can get in except for you and you are not responding!

Trusting that no one was able to get my attention, therefore nothing can be wrong just does not work. But neither does staring at a console and never being offline. There has to be a balance. It is clear that simply hoping that you will be able to be reached is a recipe for disaster. Maybe a disaster that takes many, many years to finally happen, but a disaster that is almost certainly going to happen eventually. There are just too many variables that can go wrong.

Pull monitoring and alert systems are therefore important as a means of verifying that pushed alerts are working currently or to work around known disconnects. Or to run an active monitoring site.

Alerting's systems, too, struggle with defining a successful alert. Many mechanisms like email and SMS texting will confirm only that a message has been sent or possibly received by the recipient's infrastructure vendor, but they do not give any indication that the message has made it all of the way to the end user's device, or that the end user has been displayed the message. Even if a message does go end to end, does it get filtered into a spam folder and hidden? Unless we have a human actually acknowledge an alert there is very little, we can do to have confidence that an alert has truly been seen.

Pulled alert systems, generally displayed as dashboards, are typically what we picture as a red light, green light system. It is common to display monitored systems or components graphically and to show systems believed to be healthy shown as green, those experiencing problems, but not yet indicating an outage as yellow, and those that have failed whatever sensor test we are performing as red. This does not just give the humans the ability to quickly eyeball the range of alerts, but also makes it trivial for the system to roll up large groups of alerts into single displays. As long as those are green, you know everything in the group is healthy. If it is red, you can dig in to see exactly what is wrong. At the highest level you can, in theory, even have a single big indicator that shows as green or red. Green if systems are one hundred percent good and red if any system has failed. Simply: is action needed or is it not. If you can be green most of the time, this might be exactly what you need to combine reliable monitoring with low overhead in verification.

Most alert systems will offer both a dashboard to show pulled alerts along with push notifications to improve response times and reach people who are not actively checking alerts. Pulled alert systems are often used at the core of a call center where humans on shifts watch the pulled alerts around the clock and either enact the push alerts to the concerned parties or follow up to ensure delivery of automated alerts. It is less common for organizations to open pulled alerts to many staff which can be a mistake as it can lower stress and increase alert reliability.

Similarly, the interaction between the monitoring system and the end points that are being monitored can work in either direction or both. The monitoring system may have remote access to the end points and actively reach out to them to request their status. Or an agent running on the end points may reach out to the monitoring server to push their status over to it.

In house and hosted monitoring

Monitoring and alerting are, in some ways, two separate pieces and there is software and services that do either, and those that do both. The monitoring component determines if a series of sensors detects something that is wrong in our workloads. The alerting component takes the results of the monitoring and attempts to notify the correct parties. It used to be that the two components were always merged into single products, but in more recent years with more and more types of systems needing to be monitored (more than IT systems, that is) and as alerting needs have increased and have needed to become more robust, different vendors have started to build each independently in some cases.

Today monitoring solutions come in a good variety of packages and styles. You easily may decide that you would benefit from using more than one. Choosing a good package might be the hardest part of your monitoring puzzle. Monitoring software is available commercial and free, closed and open source, and built to run on nearly any platform.

Monitoring is a function that you will generally want to host externally to your primary infrastructure as you want it be less quick and reliable, and more totally independent of your other systems compared to other workloads. You can host it yourself on your own equipment, go with third party cloud or VPS hosted infrastructure, or get a SaaS application from a provider. You can tackle this in any manner than makes sense for your organization. But rarely do you want your monitoring solution to sit on the same hardware, let alone the same datacenter, as your other workloads or you risk losing monitoring when you lose everything else.

Whether to run and maintain your own monitoring or to go with a hosted product will mostly be a question of cost and politics within your organization. Even free and open-source monitoring solutions can be robust enough for the most demanding of organizations. You will need to determine if the cost of building, maintaining, and *monitoring* a monitoring solution makes sense for your organization or if simply buying that functionality ready to go makes sense for you. In most cases this is determined by scale. If you monitor a very large number of workloads or your monitoring needs are highly unique you may benefit from building the expertise in house and having full time specialists dedicated to this project. Generally for a project like this to be cost effective to keep in house you will want to have either fully dedicated staff or heavily dedicated staff who have the time and resources to really learn the products and maintain them properly. Often monitoring and logging will be bundled together whether as a single product or under a single person or team as they overlap so heavily, and logging can be thought of as a specialty function of monitoring. The two may be operated separately or combined. Using both through the same alerting channels generally makes sense as they can leverage the same effort and infrastructure.

Like most things in this chapter, the real struggle around best practices is finding what we can distill as being *best* as guidance here is very broad and mostly ambiguous. There is no one size fits all. It seems like cheating to say, but it is true that the real best practice here is to evaluate your business' needs based on cost, functionality, support, separation from your production environment and determine what monitoring and alerting mechanisms are right for you.

RMMs and monitoring

If you work in internal IT then the term RMM might be something that you have never heard of, but if you have worked in the service provider sector then RMMs are the core tools expected to be used for customer support in that area.

An RMM, which stands for Remote Monitoring and Management, is a tool category designed around the needs of service providers who almost always need to work remotely to their client sites and to be able to quickly monitor many disparate client systems at one time. This is generally quite different from internal IT needs where often they are not remote and even when they are, their systems are typically integrated.

A rare few non-service providers still lean on RMM tools as a monitoring mechanism. Typically RMMs are very light and inflexible but are, at their core, monitoring systems much like what we are discussing here. So you can certainly consider using an RMM that you purchase or run yourself, or if you have a service provider, this might be part of the service that you are already paying for. RMMs are even available as free, open-source products. So no company, of any size can say that they do not have the resources to at least do the most basic levels of monitoring.

In some cases, traditional monitoring tools designed for internal IT teams are so robust that they actually displace RMMs in service providers. Or the two could be used in tandem, as well.

The rule of thumb is that more monitoring is better than less, hosting outside of your environment is generally best, make sure that alerts have many channels to find a way to get to a human, make sure that pull monitoring is available to at least verify that push is working, and consider having your monitoring system create actionable tickets for your support team automatically to track follow ups.

Is there an actual best practice? Yes. The best practice here is simple and broad: if a workload has a purpose, then it should be monitored. Monitoring, because it does not directly stop production from running if it does not exist, can too easily be overlooked. Almost no one gets promoted for doing good monitoring or fired for lacking it, but implementing proper monitoring is effective in separating the good administrators from the run of the mill.

Go set up some monitoring!

Summary

In this chapter we have looked at the range of key non-systems components that surround the systems themselves. Documentation, system measurement, data collection and planning, log collection and management, and finally monitoring sensors and alerting based on them. These could almost be considered soft skills within the systems administration realm.

Consistently in environments that I have taken over we have found documentation to be practical non-existent, measuring systems to be all but unheard of, capacity planning being a process no one has ever so much as discussed, monitoring often minimal and unreliable at best, and log collection while well understood, simply a pipe dream when it comes to real world implementation. Yet a single system administrator with almost no resources could, with just some time, pull together some free, open-source software and tackle each of these projects on their own with little to no budgetary constraints and could often hide the workloads somewhere within the system if it was necessary to do so.

This chapter has not been about how to make your systems run better; it has been about everything else. How do we know that they have been running better? How do we know that they are running right now? How do we know that we can pass the proverbial baton on to someone else should we win the lottery? How do we confidently say that we are doing what needs to be done to make a best effort against a malicious attack? The topics in this chapter have made us look at how to be better at all of the things that we do and not just the ones that are most visible.

Up your visibility

Too often what we do in IT is completely invisible to those on other teams, even those in management to whom we report. Maybe it is invisible because there is simply nothing to show. Or maybe what we do is too hard and complex for people outside of our realm to really understand. Or maybe we are invisible because we choose to accept being invisible.

Most of the topics in this section provide perfect opportunities to step out of the IT dungeon or closet and get in management's face(s) to do a little IT team self-promotion. From monitoring dashboard to beautiful documentation, to capacity charts, to log drill down examples there is almost always something that we can print out or show on a big screen and look pretty impressive for having implemented.

Getting the attention of management and showing that we are being proactive, that we are following best practices, this is where we can make a sales pitch for just how amazing our value is to the organization. Do not be afraid to do some self-promotion, you deserve it. Make some noise and show off how you are preparing the business for the greatest success.

Go out and make sure that all of these systems mentioned in this chapter are implemented in your environment. Keep it simple to get started, but do not skip systems.

In our next chapter we are going to move on to scripting and system automation including DevOps, which I know that you have been waiting for.

8
Improving Administration Maturation with Automation through Scripting and DevOps

I think that it is safe to say that for most of us in system administration that scripting and automation and where we naturally gravitate towards for thinking of what creates the best opportunities for overall system improvement. This might be treated, and automation is very important, without question, but it is not the end all of system administration either. It is safe to say that the more that we learn to script and automate, the more that we have free time to focus our energies on tasks that only humans can do while also developing a deeper appreciation for what developers do which is always be helpful for those of us in IT.

System automation is an area where it becomes much easier to obtain bragging rights as to what our daily task list looks like. When sitting around having beers at the proverbial system administrators cocktail lounge, we get little satisfaction over telling our compatriots how we wrote some really clean and easy to read documentation. But when we explain how we wrote a long, complicated script that takes hours of our weekly workload and turns it into a task that is run magically by the computer's scheduler we get kudos, attention, streamers, a ticker tape parade, fellow administrators buying us rounds of our favorite beverage and, if we are truly lucky, a piñata.

Automation is typically the area of administration that most of us find to be both the most exciting, and the scariest. There are more building blocks, more concepts to understand, than in other areas of system administration. For the most part, system administration is a lot like taking a history class where yes, there are real benefits to knowing more pieces of history so that you have a larger context when learning something new, but generally you can learn about any specific event without a large understanding of all of the events related to it and that led up to it and still come away having learned something valuable and essentially understanding it. You will not be lost when learning Roman history just because you did not grok Greek history first. But scripting and automation is a lot more like math class where if you fail to learn addition, then learning how to find the square root is going to be completely impossible. Scripting is a skill that builds on top of itself and to be really useful you will want to learn a bit of it.

We are going to start by looking at unscripted command line administration in comparison to working with a graphical user interface and use that as a foundation to move into automation itself.

In this chapter we are going to learn about the following:

- The GUI and the CLI: Administration best practices
- Automation maturity
- **Infrastructure as Code** (IaC)
- **Documentation First Administration** (DFA)
- Modern tools of automation

The GUI and the CLI: Administration best practices

If you are coming to Linux from the Windows world, you may be excused from realizing that nearly everything should be done from the command line, not from the graphical user interface. But really, even on Windows, Microsoft has been very clear, for a very long time, that the desktop experience is really for end users and not for system administrators and that they recommend either using PowerShell as the administration tool of choice when working on a local system directly or any number of remove management tools that connect via an API. Microsoft pushes quite hard to encourage those installing their systems for the past few generations to install their operating systems and hypervisors without graphical user interfaces at all.

Graphical User Interfaces, or GUIs as we will call them now to keep things short, present a lot of problems for system administrators.

The first issue with GUIs is bloat. During the installation of an enterprise operating system, when a GUI is available it is often more than half of all of the code that will be deployed in a system. Every additional line of code means more data that we have to store, the more we store the more we have to back up; more code to worry about having bugs or flaws or intentional backdoors; more code to patch and maintain and so on.

Next is performance. GUIs require much more compute power and memory consumption while running than do non-GUI systems. It is not uncommon for a GUI to require 2GB or more of additional system memory above and beyond what is needed for the system workloads. This might sound trivial, but especially if we are dealing with many systems consolidated onto a single piece of hardware it can add up very quickly. If we have twenty server virtual machines running on a single physical server it might not be uncommon in the Linux world for the average workload to only be between two and four gigabytes of memory. Adding two more gigabytes to each system would mean not only a nearly fifty percent increase, but forty gigabytes across the machines.

Consolidation and the age of squeezing systems

In the 1990s and 2000s, before the prevalence of virtualization, we were in an era where servers were gaining performance rapidly, but each individual system would only run a single operating system no matter how small the workloads on that system were. As systems became more powerful, much faster than software used more resources, there was a strong trend towards allowing system bloat because, at least when it came to hardware, it did not matter very much.

CPU and memory resources would normally come in discrete chunks and to have enough we would normally have to overbuy. It was rare to run a system close to its limits because it was so difficult to expand systems in those days. So a system would typically have a lot of spare resources by design to allow for a large margin of error and, of course, growth. Because of these factors, running a GUI on a server was more or less trivial.

Many factors have changed since those days. We could probably write an entire book just discussing why the industry temporarily moved from command line after decades of using nothing else and for a small blip from the mid-1990s to the early 2000s had GUIs seemingly taking over as the dominant approach in server management only to go right back to the command line by around 2005. Ignoring social trends driving changes we are concerned here with capacity concerns.

Once virtualization became mainstream, and even more so as cloud computing began to become a major trend, the availability of spare resources for operating systems ceased to be a common thing, almost overnight. This might seem counterintuitive, given that virtualization inherently gives us more flexibility and power. But it also gives us the ability to scale down very effectively, and this is something that we did not have before. With virtualization we are rarely in a position of having dramatically excess system resources, especially predictably excess resources, and so there is a huge advantage to keeping individual virtual machines as lean as possible, and that means not running an enormous and largely useless GUI process. Very small businesses that still cannot combine their workloads to reach the effective lower bounds of a single server remain an exception to this rule and still have plenty of overhead to implement GUIs unless they would be good candidates for cloud computing.

In a traditional business, where there are multiple servers, a major advantage of virtualization is consolidation and avoiding the installation of GUIs may mean that fifty or sixty workloads can be installed on a single physical server instead of thirty or forty on the exact same hardware. This equates to the need to buy fewer servers and that means cost savings not just from lowering purchase costs, but also lowering power consumption, cooling costs, datacenter real estate costs, software licenses, and even IT staff.

If we look at an example on public cloud computing, we can see the advantages of not having a GUI even more easily. Small workloads, which could include email servers, web servers, proxies, telephone PBXs, and on and on might only cost between five and ten dollars per month to run on their own. Adding a GUI will easily cause the cost of a cloud hosted virtual machine to double from five to ten or ten to twenty and so forth as the GUI creates a need for more CPU power, more storage, and most importantly, much more memory. It does not much effort to see how moving from ten dollars per month for a workload to twenty dollars will add up exceptionally quickly. As most cloud-based workloads are quite small adding a GUI to each one could have staggering capacity consequences as much as doubling the compute cost of a company's infrastructure!

The lack of appropriateness for a GUI is so dramatic that many vendors have traditionally not even provided a mechanism for attaching to a GUI in the cloud space. Amazon famously did not make GUIs possible on their standard cloud instances effectively forcing organizations to learn command line and even more advanced techniques involving management without a login. But nearly all cloud users opt for remote logins via a technology such as SSH. The cloud did more than anything else to demonstrate the risks and costs of the GUI.

Prior to ubiquitous virtualization and cloud computing system administrators, especially those in the Windows world, would argue that GUIs just did not add that much overhead and that if they did anything to make someone's job easier that they were worth it. That myth has been exposed and no one can honestly make this claim today. GUIs are clearly nonsensical in any broad sense.

Often the biggest selling point for command line management at a managerial level is about security. A GUI presents a much larger attack surface for a malicious actor to use to attempt to breach a system. All of that extra code alone makes things much easier for a would-be attacker. And, of course, the functionality of a GUI has to make for very enticing attack surfaces just by the nature of needing to have more means of being accessible. More lines of code, more access methods, more management paths, lower performance and more all come together for overall increased security risk. Taken all together it may not be incredibly major, but the increase in risk is real and is measurable or, at least, estimable.

The final major point as to why command line management has become the *de facto* standard is efficiency. Yes, the very reason that so many point to as to why they chose to keep a GUI. The reality is that system administration is not a casual task, nor one where you can effectively just poke around and guess about what settings should be wear. To do the job well, or even safely, you must have a pretty solid understanding of a large number of items from operating specifics to general computing and networking understandings.

The GUI in management was traditionally promoted as a tool for those that were not used to an environment to be able to be effective quickly with less training and knowledge. A great concept if you are talking about a janitor. In system administration the last thing that we want is someone without deep knowledge and experience being able to act like they know what they are doing. This is dangerous on many levels. GUIs, sadly, actually make it much harder for many organizations to evaluate which candidates are even minimally qualified for a technical position.

Not only does a GUI pose a risk that someone without proper knowledge will start poking around, but for someone who knows what they are attempting to do the command line is vastly faster. It is faster for performing simple tasks, for performing most complex tasks, and it is far easier to script or automate. Command line management flows so easily directly into scripting that people often fail to be able to tell the two apart. If you ever truly compare tasks, it is not unheard of for command line work to take less than ten percent the amount of time that it takes to do the same task with a GUI!

Command line is not just more efficient for a system, it also makes multi-system management much easier as commands can be duplicated across systems in ways that GUI actions cannot be. Command line management can also easily be recorded, catalogued, searched, and so forth. It is plausible to do the same with a GUI but it requires long video recording which results in large amounts of storage needs and no simple way to parse or turn into documentation, and so on.

In the more modern era, we have also begun to face the problem of needing to perform most or all system administration remotely. This inadvertently played right into the command line's hand. The amount of data that needs to be sent, and the sensitivity to network lag for the command line are far smaller than for a GUI. A remote GUI session to a server generally uses a noticeable amount of network traffic. Remote GUI sessions are video streams, generally in pretty high resolution. In some cases, even a single user can cause network issues, especially if the server exists in a location with bad Internet access. The standard method for remote command line management is SSH.

SSH remote sessions will work just fine even over an archaic dial up Internet connection. And even the slowest modern Internet service is enough to handle scores, if not hundreds, of simultaneous SSH users. This is something that you generally cannot do with remote GUI sessions. Command line is nearly as effective over a tiny Internet connection from the other side of the globe while GUI remote management suffers noticeably from any network blips, limitations, or distance.

Command line is here to stay, but it is important to really understand why. It can be easy to forget that it is far more than just one or two small factors. There are really good reasons why you should be using command line whenever possible. Moving back and forth is not conducive to learning to be more efficient in either approach, as well. From a personal level it is expected that you would want to avoid the use of GUIs as much as possible so that you can focus on learning command line skills. Using the command line consistently is needed to really become efficient.

Now we must acknowledge that there are a number of different command line options for Linux. We can use BASH, Fish, zsh, tcsh, PowerShell and more. Linux is, as we know, all about options and flexibility. This is a situation where less is probably more. Some of these shells are very nice and useful but, we must remember, that we are system administrators, and we need to make sure that we are totally familiar with the tools that we are likely to have access to in an emergency. Moving between shells is not particularly hard, especially in the Linux use case, but we should still be wary of spending time learning the nice keyboard shortcuts and auto-completion and other perks of a shell-like Fish or zsh because we may not be able to use those skills in the next job and that always has to be a consideration. And, in the case of an emergency if you were to get called to work on a system that you have not had a chance to set up previously you may be stuck with no option except for BASH. For me, this means that BASH is the only tool that I want to be learning.

And there you have it. All of the logic and reasoning so that you can go back to management and explain why you need to be working from the command line, why you need staff that works from the command line, and why your systems should rarely get installed with any GUI environment at all. In our next section we are going to talk about maturity levels in automation for systems.

Automation maturity

While there is no formal system for measuring automation maturity levels, there are some basic concepts of automation maturity that we can discuss. The idea here is that organizations sit, more or less, along a continuum from having no automation to being fully automated with most organizations sitting somewhere in the middle, but more likely to be towards no automation than towards being fully automated.

Not every organization needs to be, or even should be, completely automated. But in general, more automation is better when the cost to implement the automation is low enough to do so. Automation is not free, and it is quite possible to find organizations investing more in automating a process than it would cost to perform the duty manually over the lifespan of a system. We do not want to automate blindly only for the sake of automating.

Typically, however, what we find is organizations skipping automation in nearly all cases and uses manual labour with all of its costs and risks instead. There is a natural tendency towards this because in the moment, any task will be easier if done manually. If we do not look ahead and invest, we would simply never automate, and this is often how companies view IT needs. If a task takes one hour to do manually and three hours to automate, that's the time of three tasks and hard to justify ignoring the fact that the same task will happen monthly and in four months would not only have been less effort to have automated, but the automation would make the task more reliable and consistent.

Nearly any organization will benefit from automating more than they do today. There is no need to look at automation and feel that it is an all or nothing proposition. Automate what you can, skip what you cannot do. Be practical. Hit the low hanging fruit first. The more than you automate the more time you have to automate other things in the future. You will improve your automation skills as you practice, as well, making each new automation something easier than the last. Automation is a great example of that kind of thing that is really hard to do the first time but gets progressively easier and easier until it is just the obvious way to approach things and becomes second nature.

Automation maturity is not exactly a direct continuum with each step *more mature* than the last. For example, if we look at scheduling tasks and scripting tasks, each of these can be done independent of the other. Both are useful on their own. We can script complex operations and run them manually. Or we can schedule simple, discrete commands to trigger things without human intervention. We can then put the two together to automatically kick off complex scripts that do many things at once. Which one do we consider first and which second is just arbitrary.

Local and remote automation

In case it is not overly obvious, we have the choice of implementing automation either locally with tasks scheduled or triggered to run on the server in question, or we have the ability to push our automation from an outside source which can give us a sort of centralization of automation. There is a sort of hybrid approach where a local scheduler or agent reaches out to a central server to request automation information which is technically still local, just with a centralized store to mimic control.

Often overlooked is the advantage of local automation being able to run even if remote systems become unavailable even to the point of the local system completely losing networking. My favorite task to keep local regardless of other automation decisions is system reboots. While it would be convenient to centralize reboots and I have seen organizations opt to do so I very much appreciate having a local, forced reboot that happens at least weekly and sometimes even daily. This gives me great peace of mind that even if something completely crashes on a server if it is still functional in any way that eventually a reboot process is going to make an attempt at restarting the machine and hopefully bringing it back online. A very niche need and one that may never be important to you, but I have witnessed systems become inaccessible for remote management while still providing their workloads and an automated, locally scheduled reboot brought them back online and making them accessible again.

An increasingly popular happy medium approach is to have a central control repository that contains all of the desired automation which is then pulled by an agent on the end points being automated. This repository contains both the automation itself, such as scripts, as well as the scheduling information or triggers. Then the information is actualized by a local script that is able to keep functioning independently even if the remote repository fails or becomes unavailable. In this way you only really risk losing access to update the list of scheduled tasks to make changes to them or to the schedule. As long as you do not need to send out new updates to the automation you do not have to worry about your repository being offline.

Command line

Not so much a legitimate maturity level but more of a basic starting point, so we could think of this as a level zero, is moving to the command line and the use of a proper shell environment for interactive (that is: non-automated.) As we just discussed, being on the command line and learning command line syntax and tools is the fundamental building block on which all subsequent automation is going to be based. Understanding how to do tasks at the command line, how to manipulate text files, how to filter logs, and other common command line tasks will build quickly into obvious automation capabilities.

Scheduled tasks

The best and easiest place to begin with automation is the scheduling of tasks. This may sound like the simplest and most obvious step, but surprisingly many organizations never even get this far. Linux has long been a stronghold of reliable, easy to manage local scheduling with `cron` having been built into not only Linux, but essentially all UNIX systems for almost half a century as it was released in 1975. Cron is fast and efficient, ubiquitous and well known. Any experienced UNIX admin should be able to at least schedule a basic task when needed. Cron even handles tasks happening at boot time.

Simple tasks of all sorts can be run through `cron`. Common tasks used in most environments could include system updates, data collection, reboots, file cleanups, system replications, and backups. You can schedule anything, of course, but these are some good ideas for first time automaters looking for obviously recurring system needs.

Another common area for simple scheduled tasks are code updates via a repository like when we pull new code via `git`. Tasks such as code updates and subsequent database migrations can all be easily scheduled.

Scripting

When we say automation everyone always immediately thinks about scripting. At the end of the day, nearly everything in automation is scripting either directly, or under the hood somewhere. Scripting delivers the power when we want to move beyond the simplest of tasks or just calling scripts that someone else has made.

We cannot possibly teach scripting itself here, that is an entire topic in and of itself. Scripting is the closest that IT comes to crossing paths with the software development world. Where does combining IT command line tasks together turn into programming? Technically it is all programming, but an incredibly simplistic form of programming focused purely on system management tasks.

Typically, on Linux we write scripts in the BASH shell. BASH is a very simple language designed to be primarily used interactively as a live shell, BASH is how we assume all command line interactions with Linux will be performed. BASH is relatively powerful and capable and nearly any script can be written in it. At least when starting out, most Linux admins will turn to the BASH shell that they are already using in their command line environment and add scripting elements organically to move from a single command, a few basic commands strung together, and into full scripting just a little at a time.

Any shell, such as `tcsh`, `ksh`, `fish`, and `zsh`, will allow you to script and in many cases with more power and flexibility than you can with BASH. Traditional shells, like `tcsh`, `ksh`, and BASH, can be very limiting and cumbersome to attempt to use for advanced scripting. Apple for its macOS UNIX operating system has recently moved to `zsh` to modernize it compared to other UNIX systems. Typically, a Linux system is not going to have a more modern, advanced shell installed by default, even though they are easily available on essentially any Linux based operating system.

You may work in an environment where an alternative shell is consistently provided or offered, or you may have the option of adding it yourself. If so, and especially if you will be doing cross platform scripting with macOS you might consider using `zsh` instead of BASH, or if you are doing a lot of Windows scripting its native shell PowerShell is also available on Linux.

PowerShell on Linux

One of the weirdest things that you may ever encounter is the idea of running Microsoft's PowerShell on Linux. Many people are confused and believe that it does not even work. PowerShell does actually run just fine on Linux. The problem with PowerShell on Linux is that PowerShell users on Windows actually spend essentially no time at all learning PowerShell and nearly all of their time learning a range of CommandLets or small programs that can be called by PowerShell and easily combined with other little programs to give power to the system.

On Linux, of course, the same thing happens. If you are scripting on Linux you will surely be using tools like sed, awk, cut, head, tail, grep and so forth. These tools are a lot like CommandLets, but are actually just every day system executables. If you were to port BASH or zsh over to Windows you would find that the tools that you are accustomed to using on Linux were still not available. That is because they are tiny programs that you are calling from BASH, not part of BASH itself. BASH is just the programming language.

The reverse is also true. If you run PowerShell on Linux you still use sed, awk, cut, grep, head, tail and on and on. It is the language that has changed, not the operating system's suite of tooling and components.

So, while there can be value in learning one scripting language and attempting to use it over and over again between different operating systems, there is not the value that one might assume. You will likely spend far more time tripping over integration quirks, misunderstandings, and poor documentation than you could ever recuperate from language learning efficiency. Books, online guides, example code and so on will never work for you if you try to use PowerShell on Linux. There will assume that you are trying to do Windows tasks with access to Windows tools, always. PowerShell is, at its core, designed to be a truly modern shell that uses operating system objects to do its heavy lifting. BASH instead is focused on text processing and manipulation as Linux is traditionally built on text files and needs a scripting engine that will easily accommodate that.

Using something so foreign as PowerShell on Linux is a great tool for exposing where different components that we often see simply as *the command line* or *the shell* for what they are. If we use zsh on Linux, nearly everything that BASH has built in is replicated in zsh, and they both conventionally use the same operating system tools. PowerShell has few, if any, replicated built in commands and no shared conventions making it painfully obvious what was coming from the shell and what is part of the operating system outside of the shell.

In general, however, it is most advised to do all scripting in languages that are well supported in your environment. Just like we have said in other areas of system administration, it is important to use tools that are ubiquitous, well understood, and appropriate for the environment. For most that means BASH exclusively. BASH is the only scripting environment that is going to be absolutely available on every Linux system that you ever encounter. Other shells or scripting languages might be common, but none other are so universal.

When BASH proves to be too limiting for more advanced scripting it is uncommon to turn to another shell, such as `zsh`, as other shells remain very uncommon and generally lack in the extensive power that you are likely looking for once you are abandoning BASH. Traditionally it has been non-shell scripting languages that are used as BASH alternatives for advanced scripting such as **Python**, **Perl**, **Tcl**, **PHP**, and **Ruby**. Ruby has never gained much favor. PHP, while very common for certain tasks is pretty rare as a general system automation language. Perl and Tcl have fallen out of favor dramatically, but at one time Perl was the clear leader in system automation languages. That leaves Python as a very clear front runner for advanced scripting needs.

Python has many advantages overall. It is decently fast. It is available on nearly any platform or operating system (including all Linux, alternative UNIX, macOS, and Windows.) It is quite easy to learn (often used as a first language for new programmers to learn.) It is very often already installed because a great many applications and tools on Linux depend on Python so you will regularly find it already installed even when it is not intentionally installed as a standard. Because it is used so commonly for these tasks, it has become increasingly well suited to the role as documentation in how to use Python in this way has grown and other tools written around it have sprung up.

At this time, nearly all system scripting for Linux is done in BASH when possible and in Python when more power or flexible is really needed. All other languages are really niche use cases. This means that BASH and Python also have additional reasons that we should be strongly considering them when choosing languages for our own scripting: standardization.

System automation is different than general programming. With broader programming developers spend years learning multiple languages, language families, constructs, and spend all of their time in their programming environments. Moving between languages, learning a new one, adapting to language changes and so on are all part of the daily life of a developer and the overhead to move between languages is very low. For a system administrator this is a bit different. In theory we spend very little time learning programming and rarely are exposed to any real variety of languages. So for administrators, having just one or two languages that you use is important for being able to find resources, examples, peer review, and others to provide support for our automation in the future. These are certainly not the only acceptable languages, but they do hold rather sizable advantages over most other options.

Of course, scripting is a very general topic and scripts can run from being just a few lines of simple commands run sequentially to giant programs full of complex code. Growing your skill in scripting is a topic all to itself and well worth investing significant time into. Good scripts will generally include their own logging mechanisms, error detection, functions, reusable components, and more. You can essentially invest indefinitely in greater and greater programming skills to apply to system automation scripts.

Developer tooling for script writing

Whether you are just writing a very simple script or working on a masterpiece of automation to be passed down from generation to generation of system administrators in your organization, it may be worth taking an additional step to learn something about tooling used in software development to potentially aid in the script writing process.

On the simplest side are tools like integrated development environments or IDEs that can make writing code faster and easier and help you to avoid errors. Developers nearly always use these tools, but system administrators will often overlook them as they feel that they write scripts so little of the time that learning another tool may not be worth it. And perhaps it is not, but the more tooling you learn the more likely you are to use it and to write more scripts. A good IDE can be free and quite easy to use, so is a good starting point as you can integrate one into your process without really spending any money and only a few minutes to download and install one.

The other truly enormous toolset that developers almost universally use and system administrators rarely do are code repositories and version control tools like Git and Mercurial. With tools like these, and a central hosting of your code which is often associated with these tools, we can really leap forward in our script writing and management. These tools are also really useful for the management of other forms of textual data in our environments. Linux especially uses text-based configuration files which can be treated just like scripts and kept under version control and stored in version control systems. An excellent use of cross-domain skill sharing.

Version control is certainly the most must have technique from the software development world for use in our own scripting. Version control allows us to track our changes over time, to test code and have a simple ability to roll back, it allows for integrating multiple team members into the management of the same scripts, it tracks changes by user, empowers code review and auditing, simplifies data protection and deployment and so much more. If you use only one major development technique, this is the one to use. At first it will feel cumbersome, but quickly it will become second nature and make so many things that you do so much easier.

The development world has many other tools that we might potentially use like continuous integration, automated deployments, and code testing that all might provide to be useful depending on the scripting that we do, but nearly all of those are niche and completely optional even in a very heavily automated environment. Learning about these tools can expose you to options that may or may not make sense for your workflow, and will also give you great insight into how your potential development teams may be working.

Look to software engineering as a source of ideas about how to better approach your own script writing, but do not feel that you need to or even should adopt every tool and technique. Scripting for automation and product development do have overlap, but are ultimately different activities.

There is no secret to scripting other than just doing it. There are many good books available and resources online. Start with the simplest possible projects, look for opportunities to do scripting where you might have done work manually before. System tasks such as deployment or system setup checklists can be a great place to start. Or scripts to deploy a set of standard tools. Or perhaps a script to collect a specific set of data from many machines. I often find myself scripting data processing tasks. Once you start looking you will likely find many places where some new scripting skills can be put to work.

One of the best places to start using unscheduled scripting is for basic build and installation tasks. Using scripts to perform initial system setup and configuration including installing packages, adding users, downloading files, setting up monitoring, and so on. These tasks generally offer large benefits at relatively little effort and can serve as being essentially a form of documentation listing any packages and configuration changes needed for a system. Documentation done in this way is highly definitive because it truly documents the actual process used rather than one that is intended or expected.

Documentation first

In software engineering circles there is a concept of writing tests to verify code. While not perfect, running tests makes it far less likely for software to have bugs because there are tests that look for expected behavior and ensure that it is happening. We can still have bugs, this is anything but a guarantee, but it is a great step. After decades of writing tests for code, the idea that it would be feasible to write tests before writing code was floated and in research it is sometimes found that in doing so not only are bugs reduced but code writing efficiency can actually improve simply because test writing encourages thinking about problem solving in good ways. Test-first coding was considered a breakthrough in approaching software development.

This concept can carry over to the system administration world, in a manner of speaking, with the use of what I call documentation-first engineering. In this concept we start by writing documentation and then using that documentation to build the system. If it is not documented, we do not build it. Like test-driven coding, this approach forces us to think about how we want a system to work ahead of time which gives us another opportunity to make sure that what we are doing is well planned and sensible. And it allows us a chance to verify that our documentation is complete and sensible. When we write documentation after the fact it is far easier to make documentation that cannot really be followed for completing a task.

In some cases, such as those with low automation levels, this could mean simply documenting what we can in a wiki or word processor document and working from that as we deploy a system. If we have higher levels of automation then we may actually write code as documentation that builds our systems for us. Since the code itself functions as the documentation it is not just documentation-first, but the documentation actually does the work which absolutely guarantees that the documentation is complete and correct!

It is an intrinsic nature of automation to encourage better documentation and to move from documenting after the fact to before the fact and on to using the documentation itself as the build mechanism. This also means that we can potentially see double gains in efficiency as version control and backups and other mechanisms that we want to use for both documentation and scripting can be automatically applied to both.

Using advanced tools for our scripting may also be considered a higher step of automation maturity in a way.

Scripting combined with task scheduling

The hopefully obvious next step is to take task scheduling and our new found scripting knowledge and combine the two for even more power. Making complex tasks and making them able to run automatically without any human intervention.

Common tasks to automate in this manner will often include software updates. Having a script that looks for the latest updates, downloads them, prepares the environment, and deploys them all automatically on a schedule is very handy. Nearly any complex set of tasks that should be performed together can be scheduled in this way whether it is every minute or just on the third Tuesday of the month. Scripts are also very good for dealing with conditional situations where actions should only be performed under certain conditions such as only if storage is beyond a certain level or if certain people are logged in.

Almost a special case, and therefore well worth mentioning I believe, is using scheduled scripts to manage backups or replication.

State management

One of the most amazing changes that we have experienced in system automation is the introduction of state machines and state management for systems. State can be a difficult concept to explain as this falls far outside of the normal thought processes in IT and system administration. State is often seen as the future of systems, however.

In traditional systems administration and engineering we talk and think about tasks: how do we make a system get from point A to point B. In state theory, we do not talk about the *how* of managing systems. Instead, we focus only on the intended results or *resultant state*.

To think of it in another way: we start focusing on ends, instead of focusing on means. We move from process oriented to being goal oriented.

This approach forces us to really change nearly everything that we think about and know about systems. It is a game changer in every sense and frees us, as humans, to focus far better on what we are good at while allowing the computer to do far better what it is good at.

All of this magic is done by what is called a *state machine.* In the context of system administration, a state machine is an agent or portion of code that is given a document or series of documents that dictate the desired state of the system. State can refer to nearly anything about a system such as what packages are installed, what the current patch level is, the contents of configuration files, the ports open in the firewall, and the list of which services should be running.

A state machine will take this documentation as to the intended state of the machine and guarantee (or at least attempt) to ensure that the system is in the state desired. If a package is missing, it will install it. If a service is not running, it will start it. If a configuration file is incorrect, it will correct it. The opposite is also true, if a program appears that is not supposed to be installed it will be removed. If a service starts that is not supposed to run, it will be shut down.

The state machine typically runs every so many minutes or seconds and scans its state file and determines how the system should be, then scans the system to verify that everything it knows about how it is, and how it should be, match. It then takes whatever corrective action is required. Of course, under the hood, this is all done by complex scripts and system tools, that when assembled provide the power to enforce state. The degree to which corrective action can be taken by the state machine is determined by the power of the scripts that it has access to use. It is not unlimited, but generally on Linux a state machine will have enough power to do whatever is effectively needed in a real world, non-attack scenario and will remain highly effective at slowly or thwarting many less intensive attacks as well.

In this way, in theory, a state machine keeps a system in a nearly constant state that we desire. With state machines we spend our time writing the documentation of how we want a system to be and we let the state machine itself worry about how to make the system behave in the desired way. This includes the initial setup of a machine to take a basic, vanilla operating system install and turn it into a functional component of a specific workload. State machines work at the hypervisor and cloud levels as well, allowing us to maintain a standard conceptual approach not just inside of an individual system, but at the platform level providing the systems in the first place.

The end of the login

The nature of state management is to encourage, if not enforce, the end of the concept of logging into a server for administration purposes altogether. Before we think of eliminating logins, however, state management systems serve to improve traditional remote management through state.

Traditionally the biggest risks in remote administration are the needs to open ports and to have those ports open flexibly to allow for management from whatever location is necessary at the time. The idea of opening very few ports and locking them to a single IP address sounds like great security in theory, but is all but useless in real world practice. Having the flexibility for an administrator to log in quickly, from wherever they are at the time of an emergency, either requires too much exposure or far too many steps to limit access.

Enter state management. With state management a system can be instructed via the state definition file stored centrally in a repository to enable the SSH service, open a random port that gets used for that SSH service, and to lock it to the current IP address of an administrator or group of administrators. And the version control system will easily track that the change was requested, when it was requested, and by whom. In theory there could trivially be an approval step included as well as part of the mechanism. Once recorded the system would authorize access for the specified administrator(s). And once they were done, or on a set schedule, the state management system will revert the changes, after it has all been documented, and completely close off all avenues of access. The potential for enhanced security is incredible.

But that is only an interim step. With full state management we should, in theory, never need to log into a system at all. We should be able to do perform any management steps necessary via the state system itself or, even more appropriately, those steps should be being performed automatically by the state engine to ensure that proper state is maintained.

To fully enable a login-less mechanism of this nature we have to combine something like state management with concepts that we talked about in the last chapter such as having remote log collection and alerting so that even for tasks like capacity planning that there is no need to be physically logged into an individual system. To traditional system administrators this often sounds like blasphemy and all but impossible, but this is how many companies operate today and it is entirely possible with the right work being done up front.

For systems running on physical hardware inside of the office this might sound like overkill, and perhaps it is. For systems running on cloud servers this is highly practical in many cases. Human intervention should not be needed for a properly tested, documented, and running system. Manual management is difficult to document, very difficult to repeat, and highly error prone. Of course, human intervention can always be saved as a last resort, but the ability to remove it completely and depend on redeployment as a last resort is very obtainable today.

Automation maturity models give us a sort of roadmap of how to get from where we are to where we hope that we could be. Certainly not every organization has to get to the point where state management is handling all of their needs. Not every environment even needs to be scripted! Most organizations will continue to benefit no matter how far our maturity level is taken.

The final level of the maturity model we are saving for its own section. Taking what we have learned and applying the techniques we arrive at...

Infrastructure as code

Taking concepts that we have discussed here and looking at them another way, we discover the concept of *infrastructure as code.* Meaning that we can write code or configuration files that represent the entirety of our infrastructure. This is powerful and liberating.

It is easy to confuse infrastructure as code concepts with state machine concepts because they will, in many cases, overlap quite extensively. There are critical differences, however.

Infrastructure as code can go hand in hand with state machines, but state machines do not allow for imperative system definitions. Infrastructure as code can be used to define state, also known as a declarative approach to infrastructure as code, or an imperative approach by which operations are defined rather than final state, making it feel much more like traditional systems administration where we focus on the means rather than the ends or the *how* rather than the *goal.*

Platforms and systems

Infrastructure refers to both the systems that we run, that is the operating system containers, as well as the platforms, that is hypervisors and physical machines, on which they run. For most of what we are looking at in this section and certainly concepts like infrastructure as code, the applicability is equal to both aspects of infrastructure.

The tools and techniques that we apply at the system level will work with our platform level and vice versa. That means that we do not just get to rely on these awesome tools and techniques for configuring our operating systems and applications, but they can be used to actually deploy and build the virtual machine containers (both full virtualization and containerization) in both cloud and traditional non-cloud environments.

Applying the same or similar tooling across these two domains means greater overall power through conceptual integration and a more complete picture of our infrastructure as a whole. A key benefit in much of modern computing is moving from purely seeing operating systems as the building blocks for workloads to also seeing the hypervisor as playing a direct role in workloads as well. Instead of the hypervisor simply providing a space for an operating system to contain a workload, the hypervisor or hypervisor cluster can serve as a platform that is workload aware and act as a tool in dividing up resources to provide as a workload component level.

For example, the hypervisor or even hypervisor cluster level management will be aware that it is providing workload containers for application servers, proxies, processing nodes, storage, backups, databases, and so forth. Because the platform level is workload aware, it can then make intelligent provisioning decisions as to not only what kinds of resources will be needed, but also onto which nodes a workload should be deployed. If we have a three node hypervisor cluster and we provision three application virtual machines our provisioning should know to spread these out with only one virtual machine per node to increase redundancy; and it should know to do the same with the accompanying database that feeds those application servers. But it should also know to keep one application server on the same node as one database and to configure those to talk to the local database instance rather than a randomly non-local one. Bringing workload level awareness all the way down from the application, through the operating system, and down to the platform means better performance, with less effort, and more data protection.

Typically, as we move into infrastructure as code, we naturally begin to merge systems and platform administration groups because it just makes sense to see these as two parts of a larger, holistic infrastructure vision.

Imperative infrastructure as code design gives us many of the upfront benefits that we also saw with state machines, but require less work to set up initially and much more work to maintain over time. Because of this, imperative systems were considered normal when infrastructure as code was first introduced but as the market matured and declarative (stateful) tool sets and pre-built imperative structures were made available, the shift to declarative infrastructure as code was inevitable.

Under the hood, of course, all systems like this are going to be imperative. Any declarative system ultimately uses pre-defined imperative steps to arrive at the desired state. So in order to have a declarative system for us to use, either we or someone else has to wrote imperative scripts and tools that get us from many different starting points and result in the same ending point. It takes a lot of time and testing to build these components even once a base state engine has been designed.

These scripts have to exist for every task. If you want to define that a file must exist then we need tools under the hood that check for the existence of the file, logic to determine what to do if the file does not exist, how to find it, how to copy it, where to put it, what to do if a file copy fails, and so forth. This is probably the simplest use case and you can easily imagine how much more complicated it is to deal with any other task. Even a simplistic declarative system is going to be made from a myriad of imperative scripts that the administrator may never know about. Or it could be built by the administrator for very specific needs unique to the systems in question. The concept is flexible, but complex.

In theory we can have declarative state management without going as far as to have infrastructure as code. As odd as it may sound, it is actually somewhat common for those starting out with state systems to do so almost manually, attempting to issue stateful commands in a nearly imperative way to inform the system of desired state without documenting all aspects in code before doing so.

It is also common for the tools for these techniques to be deployed, and even used, but in a very incomplete way. This can be due to frustration, a lack of planning, internal corporate politics, you name it. Because working completely in this mode is so intensive and requires great planning up front it can be very difficult to obtain adequate time and buy-in from the powers that be to allow for complete implementations. Because of this we often see these tools being used to roll out basic, well-known functions using pre-built third party scripts, but complex configurations often unique to an organization still being done in a traditional way. The rush to get systems deployed will often drive this behavior.

This brings us to what is really the key best practice around infrastructure as code. What matters most is getting broad organizational buy-in and investing in proper documentation and code completeness prior to workloads being put into production.

Like any code that we would talk about in software engineering circles, our infrastructure as code requires testing before being used. Testing system administration code is often far easier than other types of applications, however. Creating our code for a new workload then attempting to deploy that workload from scratch is quite straightforward and can be attempting over and over again until results are perfect. Then system modifications, breaks, and other potential scenarios can be tested to ensure that the system will respond well under potential real-world problems.

One of the great benefits of this type of testing is that the function of our code is to create infrastructure out of, roughly, nothing. So all we have to do is start with a blank slate and, if things are working correctly, our infrastructure will create itself and we can test from there.

It is true that even if we cannot go as far as we would like by creating an entire infrastructure that is self-creating, healing, configuring, and destroying we can at least use these tools and techniques to go part of the way. Starting with simply building the infrastructure from scratch and not maintaining it or decommissioning it is an excellent step.

Infrastructure as code as disaster recovery

We have touched on this concept elsewhere and will really dig into it in the next chapter on Backups and Disaster Recovery, but it is so important that we have to discuss it whenever we talk about a constituent component of disaster recovery in modern systems. As we have talked about infrastructure as code we keep talking about the automated creation of systems where none existed previously.

This is exactly what is needed in a disaster recovery scenario. Our systems are gone and we need to bring them back. Having our systems, their standard files, their configurations, and more all stored as code in a place where it can easily be recovered or even better, not lost at all during a normal disaster, means that we are ready to build a new system, anywhere that we want, at the drop of a proverbial hat.

Because these types of build systems spend most of their time building workloads for testing and production deployment, they are easily justified for being fast, efficient, easy to use, well understood, and heavily tested. A system built in this way will be built identically during testing, the same in production, and the same during an emergency disaster recovery. We do all the hard work to make the system fast and repeatable up front so when something terrible happens we do not have to deviate from the established process.

Traditional system restores after a disaster require building systems through a process that is unique and nothing like the process through which the systems were built initially. This is highly error prone for so many reasons. It is a process that rarely gets good testing or documentation. It is typically done without proper planning and under a high degree of stress. This is the worst possible time to be experiencing these conditions. There is so much to go wrong at a time when there is the highest pressure to get everything right.

Avoiding those problems by creating a system that builds itself the first time, the second time, every time, almost instantly and completely identically is truly a big deal. If anything justifies the benefits of automation and infrastructure as code, it is this. Having the confidence that your systems are going to come back fast and correctly is something that most companies do not have today. The fear that backups are not going to work, that the knowledge of how to get a system up and running correctly configured for a workload is missing, that licenses or system details are not documented, or that necessary packages are not readily available is dramatic. Instead of overlooking those problems and hoping for the best we have a modern approach that makes these issues simply go away.

Getting our organization to the level of truly implementing infrastructure as code is a tremendous step, but this is the future. Companies that do this have faster builds, faster recoveries from disaster, better security, are more agile, scale faster, and all of that means have a better chance of making more money. At the end of the day, our only job as system administrators is to do our job in such a way that we can increase the bottom line of the organization through our efforts, no matter how indirect and impossible to measure that they may be.

Now that we know what the techniques are it is time to talk about the actual, already existing, ready to be tested tools that make things like state machines and infrastructure as code possible.

Modern tools of automation

All of this power comes primarily from modern tools that have been being introduced into the realm of system administration over the last fifteen to twenty years. The Linux world has been very fortunate to have been at the forefront of this movement since the very beginning. This comes naturally both because the Linux community tends to be one that thrives on and focuses on innovation, but also because the intrinsic nature of a system built around command line interfaces and simple text files for configuration and software repositories all make for vastly simpler automation. The design of Linux may not have been intentional to encourage automation, but nearly every major aspect of both the system implementation and the behavior of the ecosystem have led to it having the most ideal combination of factors to nearly always make it the leader in new automation tools and strategies.

Configuration management systems

New tools are always arising and techniques do vary over time so making a definitive list here is not possible, but there are some important tools that have managed to make a name for themselves to a point that they are worth mentioning as starting points for an investigation into the tooling that will likely make sense for your environment. The biggest tools for this type of automation, for infrastructure as code, at the time of this writing include **Chef**, **Puppet**, **SaltStack**, **CFEngine**, **Ansible**, and **Terraform**. The oldest of these, **CFEngine**, is so old that it was first introduced in its earliest form less than two years after the first Linux kernel was introduced!

All of these tools, often referred to as configuration management systems, share the common approach of allowing you to write code and configuration files to define your infrastructure environment and they manage the automation of the infrastructure based on that code. Most of them offer multiple functional behaviors such as the option to function either imperatively or declaratively. Most also offer the option to either pull configuration from the systems being managed or to push configuration from a central management location. So you get the range of functional options plus a range of product options.

The biggest differences between these products really come down to the style of coding that is used to document your infrastructure and the cost, support, or licensing of the products. Most of the products in this space either started as or moved to being open source. Open source makes a lot of sense as the space has matured quickly and management tooling somewhat naturally tends towards open source as it is mostly made by end users or those that came from open source communities and is used only in technical circles. Closed source products naturally lend themselves towards more customer-visible products where managers, rather than engineers, are choosing them. A key strength to most infrastructure as code tools is that they are free, open source, and often included in the Linux distribution so engineering and administration teams can generally choose to test and deploy them even into production without management needing to approve or even be aware that they are doing so. Because of the licensing and use cases it is little different than needing to deploy OpenSSH or any other standard component used in day-to-day system administration.

Of course the easiest approach here is to simply read about a few tools, download and install a few, and see what you like. Working with more than one is not a bad thing and most of the concepts will carry from one to another, even if the style of scripting and documentation vary. Many of these tools are cross platform and while they are generally designed with Linux as the primary platform both for deployment and to be managed, it is not uncommon for other platforms, especially BSD but also Windows, macOS, and others, may be able to be managed as well. Consider the possibility of choosing a platform that will be expandable to meet all of your organization's needs well into the future. If you are a heterogeneous shop you may want to invest your technical know-how into a platform that could be managing everything across the domains rather than only managing Linux, even if Linux is where you start.

Desktops are servers

Thinking of desktops as a form of server can be a bit confusing, but in a way, they are. They are simply one to one end user GUI servers that are generally deployed to desktops or homes rather than in the datacenter. This is a trivial bit of semantics until we start trying to understand how desktops may or may not fit into our greater support strategies.

But when we consider that desktops are truly just a special class of servers it quickly becomes apparent that it makes sense to potentially group them in for common administration as well. Of course if we are using a Linux distribution for a desktop this is much more obvious than if we are using Windows, for example, but the truth remains the same. If we are going to potentially support Linux and Windows servers both using the same tooling, there is really no barrier to doing the same for desktops regardless of the operating system(s) deployed there.

Traditionally, but for no reason of which I am aware, servers and desktops have been managed using very different toolsets. I can hypothesize many reasons why this might be. Typically, two separate teams provide this management and each chooses its own tooling independently. Servers grew up in one world while desktop support grew up in another. Vendors want to sell more tools and catering to hubris always makes sales easy. And the most likely factor: servers are typically administered from the command line and most desktop support teams expect to work purely from a GUI.

When we step back and look at desktops (and other end user devices) as if they are servers it is easy to see that the same tools to manage, document, and monitor servers work equally well with desktops. There is no real reason to treat them differently. Of course desktop support teams will always require the ability to remotely see, and probably interact with, the end user's graphical desktop to be able to assist the end user's issues directly, but that is a different need entirely from administration duties.

Desktops can be, or even should be, managed using the same advanced tools and techniques like infrastructure as code and state machines as we would with any other server. That is correct, when we understand that they are actually a server, then rules that apply to all servers also apply to desktops. Good semantics make everything easier to understand.

In fact, there is even a good argument that end user devices are among the most valuable to apply these techniques to because they are the most likely to need to be rebuilt, modified, share profiles, undergo changes, be compromised, or need to be managed while not in communication with central services. This last point is especially valuable because state machines that keep functioning with their current state will continue to provide security and self-healing characteristics for machines that are off of the network and can enforce policies independently. This is possible to do in more traditional ways but is harder and less likely to be done.

Because of the nature of how infrastructure as code systems tend to be deployed it is far more likely, as well, for these systems to keep functioning when a desktop or laptop is off of the corporate LAN compared to traditional management tooling that is generally built entirely around the LAN concept. Because end user devices traditionally have a high frequency of either moving on and off of or simply never existing on the LAN having tools that are not LAN-centric is typically much more important in the end user space than in the datacenter space already.

In some ways, mobile device management, or MDM as it is generally known, is an attempt to make tools that work more likely infrastructure as code and state machines, but that are presented more like traditional tools and sold through more traditional channels with a focus solely on end user management. These tools have been successful, I feel, because they are copying much of the technology and common approaches from this space and once we work with these tools we typically find that mobile device management tools do not make sense unless we are lacking these capabilities in our organizations.

Many of the benefits of any new system, of course, come from convention rather than strict definition. One of the largest conventions in the infrastructure as code space is the move from LAN-centric to network agnostic system management deployments. It is common, and nearly expected, that the tooling for infrastructure as code systems will be hosted in some public form whether on cloud, VPS, or colocation, and kept external to any LAN(s) that may exist for your organization.

Hosting management infrastructure outside of the LAN means that any LAN-centric ties that we might otherwise make accidentally or casually are no longer possible unless we deploy a LAN extension technology like a VPN. This convention naturally moves us away from deploying technology that only works inside of a LAN or that uses the LAN boundaries as security features. Eliminating the LAN boundaries frees us to manage multiple sites, mobile users, even multiple organizations transparently from a single platform, as long as those systems use the Internet. Traditional systems break under so many circumstances, while also having common security vulnerabilities both because of the tendency to break easily or have gaps in utilization, and because LAN-centric thinking is not good security practice.

Version control systems

The other category of tools that we should really address within the topic of automation is code repositories and version control systems. These are technically two separate things, but almost always go hand in hand.

At their core, version control systems simply keep track of the changes made to documents that we have so that we can track essential data such as who made a change, when it was made, and what the document looked like before, at the time of, and at a later date than the change. This alone is quite powerful, but most any system that does this today also serves to distribute the code and the version control so that it can be used by multiple people, in multiple places. And in being able to do that, can also be used to populate a central repository which can be treated as a master location for storage so that no single person's endpoint needs to be considered vital for the purposes of data protection. That central location becomes a location for backups and recovery, as well!

Version control systems influence many modern document systems and we covered both in our earlier chapter on documentation, but in this context we could look at real world products such as Google Docs, Microsoft Office Online, and Zoho Docs, all of which present traditional document file types or interfaces, but all provide version control of those documents. These are very clunky to use for the purpose of coding and code management, but all will serve in a pinch if you are looking to just get started quickly using what you already have deployed. These systems are essentially copying the mechanisms of traditional code version control systems and applying them to spreadsheets and word processing.

Since these office document types are so well known, it is almost easier to picture these as the standard (they are not) and thinking of code version control systems as being a code-specific modification of those document tools (they are not, they came first.) These systems generally (and by generally, I mean every situation that I am aware of) work with standard text files and so can be used with any text editing tools whether you work with something basic like *vi* or *nano* directly on the Linux command line or you work with robust tools like *Atom* or *MS Visual Studio Code* that provide fully graphical coding environments with deep editing awareness and features, you can use version control systems. Some advanced environments will actually have version control integrated directly into the applications so that you can automate the entire process from a single place and make it look and feel much closer to the office style tools!

In a practical sense, at this time, two protocols for version control have risen so far to the top that it almost feels like there are only two choices really left on the market: *git* and *mercurial.* In reality, there are many, but only these two require mentioning. Feel free to research other tools and protocols, but make sure that these two are included in any research short list that you might have. Both are free and work similarly and allow for all of the features that you often expect today including central repositories, copies on end user machines, automated deployments, version meta data, and so on.

Beyond the protocols that are used, much of the power stemming from version control systems today come from the online repository services that power them. Of these there are more and you can run your own in house as well. The two key players are Microsoft's GitHub and the open source GitLab. Both are hosted services with extensive free offerings, and GitLab also offers their software for free that you can host yourself if this is a business or technical requirement for your environment. These two services, and others like them, provide central git and Mercurial repository locations, a centralized location for backups, a simple web GUI for code manipulation and management, and a raft of processes, tools, and services around code automation. Much of which is likely overkill or useless in a system administration environment, but much of it does have a potential use. You can certainly get the benefits that you need without these types of services, but it is far harder to do so and nearly all successful environments have been relying on them for years. They are almost always free for the needs of system administration, do not avoid them. As hosted services, their use is yet another means of *breaking free* from LAN thinking, as well.

There is not much of best practices or even rules of thumb to discuss when talking about tools. Testing multiple tools, keeping up with the market as to what is available and what is nearly developed, evaluating the tools that make sense for your organization, and learning your chosen tools inside and out are all standard good approaches.

Rules of Thumb:

- Management tools should almost always be open source. This is an area where security is of the absolute utmost importance and where licensing limitations create security risks on their own. So open-source matters more here than in most areas.

- Deploy management tools in a network-agnostic way. This means deploying them on the Internet in a place that is accessible to reasonably any machine located anywhere. Avoid any semblance of requiring or relying upon traditional LAN networking for connectivity or security unless absolutely necessary.

Best Practice: Keep code of all types under version control and in a repository.

Now we have covered the techniques and talked briefly about a handful of real world tools that you can use to initiate your investigations into tools that you want to try to deploy and learn for your own environment.

Summary

In this chapter we have looked at why automation is important. We investigated how we should approach automation and where to look to get started. We discussed maturity modeling. We delved into the rather complex topics of state machines and infrastructure as code. And finally we tackled actual tools that you can download and learn today to take your coding to a totally different level, entirely.

In our next chapter we are going to be going into one of the absolutely most important, and most commonly avoided, topics in system administration: backups and disaster recovery. Do not try to skip over this coming chapter, if there is one thing that we need to get right in administration, it is our ability to avoid or recovery from a disaster.

9
Backup and Disaster Recovery Approaches

I have said it already earlier in this book and it bears repeating no matter how many times it takes: nothing is as important in what we do as system administrators as maintaining good backups. This is our utmost priority. It is so important that many organizations maintain an independent system administration team that handles nothing but backups to make sure that it maintains constant attention.

Backups are not glamorous, and they are rarely exciting. This does not just make them a challenge for us in the technical world to want to spend time thinking about them when we could be implementing new automation or something else admittedly more exciting, but it also means that management often does not prioritize budgets or prioritize around backups. This creates a potential danger for system administrators that our careers can be slowed if we focus on critical functions like backups instead of doing flashy, high-profile projects to keep the interest of management while at the same time often being at very high risk of being punished if the backups that they do not prioritize do not work flawlessly when needed.

Backups are moving more and more to the forefront of our consciousness, however, as they have moved from primarily protecting against catastrophic hardware failure to being on the front lines of security concern. Backups have evolved in recent years and now present opportunities to shine politically within our organizations and opportunities for renewed technical interest. The boring backup and restore landscape of the 1990s and 2000s is, quite literally, a thing of the past and today we have so many approaches, options, and products that we really must approach backups as a broader concern than we have in the past.

In this chapter we are going to cover a lot of different aspects of backups. We will start with a broad overview of how we take backups and what components exist within backup systems. Then we will look at similar technologies that are often used in conjunction with and might be sometimes confused with backups. Then we will cover in depth what we have mentioned throughout this book, the idea of a modern *DevOps centric backup approach* and talk extensively about that. Then we will explore agents and the issues with crash consistency of our data in backups. Finally, we tackle triage and what we can do to make restores better once a disaster has happened.

Without further ado, let us begin what is assuredly our single most important chapter in this book. Of any book, I would assume.

In this chapter we are going to learn about the following:

- Agents and crash consistency
- Backup strategies and mechanisms
- Snapshots, archives, backups, and disaster recovery
- Backups in a DevOps world
- Triage concepts

Agents and crash consistency

In the next session we are going to look at mechanisms for taking backups. Before we do that, I want to look at why backups are so hard to do well in the first place. In order to do that I will work with two examples. One is a text document that we create using our favorite text editor. I will assume that you are a normal person and love the vi editor as much as I do. And we will compare that to another common use case, the data file of an enterprise database system.

When we talk about backups, we are talking about taking data that is stored on a physical medium and replicating that data somewhere that is separate from the original system in such a way that it is able to survive in many cases when the original system has failed. That is a very high-level view of the goal of backups. It serves our purposes. Therefore, in order to perform a backup we must be able to take the data that the system has, read it, move it, and write it.

Of these steps it may seem like moving data to another location or writing the data somewhere would be the biggest challenges. This is incorrect. Being able to read the data is actually where the real problems tend to exist. Primarily because what data we need to read is not always clear.

How do computers store data on disk in the first place? Not every computer system works the same way, but there are certain basics that exist for which we know no alternatives. While working with data, computers hold the data in their random-access memory. While there it is completely volatile, but it is very fast. When the computer is done with a file it takes the data that is in its memory and then writes it onto the storage device, presumable a hard disk or SSD.

If we attempt to take a backup of a system while data is in the computer's memory and not yet written to disk our backup would contain none of the data in question. This may sound silly to say, but it is necessary to remember. It is common for people both technical and management to assume that data is always stored somewhere, even when it has only gotten into memory and not had time to be stored anywhere yet.

Once data is written to the disk, and then a backup is taken, the data from the disk should be copied to the backup location. All is well with the world.

This is all well and good, but computers deal with situations far more complicated than just receiving files, and saving them to disk. In the real world, computers nearly always are reading existing files off of their storage mechanism, holding it in memory while manipulating the data, and then saving it again back to disk with the new changes incorporated into the data set. This is where things start to become tricky.

We should start using our examples now. First the case with a text file editing in `vi` (if you insist, you can edit in Atom or something else.) We will also assume that we are not creating a new file, but rather editing a configuration file that already exists on the disk. A great example is the `hosts` file (found at `/etc/hosts`.)

If we take a backup while the file is being edited, we would get a copy of the old data before any edits because any edits that are happening exists only in memory, not on disk. The fact that the file is open at the time does not mean that data is being written to disk. The old data is still on the disk while the new data exists in memory.

When the file is saved, the data is written to the disk. Once written, we can make our copy. Of course, there can be a short moment while the data is being written where what is on disk is neither the original version nor the new version but something in between as the data is not finished being written.

We can attempt to lock the files, but even with a file being locked we have a complicated situation. The backup mechanism, whatever it is, has to decide if it is going to skip over a locked file, wait for the lock to be released, or ignore the lock and take the backup anyway. No situation is very good. We either fail to backup all of the files in any particular backup process, we risk a long wait for an event that might *never* happen, or we risk getting a file that is out of date or worse, corrupt. There is no completely simple answer.

Locking mechanisms in Linux

When processes are using files on Linux, and more or less on any operating system, there are three essential strategies that we normally see in action. The first, and most obvious, is to do nothing. That is correct. In many cases there is no locking performed whatsoever. The contents of a file are read and the file itself is not marked as being open. It is just read. This approach is great in that any other process can continue to use the file in question in whatever manner they desire, but it also gives us no information as to whether or not the file is about to be updated in some way.

The primary alternative mechanism on Linux is called an advisory lock. With an advisory lock, the operating system marks when a file has been opened by a process. The locks are so called advisory because the operating system advises other processes that the file is in use. Another process can opt to ignore the lock and continue to read, or even to overwrite, the file that is locked. This is handy in that we can lock a file and not worry that we are completely blocking others from accessing it. The risk is that our lock is not honored, and race conditions are encountered where data is changed out of order or data that is believed to be saved gets overwritten. Ignoring an advisory lock to peek at a file and check out some old data is pretty safe. Ignoring an advisory lock to overwrite the file with changes without the original process knowing that it has happened is dangerous. Processes that fully support and honor an advisory lock are called cooperative processes.

The third option is a mandatory lock. As the name implies, processes have no option but to honor a mandatory lock. Mandatory locks are managed by the Linux kernel and only exist if a filesystem is specifically mounted with this form of locking enabled. Mandatory locks on Linux, however, suffer from implementation problems that make them unreliable and subject to some uncommon race conditions which effectively defeat the purpose of the lock. Because of this mandatory locking is almost always ignored on Linux systems.

Locking is conceptually hard. It carries risk and system overhead. The best way to respond to a locked file is not always clear and may vary by many different use cases of which the operating system is not aware. Because the operating system does not know how or why a file is being accessed in the first place, it has little means of usefully enforcing limits on secondary access. There is nearly always a reason that we might want to read a locked file and sometimes a good reason to even write to a locked file. Trusting that we will only run processes that will behave as needed for our use cases is our best course of action in nearly all cases. When more complex access is required, such as is the case with most database files, it is necessary to move to single process access with that process handling additional data access from higher in the application stack where more knowledge of the intended use of the file is know.

Other than locking, the major concept that we have to understand is quiescence. Quiescence simply refers to the storage reaching a state of dormancy, a lack of fluctuation. When we say that we have achieved quiescence in our storage we mean that all of the data that is currently *in flight* whether actively being used by an application or just being held in one of any number of different cache layers has been written to disk. In some ways, we can think of locks as being a mechanism meant to warn us (or a process) as to a system not being currently quiesced.

Unfortunately, there is no universal mechanism to enforce or even ensure quiescence on Linux. There is none on Windows either, dangerously contrary to many claims that VSS (the Volume Shadow Service) does this. VSS is the standard LVM on Windows and as such is often used in many storage operations. It is commonly said that VSS guarantees that all files are quiesced, but this is false. VSS has special hooks into common applications, primarily from Microsoft, such as SQL Server, Exchange, and Active Directory so that they are able to communicate effective with the storage layer as to the state of their quiescence. This is an amazing feature and incredibly handy to be integrated with the operating system automatically. It does not address third party applications which rarely have VSS integration leaving them dangerously unquiesced while many Windows administrators believe that VSS magically manages data flushing from all application layers that it does not, and cannot.

At the operating system level, where we typically have to work as system administrators, our primary tools for quiescence are snapshots from a logical volume manager (such as those provided by LVM or ZFS) or to freeze the filesystem itself as is possible with XFS. The higher up the stack (higher means closer to the application itself) we freeze the system, the less likely that there will be some artefact still in flight and not written to disk. At the end of the day, however, unless an application itself guarantees that it has flushed all data down to disk and left nothing in a cache or in process of some sort, the best that we can hope for is a best effort situation. This is true regardless of the operating system. It is a universal challenge. Computers are simply really complex things.

In the majority of cases, nearly all cases in fact, one or more of these quiescence methods will work just fine to get data safely to disk. In fact, most backups and most businesses rarely even rely on this much protection and typically rely on best effort file locks and multiple backups to provide working copies of a sufficient number of files on any given system. Most workloads write to the disk infrequently making relatively few into high-risk scenarios. But we cannot overlook at all of these methods leave us with no ability to truly ensure that a system is completely quiesced and we are always taking some degree of chance that critical data will be in flight in some form. If we are talking about a web server that only serves out static websites the risk might be so low as to simply ignore it. If we are talking about a critical enterprise database used for financial and accounting data the risk might be enormous and one that could never be undertaken no matter what. So, if we want to take backups at a system level, we need to consider our workloads, their quiescence, and what risk that creates for us.

When we take a backup using one of these mechanisms that does not coordinate end to end with the application workloads to ensure that the data on physical media is complete and quiet, we refer to our backup as being *crash consistent*. This is an important term and one that is used often. Crash consistency does not mean that the system has actually crashed, but rather it refers to the state that storage is in after the abrupt crash or loss of power in a computer system.

Crash consistency is such a critical idea that we need to really understand what we mean when we say it. For some it is said as a scary warning of impending doom. Others use it to imply a reliable system state. So, what does it really mean?

When any computer crashes completely, meaning that there is no ability for recovery or protection after the event has occurred such as when there is catastrophic hardware failure or power is suddenly unavailable to the system, there is absolutely no mechanism to ensure that the data that is on the disk is accurate or complete. As any computer user is aware, sudden loss of power to a computer is generally not a problem, except that any unsaved changes to a file or our latest progress in a video game or whatever will be lost. This feels obvious and we do not normally think of it as being a significant problem of system design. Once in a while, we will find that a program or data set has become corrupted. We all fear this when a system returns from a crash, but rarely is it actually a problem - at least as an end user.

This state of a system after an actual crash is essentially identical to the state of a system where a backup is taken without any quiescence. Any data that is currently being modified will be lost. Once in a while data that we thought was already saved to disk will be corrupted. By and large, everything will be there. This is why we call backups or any copy in this state crash consistent, meaning in the same consistency as a catastrophic crash on a physical machine.

When talking about desktops it is easier to illustrate the point. The same risks exist on servers, but to a much more serious degree. With a desktop we tend to have a single user who has a very good idea of all data that is likely to be currently in flight and storage is pretty straightforward. In a server environment not only do we likely have lots of storage complexity with different possible and possibly risky mechanisms at every turn, but we likely have larger and more numerous cache situations throughout the stack and one or more multi-user workloads sitting on top that will rarely be interacting with users meaningfully at the time of a crash.

Consider an example web server, a common example, with data stored in a database. A user has some interaction with the web page maybe entering contact details for sales or uploading a form or processing some banking transaction or placing an order. The user believes that the transaction completes as they receive feedback saying that their submission is complete. The user will likely not refresh the page or return as to them, the transaction is completely. But is it?

The web server may have queued the transaction in memory to be handled a few seconds later, or left part of the data in a cache. Or it may have written the transaction to a database, and that database might still be holding it in a cache. Maybe the database has written the data to its file, but the filesystem has the data in a cache. Maybe the filesystem has committed the data, but the logical volume manager has the data in a cache. If you have hardware RAID, there is likely a cache there and before data gets written out to physical disks it will sit in that cache. The physical disks themselves can have caches which, one would hope, would be disabled in this situation but may not be and represent yet another place where data might be cached before the drive head actually puts the data physically onto the disk.

Of course, lots of mechanisms probably exist to protect against many of these failures. Disk caches should be disabled, RAID caches are often protected by being non-volatile RAM or have a battery to allow them to flush to physical disk even after the system experiences a power loss, databases often log transactions before doing their final write so have the ability to roll back if they cannot roll forward and applications generally do not report that all is well to the end user until at least some degree of certainty exists that the transaction has been recorded. All of those are situations that we hope exist, however, and nothing guarantees them except for a combination of application and system design end to end. The best designed application can still be fooled by a hypervisor that presents volatile RAM as if it were a hard drive and reports quiescence where there is no storage written at all. So, trust is required.

Dropping data in flight is only one concern. The other is corruption. Many layers along this path have a potential to become corrupted by an incomplete write operation. An individual file is often where we see corruption occur, this is the most common place. Twenty five years ago, we saw filesystem corruption regularly as well. Today this is not a common worry, but it does exist as a risk, still. Different filesystems carry different levels of risk for this type of scenario. In theory a logical volume management layer could hold some corruption risk. Certainly a storage layer of RAID or RAIN can become corrupted and sometimes, such as parity RAID, have some potential for total data loss during a corruption event.

All of these steps are unlikely, but all are possible. Missing or corrupt data will generally not happen during a normal crash. How likely a disaster is to happen depends on many factors. We should neither look at crash consistency level solutions as useless or total loss waiting to happen, but neither should we pretend that they do not ignore basic risks of not having any system in place to ensure consistency. In all cases we are simply rolling the dice and hoping for the best. I have seen a great many small businesses lose data with what they had thought were completely reliable backups because they were depending on crash consistency and a vendor had pretended that data corruption was not a concern. Because crash consistent mechanisms are cheap and simple, they get a lot of attention from vendors.

The alternative approach is called **application consistency**. We call it this because it refers to the potential state of the system when the workload application(s) are able to confirm that they have flushed all of their data to disk and have nothing currently in-flight. Applications may do this simply by confirming that they are quiesced currently, or they may have mechanisms to force such a state to occur. In either case, they avoid the problems of crash consistency by applying intelligence from the very top of the application stack and verifying that data is ready to be backed up.

This approach requires the application support providing this quiescence and if we want to coordinate this action with a backup mechanism, of any sort, we have to have some way for the backup mechanism to call the application to request quiescence, or for the application to call the backup mechanism. In the Windows world there is a standard interface in VSS as we discussed earlier than any application can support if they chose to do so. Linux lacks a similar standard at this time.

Because all of these mechanisms are complex and non-standard it presents major challenges when working with anything outside of a limited scope of popular services that are well known and supported. In reality, Windows is also similarly limited but the use of a smaller set of very standard services is more common there.

Many applications choose to tackle this problem by incorporating their own backup services rather than attempting to depend on some other mechanism. Some may build an extremely robust backup mechanism that directly supports many different options such as built-in schedules, multiple backup levels, and storing directly to many different types of media. This is not very common but certainly exists.

The most common option used by everything from the simplest of applications to enterprise databases is to simple write a backup file from the application directly to the local filesystem. The is extremely common and very effective. This produces a simple file or set of files that can be designated as a backup and kept from being used other than for the purposes of another backup mechanism using these files to then copy or send to another location.

A really common example of this is MySQL and MariaDB databases. Databases are our hardest type of application to safely backup and so nearly any database system will incorporate some means of safely protecting the data without having to resort to a drastic, brute-force step such as shutting down the databases, coping files, and starting it back up when completed.

MySQL example with mysqldump utility

Almost certainly the most well-known application level backup tool in the Linux ecosystem is `mysqldump` that comes with MySQL and MariaDB. This simple command line tool connects to a running MySQL or MariaDB database server, locks it and quiesces all of its data in RAM (it has no need to flush to disk), and then saves a copy of this data to disk. It is as simple to use as any utility could be while also being outrageously power. It has almost no impact on a running system and because it does not need to do any complicated flush to disk before taking a fully application-consistent copy to disk it does not need to stop read operations to the database and only needs to halt write applications for the briefest of moments.

I will show here the simple, single command to take an easy backup of every database managed by a single MySQL or MariaDB instance:

```
mysqldump --all-databases > mysqlbackup-`date +%H%M`.sql
```

This example is about as basic as you can get, and yet it is so effective. You can run it anytime and because it takes the current time during the process it will not overwrite another backup file, even one takes just a minute earlier. You can make it write anywhere on disk that you like. You can use standard utilities to compress the resulting backup file while it is being written or at a later time very easily. Keep in mind this is a backup file, a file for the purpose of making a backup, but not an actual backup yet at this point.

Once this file is stored to disk, it is safe and any backup tool can now take this file without worry than an application is using it or modifying it and place it wherever it makes sense. Total flexibility and simple application-consistency.

At this point I think we understand consistency, how locks and quiescence play a role, and all of the problems that we may face while trying to make a backup of even a single file or application (remember that not all backups involve data that is actually in file format on storage anywhere), let alone trying to back up entire filesystems or systems. Appreciating the challenge is key to understanding why we are concerned about different aspects of different types of backup tools which we will look at next together.

Backup strategies & mechanisms

Backups are actually far more complex animals than most people imagine. So often when dealing with backups we are simply told to *take a backup* as if this is a straightforward activity with few variables. In the real world we do have some stock approaches that meet the majority of needs, if only minimally. There are cases, however, where to do effective backups requires a lot more thought and deep understanding of our workloads and infrastructure to be able to get correct.

In the good old days, you know like the 1980s and 1990s, backups were almost always the same. They involved a simplistic agent of some sort, like the standard Linux `tar` command, that would run on a schedule (that we probably had to set manually with something like `cron`) that would take all of the files in a directory or, more likely, the entire system and package them up as a single file and place that single, large file onto a tape device. That tape device would then require a human to remove the tape and transport somewhere for safe keeping.

Over the years new technologies would come out and backups slowly became more robust and more complex to discuss. Tape became less popular and other backup targets ranging from swappable hard disks to constantly online storage emerged. With backup media evolving the mechanisms that took backups had an opportunity to move from discrete *one tape per backup* schedules to more flexible or complex designs. With all of this came complex backup applications and the backup world went from simple and basic to quite complex.

Because there are so many variables today, simply taking a backup is no longer a straightforward concept. People all have different views and ideas of what a backup entails based on their desires and experience. It is so varied that we even risk tunnel vision and thinking that all people see and experience backups similarly when, in reality, there are so many different ways being used in the real world to handle backup needs.

Types of backups

First, we will discuss the actual backup mechanisms that may be used. These generally fall into two categories: agent based and agentless (I really hate both of these terms) but people normally think of them as three categories with a third being ad-hoc scripts. All of these ideas around their identities are totally wrong, of course, as are so many things that people tend to say. The terms have made inroads and you will need to use them regardless of their inaccuracies, however.

System level backups

System level backups are the so-called agent-based backups. These get this name because typically a software agent that runs on top of the operating system is installed and is visible. It can be seen in an installed software list somewhere on the system. The agent then runs and grabs files from *inside* of the operating system context and send them off somewhere else to be stored. There is often additional processing steps such as packaging, compression, deduplication, and encryption, but all of those are technically optional.

Everything is an agent

When we say that we have an agent-based backup solution it brings to mind software that must be purchased, downloaded, and installed onto our computers. Certainly, this is a very common thing that people will do.

The idea of an agent is, however, far broader. Backup agents may be complex backup utilities that are built into the operating system rather than obtained separately. In the Linux world this is just as likely to be the case as not because so many powerful backup options are included in the repository ecosystems of the typical business-class distributions.

Agents get even broader still. Classic utilities like `tar` and `rsync` are, or at least can be, backup agents. They are installed software components that can be used to carry out a backup from inside of the operating system.

We can keep going. If you write your own script to move blocks or files around that, too, is an agent. The idea of an agent is incredibly broad and, logically, some component is always needed to perform a backup job and that component is always an agent.

By working from inside of the operating system, an agent located here has the ability to utilize obvious access methods to the data that we want to back up. It can access the block devices directly either by talking at a block level with a tool like `dd` or by using a slightly high-level logical volume tool like LVM which can provide robust block level handling; or it can use the filesystem layer to request files one by one; or it can use APIs to talk to running applications that can provide some type of higher level managed data access. In one or more of these ways an agent can access the data necessary for a backup.

All of this is to say that agents have the task of talking to the operating system in an attempt to trigger some level of quiescence from the storage subsystem to increase the chances of getting a consistent backup. Some agents talk extensively to many applications and some talk to none.

Agents do all of the work and offer us many ways to talk to our data. But they have the limitation that they can only access what the operating system can access and they can only be used when the operating system is up and running. This last point means that any resources that the operating system itself is using will be in use at the same time that the agent needs to access them. This makes for a real challenge in system-wide consistency. However, we rarely are truly concerned with consistency on a full system scale. Of course, we would prefer it, but the majority of system files are of little consequence and easily replaced if lost or corrupted. Typically, only critical application data or maybe installed applications themselves are of major concern and the rest is just useful in case of a need to restore quickly.

Typically, but not universally, with system-level backups it is necessary when performing a restore operation to install an empty operating system and then install the restore agent and allow the agent to run against the stored backup and put files back into place inside the operating system. This often means that system-level backups are excellent at taking backups of individual files and being able to restore individual files but find it more difficult to deal with having to restore entire lost systems.

There was a time when essentially all backups worked this way and the general assumption was that nearly all restore operations were to retrieve single lost files rather than to retrieve entire failed systems.

People just do not lose files like they used to

In writing this chapter I was struck by a massive, fundamental change in the computing experience for most people over the last thirty years. Through all of my younger years in the industry and even before simply as a computer user, the big concern that we always had was losing, deleting, or having an individual file become damaged in some way. We really all pictures the concept of backups as just being lots of individual files being copied somewhere, and we envisioned any restoration to be a process of copying those files back once we had a place to put them and more often than not, needing to do so when there had been no failure whatsoever.

This last point is what is most interesting to me. For decades it was assumed that people were going to accidentally overwrite or just delete critical files that they needed with great regularity, and it really happened. Needed to track down and restore an individual file that an end user deleted was a completely common task that was done so often that most large companies had teams dedicated to doing nothing else.

Computing has changed such that today, this is rarely the case and the idea that an important file would be lost on its own does not seem impossible, but certainly seems unlikely. I assume users have become better educated and far more diligent with how files are managed, and of course a huge percentage of work that used to be done as individual files has moved to some sort of database of data stored behind an application front end that protects end users from themselves (this could range from Google Docs to Flickr, but even end users working with their own file management has decreased significantly.) Most operating systems, and even some applications themselves, now implement *trash* features that hold deleted files until they are explicitly disposed of to give end users lots of time to change their minds or find what was deleted accidentally. And finally there are features like self-service file restores included in operating systems or from simple add-on applications that allow files to be brought back from other media without needing to engage traditional backups or backup teams as the data is still stored somewhere locally (not to mention online backup systems that can be used by end users without intervention.)

It is simply interesting to note that something that had become such a common problem that it drove most of an industry not that many years ago is effectively gone today. The days of computers being seen as file management and manipulation devices are gone and have given way to being online data consumption devices where most users are not even file-aware any longer. Many younger users can even be confused at the concept of file and storage management today.

Platform level backups

The alternative approach is to take a backup from the platform level, that is from the layer underneath the operating system. In practical terms this means the virtualization layer which would entail typically the hypervisor, but in some cases can be limited to external (to the operating system) storage virtualization, this would be especially common if using a SAN device, for example.

This approach became all the rage with the advent of virtualization. New backup vendors entered the market with new technology aimed at making extra-operating system backups easier and faster. Backups at this level, of course, do not have access to things like the filesystem or application APIs to communicate with those components or to have knowledge of the data layout. So backups taken at this level must do so by talking directly to the block devices or to hypervisor level storage abstraction layers, which are essentially block devices themselves, leaving us generally blind to what the data is that we are actually getting. At this layer we might know, but only maybe, about the physical separation of devices presented to the operating system. But we have no visibility into filesystems or deeper. So all of the concerns that we have with system-level backups and consistency are potentially magnified many times over. Not only do we have no way to know if an individual application workload has quiesced, but we cannot even tell inform the system-managed filesystem, file locks, or logical volume manager that we are attempting to read the block devices. So if we take a backup at this level alone, we are totally blind and simply reading blocks off of a virtual disk with no reason to believe that it is in any sort of a consistent state and nothing tells the operating system that anything is happening at all. If we do this alone, the operating system literally has no way to know that it is happening.

It does not take very much thinking about how this mechanism has to work to realize that platform-level backups without an agent are all but impossible. We could shut down a virtual machine and take a backup of its storage layer while the machine is powered down, of course. The shutdown process of a total operating system is effectively the best way to ensure full quiescence and consistency. How do we shut down a virtual machine from the platform level without just pulling the virtual plug on it and putting ourselves into a state of crash consistency? Surprisingly, with an agent.

This is the big secret of platform-level (aka *agentless*) backups: they use an agent! They have to. In the real world taking a backup directly from any lower level service would be completely unreliable. But I hear you murmuring *I've done this, everyone has, and we don't install any agents.*

What makes the backup agents in virtualization environments hard to identify is because we install them automatically and universally so that we do not think about it. In some cases, these agents are built right into the operating system. They have different names depending on the platform that you are using. The most famous is VMware Tools as VMware uses this and the installation process is well known. KVM, the primary hypervisor in the Linux world, has the guest agent built not just into most Linux-based operating systems, but actually baked right into the Linux kernel itself! So you never actually need to see the agent, even though it is almost always there. Many Linux distributions do the same with Hyper-V's equivalent agent called Linux Integration Services, which are not built into the kernel but are often included in the operating system automatically.

In these cases, the agents from the hypervisor vendors that I have used as examples are not strictly backup agents, they are general purpose agents used to coordinate activity between the hypervisor and the operating system, but in reality, that is simply a more advanced agent function and nothing more. They are agents in every possible sense of the word, they are software that has to be installed into the operating system to work, and just like more traditional backup agents they are sometimes included with the operating system and are sometimes third-party add-on packages.

These agents are generally part of a set of high-performance drivers needed to make full virtual machines work efficiently and so are rarely left missing. They provide a critical channel that allows the underlying platform layer to communicate up to the operating system layer or even higher. This communications channel is what is used to issue graceful shutdown commands to a virtual machine as well as to inform it that it needs to quiesce a filesystem or take a snapshot of a volume. In theory this agent could even hook into a specific application, although this is mostly only theoretical in the Linux world.

There are important advantages to taking platform level backups. If we ponder the workings of these backup systems, we easily see that the expectation is going to be a block-level image taken of an entire file system container or block device. The real advantage to this is in its completeness. If we take an image level backup, rather than a file backup, we have an opportunity to have the entire device, not just portions of it, that will presumably allow us to restore a system completely rather than having to rebuild and then restore partially.

Disk images

When we talk about file level backups the essential mechanism, we are discussing is ultimately a file copy of some sort. File X exists on the filesystem that we want to protect and in order to protect the data that is in that file we copy that file to another location so that the data can exist in more than one place at the same time. Easy.

With full block device copies at the platform level, we refer to the resulting copy as an image or a disk image. In theory we could be copying from one physical block device to another or, at the very least, from a virtual block device to a physical one. In nearly all real-world cases, especially those involving backups, what we actually do is copy the block device to a file. That file is called a disk image and is sometimes referred to oddly as an ISO file.

We call this an image because it is essentially a picture of the entire disk as it was in any given moment. We also use the term snapshot to refer to this same operation. In regular English a snapshot is an image. The words are nearly interchangeable. The same is true in computing. Sometimes one is used to imply something different from the other, but there is no accepted definition that makes them not completely overlap.

In every day usage, images are used to refer to images stored as regular files, such as ISO files and contain a complete copy of the original filesystem. Snapshots are used to refer to a nearly identical scenario but where a logical volume manager creates a partial image file that is in some way linked to the original file and may contain only the differences between the two. But that file is often used to make a standalone image file, which is indistinguishable from an image file otherwise, but is still called a snapshot. Which leads to the inappropriate situation where two identical files can have different designations based solely on untraceable and unknowable histories. Obviously, there are common misconceptions in what these files are that lead to these different uses of names for the same thing. Nevertheless, images and snapshots are overlapping concepts.

Because a disk image contains the entire contents of a block device it can be used to directly restore the entire block device, even onto hardware that has no knowledge of what that block device should contain. If you backup a Linux server and restore it, the entire system is restored, even if it was encrypted. An image will not bypass the encryption, but the encryption will do nothing to alter the imaging process.

Most virtualization platforms, even when using external storage devices like a SAN device, will storage the disk(s) belonging to a virtual machine as one or more disk image files (often in an advanced format rather than a raw ISO file.) This demonstrates that the image file is a full block device (virtually) and so can be used anywhere that any other block device is used. This is very important when we want to talk about restoring our data. If we have a complete image file, no restoration is needed if it exists in a place where we can access it. We can mount it directly under the right circumstances.

Image backups have some big advantages because they can be taken *closer to the hardware* when it comes to performance. There are also big disadvantages, so it is not all roses.

Advanced snapshot technologies that allow the system to essentially freeze a moment in time and work with it while the system is still under active use is a big deal for allowing major backups of active systems to have effectively. Taking full system images is by far the easiest process for handling backups because we do not have to think about what we are backing up, we just grab everything. Not only can we grab an entire operating system, but we can grab all of every operating system that exists on a single hypervisor. We can even take backups of virtual machines that are powered down as well as those that are actively running. System level backups have to happen per system, can only be done when the system is running, and only in very limited situations can they effectively take a complete backup and even more rarely can they find a way to do it that provides for a complete block level recreation of the system that can be restored in the same way.

Platform level backups have been all the rage for over a decade now for good reasons. They are efficient, they protect against failing to select the right data to be backed up, they can easily be handled by a different team than the standard system administrators if necessary, and they can be done at large scale with essentially no deployments needing to be done to work (other than the deployments of agents that are needed otherwise, and the deployment of the backup software to the hypervisor layer.) Platform level backups fit perfectly into how most organizations want to be able to work, and they are exceptionally easy for support vendors to provide blindly as a service without needed to do any due diligence.

Platform level backups come with plenty of caveats too, however, and we need to be aware of why we might not want to be focused on them most of the time. Being blind they generally require the most storage capacity to hold the backups as they tend to contain a lot of data that is unnecessary such as system files. This bloat also means that moving the resulting images around whether for archival purposes or to get the file(s) where they need to be for a restore will potentially take longer than would otherwise be required. Agents still need to be deployed. Mistakes are easily made because the system is almost universally misunderstood and, like so many things that are too complicated for the average user to grasp, they are seen as a panacea rather than just another tool to consider.

Quiescence is a real struggle for platform level backup. Even when all agents are properly in place, the nature of those agents is that they expose fewer options and have fewer hooks into the application layer meaning that there is a higher risk that backups taken from this layer will only be crash consistent. Operating system files will be protected because there is nearly always proper communications made to the operating system itself via the agent, but almost no application workload will be protected by that.

Platform level backup is flashy and cool. It is an easy way for vendors to capitalize on backup fears while doing minimal work; it is an easy path to big margins with any actual risks being pushed off to the customers who rarely do their homework. It makes customers feel safe because the risks are too complex for even more IT practitioners to grok, and it lets them tell themselves that they are protected. Ignorance makes people sleep well at night.

Platform level backup is a power tool and a modern marvel, but it is still just one mechanism, just one approach and certainly not one that fits the bill every time.

That was a lot of information about two very simple backup mechanisms. We needed that solid understanding so that we can move on to talking about many concepts that exist in and around the backup and recovery world.

Snapshots, archives, backups, and disaster recovery

There are many technologies that are either confused with backups, or may be a component of a backup. In this section I want to break down these basics and make sure that we understand what they are, why they matter, how they are used, and when we should leverage them ourselves.

Snapshots

In our last section we talked about taking block device images, how images and snapshots are truly the same thing, and what people tend to mean when using the term snapshot instead of the term images. Now we are going to really delve into snapshots.

Snapshots, as people tend to use the term, are amazing tools for doing some extraordinary things with storage. Snapshots are typically used to grab a momentary image of the state of a block storage device. The term snapshot is very descriptive in this case.

For most snapshot systems, and as it is intended by most people using the term, the snapshot that is taken is kept on local storage along with the original data and the two are intrinsically linked. The assumption is that the snapshot contains only changes, or the differential, between it and the original data. There are multiple ways to achieve this, but essentially the end results are the same: a file much smaller than the original data that is quick to create but is able to perfectly recreate the state, or *snapshot*, of the block device at the time that the snapshot was taken.

The obvious risk to this process is that the snapshot, presumably, contains *only* the differences between the time that it was taken and the original data. So, there is no protection here against system failure. If the original file becomes damaged or lost, the snapshot is useless. A snapshot might be an effective protection mechanism against accidentally deleting a file, malware, or ransomware at the system level because it allows a system to revert to its pre-compromised state easily and quickly. This is an important value and early snapshot mechanisms were often used expressly as a means of protecting against file deletion and overwrite mistakes. But since they are tightly coupled to the original data, the most important protections afforded us by true backups are completely missing here.

This risk has led to the mantra of *Snapshots are not backups!* You hear this everywhere. And it is true, they are not, on their own. Snapshots are a really critical component of backups, though. Overstating the mantra has caused many people to incorrectly believe that backups created using a snapshot as a source are not real backups.

Types of Snapshots

A snapshot is a generic idea that can be executed in many ways. Two mechanisms popular for snapshots are copy on write and redirect on write. Nearly any production system that you encounter will utilize one of these two mechanisms. Understanding these two give a good insight into the thinking and design of snapshots and explain why they can be so powerful.

First, copy on write (sometimes called COW Snapshots.) When a copy on write snapshot is initiated, a block device is essentially frozen in time. Mostly this is theoretical because nothing actually happens, and until there is an attempt to make a change to the storage, nothing will happen.

When someone attempts to write new data to the block device, at that time the storage system takes any block that is about to be written and copies it to a new location and then overwrites the original block with the new data. There is a performance hit while all this copying is going on, of course, and as changes start to add up the size of the copied data will grow. So while the snapshot is literally of zero size when we first initiate it, it keeps growing as long as we keep writing to the block device.

Copy on write snapshots are popular because they have so little penalty to being destroyed. To delete the snapshot all that has to be done is the extra data be deleted. The working block storage is untouched in all of this. Even a process monitoring the original block device would never know that there was a snapshot somewhere because everything related to the snapshot is external to the original block device. So there is basically no penalty to cleaning up the snapshot when it is no longer needed.

The alternative approach, redirect on write, works a bit differently by manipulating points. These points point to all of the block locations of a storage device. When a block is changed by writing new data to it, the system does not modify the original block at all, but rather writes the new data in a new location and simply points that blocks pointer at the new location instead. In this way there is essentially no impact to any given write operation. The impact is not zero, but it is extremely low, especially compared to copy on write which requires moving existing data around and rewriting it during normal operations.

With redirect on write we get some great features like the ability to maintain essentially indefinite versions of our storage device. In fact there are storage devices that simply use redirect on write for all operations and do not think of changes as snapshots but treat the entire storage system as an eternal snapshotting mechanism so that any portion of the storage can be rolled back to literally any point in time.

The caveat of redirect on write, because it sounds pretty perfect at first glance, is that you still have a growing amount of data over time and if you keep interim versions of data, rather than just a single point in time, the degree of growth can end up rather staggering. If you then need to clean that up the process of cleaning up the system of pointers can get somewhat complex and has a performance penalty.

Copy on write tends to be the best choice when we are talking about short term snapshots that are being taken, for example, just before a major system update is performed and the need to roll back to the time just before the update is critical, or a snapshot is taken in order to send data externally in a backup operation. At the end of either of these tasks, the snapshot would be destroyed and forgotten about. Copy on write is all about creating and destroying the entire snapshot quickly, not keeping it around, because the penalties of copy on write will exist only for a short time and the destruction process is painless.

Redirect on write is really powerful when we intend to keep the snapshot around for a long time and when we want multiple snapshots that build off of one another. Because there is almost no penalty during use and only a real penalty during the destruction of the snapshot(s) it plays the opposite role of copy on write.

So: redirect on write is probably the best choice for using snapshot as ongoing data protection and copy on write is typically what is used under the hood for things like backup software to originate a dataset.

Even if we do not have any special backup software, we can use a snapshot taken by our storage system and use it as the building block of a meaningful, manual backup. This is far simpler than it sounds. In the simplest of examples, assuming that we have something like a mounted tape drive on the same system to write to, we simply use the storage systems ability to mount the snapshot as an immutable version of the block device and copy that to the tape.

When we do this, the source snapshot file might be absolutely tiny, or theoretically even zero bytes, but what is sent to the tape (or any other storage media) is a complete copy of the entire block storage device in the state that it was at the time that the snapshot was taken. This is the miracle of snapshot technology. The automatic and generally completely transparent recreation or rehydration of the original block device state while using minimal, often trivial, additional storage space.

When we do this, we have the semantic challenge of what do we call the copy that is no longer on the original media. It is an image, of course, but typically we still refer to it as a snapshot as it is identical to the snapshot in every way and represents a snapshot of the block device at a point in time. So, when we do this, the statement that snapshots are not backups becomes false because we can have a true backup that is also a snapshot. The correct statement would be that a snapshot is not typically a backup. Snapshots do often get used as backups, however, and even more often as the basis of backups.

Because snapshots provide the ability to freeze the block device in a moment (or more than one moment) in time and then provide a way to utilize that frozen moment without having to interrupt the block device for future options it is perfect as a means of creating a source from which to take a backup while allowing the running system to continue on its merry way. It must be noted, though, that traditional snapshots all happen on top of the same underlying physical block device. So, while we do not need to halt the storage device in order to perform snapshot operations, we do use IOPS (input/output operations per second) from the original device. Our operations are not zero overhead, magically, but they are much lower overhead than other backup mechanisms will tend to be.

Because of all of this, snapshots are a popular tool to use under the hood and behind the scenes to make many modern backups possible. Essentially every hypervisor or storage device level backup (platform backup) is powered by snapshots. In many cases, even system level (so called agent based backups inside of the operating system) use snapshots today, across all operating systems, not only Linux-based ones.

Archives

Closely related to, but importantly different from, backups are archives. These two concepts are very often confused for each other, even in very enterprise circles. In theory, just by saying the names and asking someone, anyone, to describe what they think of a backup and an archive is enough to get someone to self-describe why they are not the same thing. But taking *I can define it* and moving that to *I can clearly articulate and internalize that the two are different* is not always automatic.

A backup, which we will define more in a moment, is a copy at a minimum. That much should be clear. Our English use of the term denotes this. We refer to things as our *backup copy*. If you articulate it well, everyone always agrees that if it is not a copy, then it is not a backup.

An archive is different. Archiving something does not suggest that there is not a copy, we sometimes even say things like *archival copy*, but nothing in the definition of archive is it suggested that a copy is required. An archive refers to long term storage, often assumed to be either lower cost, harder to access, or otherwise *archival*. Long term storage, but not necessarily close at hand.

Archival storage could simply be a second hard drive where data does not change. Maybe it is offline tape. Maybe it is cold cloud storage. Archival storage does not mean, necessarily, that the performance or accessibility of the storage is less than regular storage would be, but it is very common for it to be.

Most organizations use archival storage as a lower cost means of maintaining emergency or occasional access to data that is no longer needed on a regular basis. It is easy to talk about potential scenarios for this. Old data could be last year's financial records, copies of old receipts, video footage from material already processed, old invoices, old meeting minutes, blueprints from completed projects, old project artefacts, you get the picture. Businesses produce enormous amounts of data that they never anticipate needing to use again, but cannot necessarily delete entirely.

The assumption is that an archive, if it contains data of any importance, also needs to be backed up just like any other storage. The rule of thumb is that anything worth storing is worth backing up, if you feel that the cost or effort of taking a backup is too much then you should carefully reevaluate the desire to continue to store the data at all. Why continue to pay for its storage?

In a properly planned storage infrastructure this is exactly what happens and archives are a powerful mechanism for data retention and protection when properly combined with a backup. As the term archive means that the data is not changing it means that there should never be a lock on the data and backups are as simple as can be. None of the complexities that face the backup process exist when dealing with an archive. The need to take backups often also does not exist; a single backup might be all that is needed if the archive is static. Or if the archive is only rarely changed, maybe a monthly or annual backup might be adequate.

Converting backups to archives

A not uncommon thing to have happen is for a backup to become an archive. This happens far more frequently than you might imagine and happens for reasons that most of us can identify with. I have seen it happen in the largest of organizations with no policy to avoid it.

Under normal circumstances, all data that you want to store is live and in its proper location for use and available normally. A backup is taken of this data and, in case of disaster, the system can be restored. until there is a disaster all of the data exists both in the original location as well as in one (or more) backup location(s). This redundancy of locations is what makes the copies into backups. Technically, if the original storage location fails, the primary backup (which might be the only backup) stops being a backup and turns into the temporarily primary source location of the data. We never speak of it in this way as we all know what we mean by *restoring from backup*, but for that period of time when the original storage is gone, the first backup is now the master of that data and not a backup any more. It may have backups of itself, and generally we would, but you can only have a backup when there is data that is not a backup for the backup to be a backup of! Hard to explain, but a critical concept.

The reason that this matters is because once you take a backup of data, it can be very easy to feel that our backup protects us and that we can then delete the original source data. It feels like we can, because there is a backup already. The problem here is that if we delete the original data, the backup (or at least the first backup) stops being a backup and instead turns into an archive! And in most cases, if the assumed master source location is no longer there, there may be no mechanism to take any further backups of this archived data. There may not even be any means of locating it.

I have seen this myself in the real world. A daily backup task would run and end users just assumed that they did not need to store any data that they needed, even though it was a legal requirement to keep it and to keep it backed up, on the primary systems and so would delete the data immediately upon creation assuming that their data was protected by the backup system. There were some critical flaws in this plan, however.

First, by deleting the original, any backups that existed became the source location rather than a backup and were simply a tape-based archive. Given that there was a legal requirement for the data to be retained and backed up, this violated the legal retention requirements.

Second, the backup mechanism was staggered where some backups are kept for years, some for months, and some for weeks. So if the data happen to exist during a run that was kept for years there was a good chance that the one, singular tape that contained the files in question might be readable to able to retrieve the data, but if the data existed only during a backup run for a weekly job, even the archival version would be deleted in just a few weeks when that tape was overwritten.

Third, and unrelated to the archival situation, often the data was deleted before the daily backup job ran at all meaning that the data was deleted with no backup or archive ever existing and so was lost instantly. The result was that no backups ever existed, and once in a while a file would get lucky and be retained for a few years without backup, but most files were either never archived at all or were archived only briefly and deleted in a few weeks when tapes were reused.

Given that the requirement was seven years of retention, plus a backup of that retention, it is easy to see how far off the mark the process was simply because the end users thought that they could intentionally delete the original files because they thought a magic backup process was somehow protecting them. The backup team thought that all data was being kept live for seven years and that if any file was lost or corrupt that the end users would alert the backup team to restore it almost immediately. That the end users would themselves intentionally destroy the data and never alert the backup team that the data had been destroyed was never considered in the workflow, because why would it be?

It is easy to see how we can, through actions of the end users in many cases, accidentally convert our backup location into an archival location and potentially lose data because we do not understand the complexities of the backup processes. If data is not going to be retained permanently in a primary location, then very complex processes are often needed to ensure a safe workflow for deletion.

Archives are a powerful mechanism to lower cost and keep our mainline storage lean. Maybe we use archives only to reduce cost. In many cases, by keeping excess data away from our top tier of storage, it will allow us to invest in faster, but smaller systems for the data that we use every day.

Archives can apply to portions of a file as well, of course. Sometimes even databases use a mix of storage locations to allow them to be able to store tables or portions of tables that are accessed regularly on the fastest storage, while rarely touched tables or parts of tables can be kept on slower storage that costs less. This could be automatic inside of a database engine, or an application might use multiple database systems and move data between them at the application layer, for example.

Archives are a useful tool, but in no way a substitute for backups.

Backups

It is hard to believe that we have gone this far without actually digging into exactly what we mean when we say *backup*. As with many things, it is hard to sometimes jump directly into a definition as there is just so much that we have to consider.

At the most basic, a backup is a copy of an original set of data. We talked about this above. If the data does not become redundant, then it cannot be a backup. This is the most obvious piece of the puzzle.

Next, the data must be stored twice. Meaning the hardware must exist more than once.

In many cases we can do a comparison with paper or some other physical form of data storage. If we have a piece of paper with an important code on it, we feel a sense of urgency to copy that data somewhere - to make a backup. We know how easy it is to smudge, burn, lose, or send a piece of paper through the wash.

Probably the most confusing piece of a backup requirement is that it be strongly decoupled from the original material. Being strongly decoupled is a term that involves some amount of opinion that makes determining appropriate levels of decoupling a little bit hard. What is decoupled enough for one organization may also not be enough for another.

When we use the term tightly coupled, we are referring to situations where the original data, and the copy of the data, have a connection between them. At the most tightly coupled is something like the snapshot concept where anything that happens to the original file will ruin the snapshot a well, in all cases. That is fully coupled. Slightly less coupled would be something like storing the copy on the same physical device. The farther separated the copy becomes from the original data, the less coupled it is.

Coupling can involve more than physical connections. It can also include concepts such as being stored on systems that share credentials like usernames and passwords. Coupling can be complex and there is no single, clear-cut way to describe it. When we talk about keeping backups heavily decoupled, though, we generally means a few things such as: completely disparate hardware and media, physically separate location, and disconnected authentication.

The idea behind every additional step of decoupling is to keep any event that might happen to the original data to have little to no chance of also happening to the backup data. This could be an accidental file deletion, a failing hardware storage device, a ransomware attack, flood, or fire. There are so many ways to lose our original data, or to lose access to it. We have to consider these possibilities as broadly as possible and assess how much decoupling, and what type, make sense. No two systems are ever truly decoupled completely, but we can decouple to a practical level quite easily.

The first step is easy and very few organizations make the mistake of attempting to put a backup file onto the same media as the original data. It does happen, however. Using the same drive happens very rarely, but attempting to use a second drive inside of the same device is sadly somewhat common. Of course, as you can imagine, many events that would cause data to be lost on one drive in a computer can cause data on another drive to be lost. A fire inside the chassis, extreme shock, flooding, loss of power, theft, data corruption caused by many types of events, or malicious attack are all likely to destroy a backup simultaneously with the original data. This defeats the purpose of the backup almost entirely.

To compare to the physical world, we can compare to paper. You have a piece of paper on which you are storing critical data. You keep that piece of paper in a filing cabinet. You are worried about protecting that paper from something bad happening. You can use a single piece of paper and write the information on it twice. You can have a second piece of paper in the same folder in the same filing cabinet with the data duplicated. You can put a copy on a separate piece of paper in a second filing cabinet sitting next to the first one. Or you can put a copy of the paper in a separate filing cabinet in a different building.

In these examples, it is easy to see the progression from tightly coupled to highly decoupled. When sharing a single piece of paper, or two papers in the same filing cabinet, the risk is obvious. Almost anything that would hurt the first copy would damage the second. Having paper in a second filing cabinet at least gives a modicum of protection that there might be two different keys for the two cabinets, a cabinet falling into water *might* not affect the other, file damage to one *might* not affect the other, and so forth, but it does not take much to see that their risks are still coupled, just not as closely.

By putting a filing cabinet into a completely different building, possibly on a different floor or even in a different town, then having a flood, fire, or theft that is able to destroy both the original and the backup at the same time is extremely unlikely. There is also the chance of a volcanic event, nuclear war, or meteor strike that could still destroy both copies even if located miles apart. This is why we say strong decoupled, but never totally decoupled.

With our backups we have to consider just how decoupled our data is, and how much it costs to decouple it further. We also have to consider the efficacy of decoupling and how that impacts our business. It is a complex question. For most businesses, though, we will want our backups to have physical distance, a variety of media, a separation of authentication, and multiple copies. What our backups contain and how they will be used play into this heavily.

There is no simple best practice here, even common rules of thumb rarely apply outside of the extreme basics. Best practices dictate:

Best Practice: If data is worth storing, it is worth backing up.

Best Practice: A backup needs to be highly decoupled from the original data.

The range of possibilities with backups are just so broad that we really must evaluate the needs, across the board, for every business and, in many cases, individual workloads. In some cases, such as backups of entire operating systems, our primary concern may be around rapid recovery in case of hardware failure and the actual contents of the backup may be trivial. In another case our backup may contain large amounts of highly proprietary data and keeping that backup safe and secure, ensuring that it cannot fall into the wrong hands, takes precedence.

Why tape still matters

Inevitably when talking about backups, the discussion of tape media surfaces. Generally, you get one of two responses: tape is amazing and I always use it or tape is dead. Wildly disparate opinions.

Once upon a time, tape was essentially the only backup media option. Solid state drives did not exist yet and hard drives were outrageously expensive per megabyte and had a terrible shelf life and shock capabilities. Tape was the only affordable and durable media option, and even it was not very good at the time.

In the decades since, everything has changed, as it often does. Hard drives, solid state drives, and services that host these for you have all become very standard. The biggest factors driving us to tape are gone, we have many options that are all viable. This has led many people to focus on these newer options and not keep up with the advancements in tape, but just like all of the other technologies, tape has advanced too and quite significantly so.

Tape is a media almost purpose built for backups. It can move linear data at an extremely high rate, is very low cost per megabyte stored, and has incredible shelf durability and shock handling. Tape naturally, unless you leave the tape in the drive, becomes decoupled from the original system as simply as by just ejecting the tape from the drive. You can even have a remote tape system with automated tape ejection to make both technical and physical decoupling a completely automated process!

Tape carries the benefit of allowing each tape to be at least partially decoupled from each other. Tapes can be stored in different locations and multiple tapes can be used for multiple copies of data. In some cases, groups of tapes might be stored together in a single box, effectively *re-coupling* the tapes to some degree. If tapes are stored apart from each other, though, dramatic decoupling is possible not just between the source data and the backup, but within the backup(s) itself!

Tape is not perfect. It requires physically mounting the tape before being able to restore and it is terrible at locating single files buried deep within a backup. Tape shines at backing up or restoring large quantities of continuous data, but once you start to search for specific data the performance declines quickly. Some companies address the human component of tape management by employing robots and tape libraries, but this effectively puts the tapes back *online* and potentially re-couples them to the original data, at least partially, taking away one of their layers of protection. Straight tape, without a robot or library, has the benefit of needing to hack a human, on top of hacking computers, to be able to destroy the original data and the backup at the same time.

Tape is useful enough that even some online cloud backup providers use tape as a component of their storage solutions. Tape has an important place not just in the modern world, but in the foreseeable one.

Of course, backups may not be a single solution for a workload. A single workload or system might need multiple backups taken of it. One backup that is kept locally and moderately coupled that can be used for rapid restores in the event of an accident or hard drive failure. Remote backup to tape or immutable cloud storage for long term, highly decoupled data retention in the event of ransomware or total site loss.

Decoupling is so critical to a backup being functional or useful that we must include it in our definition of backups, even if we cannot absolutely clearly define what constitutes a significant enough level of decoupling. This is because what is adequate for one organization or situation may not be for another. For me, this means that our definition of decoupling has to be subjective. Stakeholders need to define what a backup is to protect against and define the necessary decoupling from there.

It should not be glossed over that modern ransomware has become a driving force in organizations beginning to analyze their traditional levels of backup coupling because suddenly the reach and threat of backups having any real level of coupling is dramatic. Ransomware techniques, at the time of writing, aggressively include strategies to ransom backups themselves whenever possible and techniques to hide their activities to thwart the ability of backups to protect against system encryptions. Backups remain the best defense against such threats, but ensuring extreme levels of decoupling, often requiring techniques like immutable storage and physically taking backups offline so that no computer can reach them without human intervention.

If you cannot recover, it was not a backup

I have heard this said so many times and I still love it. Simply, if your backup does not work when there is a disaster, was it really a backup at all? To a small degree this is overstating the case. A backup can fail and that failure can coincide with the moment when an original workload has failed. But this should be statistically so unlikely and if it were to happen there should be a coincidence that is outrageously obvious.

What many want to express is a need for testing backups. Backups are complex and knowing the speed, process, and effectiveness of using the available mechanisms is a very critical component to any backup process. If you have never tested your systems, you have to assume that they will not work. And even if you have tested them, it is best to have tested them recently. Some backup systems even do automatic restoration tests to demonstrate the efficacy of the backup every, single time.

Backup tests can be misleading. Like many other data protection mechanisms like RAID, redundancy power supplies, or failover SAN controllers the way that we tend to test backups in a predictable, pristine environment rarely reflects how a restore would be required under emergency conditions. It is common for backup tests to be performed when systems are idle or slower than usual and to reflect best case scenarios resulting in nearly always passing tests even on systems that would almost always fail in a real-world scenario. Consider adverse conditions such as actively failing, rather than having completely failed, systems, heavy load during operations, or an inconsistent data state (crash consistency) that will not be reflective during a test scenario.

Backups may seem like a simple thing and every backup vendor is going to present their product as eliminating the need for you to understand your data. Vendors hope for your blind trust that they will do the impossible and offer you a chance to open your wallet and simply hope for the best. As system administrators it is our task to understand our data, know how our backup mechanisms work, determine the consistency and coupling needs of our organization and workloads, and to assemble a backup solution that meets or beats those needs.

Disaster recovery

Ultimately the purpose of any backup mechanism is to enable disaster recovery. Disaster recovery is something that we all hope that we will never have to do, and yet is the most important moment in most of our careers. Disaster scenarios are when we earn our keep more than at any other time. Your ability to perform calmly and coolly when a disaster is striking, to be ready with the knowledge of how to get your workloads back online, and being able to adjust to whatever twist or surprise is thrown at you is key. Performance during a disaster can mean salary differences of hundreds of percents.

As we have dug through the many types and approaches to backups, it should be natural that there are now many ways to recover as well. When planning for disaster recovery we really must take all of this into consideration as our planning process will very so greatly.

If we use image-based backup methods, then the assumption is that we will approach restore processes by restoring an entire block device image (or images) all at once. This has some significant advantages during disaster recovery because we only have a single step: restore the entire system. In some cases, our restore mechanisms will automatically restore all systems at once, not just a single one! This is a very alluring prospect. The caveats to this method are that restores are typically slower and generally only crash consistent. Doing high speed restores with the utmost of reliability is the most challenging for this method.

Using file-based backups we need to, in almost all cases, first restore a blank operating system, either a template or a vanilla build, and then restores individual files back to it in the place where they originally went. This method is theoretically faster than the image-base complete restore, but in practice rarely is because the time to build the base operating system from traditional methods. While theoretically you could build the base system from modern methods, this generally does not happen because if you were to do so you would naturally move on to DevOps style backups. However, these start to overlap conceptually here and you can use a file-based backup mechanism in conjunction with traditional file restores. The change would, of course, be that the file backups would be isolated to the meaningful system data rather than a blind backup of anything and everything.

And lastly, a full DevOps style restore. This is more complex as there is no clear single definition. The assumption here is that rebuilds of the base operating system will be nearly instant because of any number of automated build mechanisms. And then that the data restore will be heavily minimized so that it, too, can happen at great speed.

Of course, all of this will be going on in conjunction with triage operations that we will discuss shortly. Planning and timing these operations so that they are well known, that processes are tested, and restore times are predictable under different scenarios provides invaluable information that will be needed during a disaster scenario.

Disaster recovery planning should be more than just individual workloads being tested in isolation, and it should be more than testing only data loss. Different organizations have different risks and testing multiple types of scenarios is important. We think of data restores when we think of disaster recovery, but that might not be what we are facing. If we have workloads running in multiple locations, we might be looking at how to work from a slower connection to a less than ideal data center location. A disaster recovery test might involve testing the ability to spin up workloads on backup systems, to failover clusters, or to run from a secondary data center.

The best practice in disaster recovery is to always test your scenarios and to test them regularly. Never blindly trust that our backups, our planning, our networking or data centers will work. We need to test not only the technology behind our plans, but the procedures of those plans as well.

Now that we have discussed mechanisms and terms in and around backups, it is time to really look at how the modern world of DevOps can redefine how our backups can work.

Backups in a DevOps world

In earlier sections of this book, we have talked about modern concepts impacting the world of system administration such as DevOps and infrastructure as code. You may be wondering if these modern concepts have a potential impact on the worlds of backups and disaster recovery. Good question! And if the section title has not given away the answer, I will clue you in now: yes, yes they do!

Traditionally we think of restoring data as either the very old fashioned way of just restoring individual files, or the more modern (think last two decades) way of restoring entire systems including the operating system and all of the files that go with it. We are so accustomed to thinking of restoring systems in this way that it is often very hard to think about the problem in any other context.

In the ultra-modern DevOps style world where systems are built via automation and defined in code or configuration files we have to start to think about nearly everything in new contexts. When systems can be automatically built easily and rapidly through standard, non-disaster processes, the need to restore those systems rather than starting fresh completely vanishes. Imagine if instead of fixing a car after an accident if for less money, and less time, and more reliably you could have a brand new, but identical, copy delivered to your door - you would never waste time trying to restore something broken when a pristine, perfect copy can be had.

Version control systems

When talking about backups, especially as they relate to DevOps, we should also talk about version (or revision) control systems. Version control systems, like GIT, Mercurial, or Subversion, are not themselves backup systems, but act as mechanisms to sync some of the most important data on a system to another location, where backups will often occur.

When talking about version control in the context of backups, it can be a bit confusion as in some cases we might be looking at our operating system as being a master location and a version control system can be used to replicate configuration files to another location and from there, they can be simply backed up using any number of normal mechanisms.

Because version control systems do not only store current data but also historical versions and changes to files, they become essentially immutable and therefore useful in a backup style situation since the backup mechanism does not need to deal with versions over time. The version control system holds all versions and changes to the files over their history. So a backup of the entire version control system automatically includes all of the changes to the files. Because of this, end users accidentally deleting files, bad changes to files, ransomware attacks on data all become moot as the ability to roll back to just before a bad change was made is built in.

Version control systems typically can be used to replicate files (and their version histories) not only between one end point and a version control server, but also to many additional end points. For example, a system administration workstation or jump box (which we will describe in the following chapter) might contain a full copy of the configuration files for every Linux system in a company separate from any server or backup system. Even if every official copy of the data was lost due to enormous catastrophe, a single workstation may be able to recreate the entire set of documentation.

These types of systems are already in wide use for non-backup related reasons and are considered absolute *must have* tools in the DevOps world. We may already be using them in places for system administration tasks. Even in systems that do not have a DevOps process, these tools can potential be used as if they were as far as data protection is concerned.

Even the most traditional (meaning as far from DevOps as you can get) Linux system can benefit from the user of version control for its configuration files. Organizations do not even need to stand up their own infrastructures for version control if it does not make sense for them to do so. Vendors such as Gitlab and Microsoft via GitHub provide enterprise hosted version control systems for free with extensive features and access control systems.

A small company with legacy style systems that wanted to embrace version control protection could, in a contrived example, simply got under the `/etc` file system and add it to a remote GIT repository and do a push. Voila, all of the data is protected, that quickly and easily. Set a `cron` job to run hours to push any additional changes and you have automated a robust backup system with essentially zero effort and definitely zero cost.

Version control systems are one of those poorly kept secrets of the development world that have leaked into and become embraced by many other professions. Any system that works heavily with text files should be jumping onto the version control bandwagon as quickly as possible. Version control is one of the simplest ways to take data protection to the next level. This approach carries far more advantages than that simple example might imply, however, restoring a system carries a non-trivial amount of risk that there is corruption or worse, an infection or root kit, that gets restored. A clean start removes those risks giving you the peace of mind of starting fresh. Likely the fastest path to a working system, performed in the most repeatable and predictable way, with the lowest risk. This approach is also the easiest to test. You naturally test at least part of this approach every time you build a new server whether for production, testing, staging, and so on.

When we build our base operating systems, install applications, and deploy configurations via DevOps-style automation we leave ourselves with only our system specific data that needs to be restored from backup systems, and even that only part of the time. If you consider a typical multi-tiered application, data that is unique and cannot be pulled from the initial build process is generally limited to database files or limited file storage. In an application running on multiple operating system instances, we often expect this data to only exist on one layer of those nodes, generally the database. If that node is replicated, we generally only need to take a backup of the data from a single node in the cluster (because we often have all of the data replicated between all nodes.) Through this process we eliminate many points of backup, many types of backups, and identify only the actual data that may or may not be at risk.

Every environment varies and in one we may find that data that requires specific protection to account for only one percent of all data, in another it might be ninety nine. We cannot say with any certainty what you will be expected to find in the real world as every organization and workload is so different. Universally the ability to limit the scale of backups, and therefore restores, brings advantages. The smaller the backup means the faster and less impactful the backup will be. The smaller the backup size the lower the cost to store it. Smaller means less to verify against corruption. The same smaller size that reduces backup time also reduces restore time in the same way.

In our multi-tiered example, we may need to rebuild three nodes to get our application back up and running. The top-level load balancing and proxy layer will be assumed to have no unique data and can be *restored* via the already tested build process that built it initially. The application layer should likewise have no unique data and be able to be automatically restored using the already tested build process.

Our last standard layer, the database, should again, be built with all configuration and applications being deployed completely using the build process. At this point, all of the layers of our application, all of the configuration for it, have all been restored and our backup and restore mechanisms have not even come into play. The only piece missing at this stage is the restoration of the latest data in the database. The database is in place, but not the data that goes inside of it. It is now that our restore process kicks off and puts that data back where it belongs. This might be a simple file copy from a remote location to the database location with a restart of the database after the files are in place. Or maybe a database-specific restore operation has to happen on the data to ingest it again. In any case, the restore is of a minimal amount of data, of very limited types.

This approach changes everything that we traditionally think about backups. It changes how quickly and how impactful backup operations are, it changes the tools that we need to use and potentially eliminates them altogether, it makes restores fast and even, potentially, fully automated!

If we go back to our example case and assume we are using the popular MySQL application as our database platform, and that all of our necessary data is stored in this one spot which is a reasonable assumption for many common workload designs, our need to use special backup tools likely does not exist. Nor do we need to rely on complicated or risky mechanisms like snapshots. We can use built-in tools to the database platform to reliably get a complete dataset from the workload layer (we hesitate to say application layer as that is a software engineering term that would be different here from the systems term) where we know that the workload, with all of the necessary intelligence to do so, was able to stabilize and lock the data during the backup operation so that our backup is safely *application consistent*.

When we step back and start ignoring convention, and we stop focusing on simply *buying a solution* but instead put on our IT hats and determine how best to protect our environment we can often find ways to both protect our organizations and to save money all at the same time.

IT provides solutions, vendors sell components

Broader than backups, and even broader than systems administration, the concept of where solutions come from is fundamentally something that we need to understand to do our jobs well at any level in IT. At its core, it is ITs job, and no one else's, to produce solutions for our business. In many cases we will need to turn to vendors to supply one or more components of any solution. IT is hardly going to fabricate their own CPUs and assemble parts to make servers or write their own operating systems and so vendors (which could, in this context, include those who do not get paid but provide things like open-source software including your Linux based operating system itself) are a necessary part of the solution process.

Vendors are not solution providers (although a great many will call themselves that as a marketing name, of course) but rather sales organizations. Their role is to provide access to tools that will hopefully benefit us in the pursuit of a solution. It is IT that determines which tools are right to use, selects them, and uses them to assemble the final solution to meet the needs of the business. IT, which is a department that provides solutions, is something we do, not something we buy. We cannot simply depend on a vendor to sell us a product that will protect us; it is not that simple. We have to understand the business need, the workload, the backup products, the approach and put all of this information together to make a cohesive solution that works for us. Every business is unique, every solution should be as well.

No vendor-supplied tool can take into account all of our unique backup needs today, let alone be adaptable for any changes in the future. If you can imagine it, many companies are so addicted to the vendor buying process that they attempt to adapt their workloads and internal processes (and needs) to fit the needs of a product that the vendor wants to sell! This would be like relocating to an undesirable house so that your boat salesman can excuse having pushed to sell you a boat when a car would have gotten you to work easily from your existing house.

DevOps and similar infrastructure have really exposed the extent to which traditional *just buy what a salesman wants you to buy* processes have required businesses to adapt to the purchase, rather than choosing to buy what makes sense for the business. In previous technology generations the options were so much less broad, and the differences so much smaller that it was easy to hide or dismiss the inefficiencies. Today that is not possible.

Enhanced backup and recovery processes turn out to be one of the best reasons to consider investing in DevOps and infrastructure as code engineering efforts. Typically, we can find many reasons to make DevOps attractive, but better backups is easily the best benefit.

Now that we understand how backups can be taken with a truly modern infrastructure, we are on to our last topic here: triage. Time to move on from taking backups, to using them!

Triage concepts

Planning is important and prepares you for many eventualities. When disaster finally strikes, though, most planning is going to go straight out the window. All of your assurances that your backups are good are not going to make you relax, end users are going to be panicking, management is going to forget that you have to fix things and pull you into meetings, stress is high, and nothing is quite as expected from the planning process.

Triage is hard because every workload, time of day, current situation has so many dynamic elements. We have to be ready to adjust to anything, and we have to get our systems back online as quickly as possible.

At the moment of a disaster is when things matter most and this is where system administrators really prove their mettle. Being prepared for a disaster is relatively easy, but staying cool and logical, evaluating the situation in real time, and managing the people around you all become unpredictable and very emotional challenges.

There is no simple guide to triage and not everyone is going to be good at it. The more we are prepared ahead of time, the more we understand the entire environment, and the better we know the business environment that we are working in the more adaptable we are going to be.

Triage is a skill best handled by a perceiver, in the Myers-Briggs chart. As such this is where administration, over engineering, really shines to its most extreme.

When analyzing an outage our first step is to evaluate the situation and determine the extent of the disaster. Are we only missing services? Is it all services? Is there data loss? What is the current, ongoing, and future expected impact from the outage. If we do not have the data at hand, keep people busy gathering that data for you.

Triage is especially needed to assist the business in understanding what it can do during the time of an outage. Many businesses panic or have no plan (or if they have a plan fail to action it properly.) But business behavior needs to be coordinated with technology recovery efforts. As we begin to approach our recovery we need to be working to also keep the business working as best as possible. Of course, we hope that corporate management will step in and guide operations, based on ITs assessments of what impact is and what recovery is likely to look like, to make them as efficient as possible. Often IT needs to be ready to provide guidance as well.

In any triage operation with have to determine the criticality to our workloads as well as the potential for restoration. The most critical workload will still take a backseat if it will take weeks to restore when many minimal services can be up and running in short order. We need to consider loss of productivity, loss of customer confidence, failure to meet contracts, and similar concerns when deciding how to approach our disaster recovery.

Recovery planning, even in the moment, needs to be coordinated with the business. What can the business do to assist in recovery, what can technology do to enable that? With good cooperation different businesses may find many different paths at their disposal to make the damage of an outage minimized.

Businesses can take actions as simple as sending staff home to relax to get people out of the way and allow IT more freedom to undertake restoration. A staff with a surprise vacation day or two might be refreshed and excited to return to the office and attempt to at least partially make up for lost time. Instead of making the team frustrated that they cannot be productive, why not reward them for their hard work?

Shifting communications to systems that are not down is important as well. Can staff move to phones if email is down? Email if phones are down? Instant messaging? Voice chat through some other platform? Maybe this is a good time to visit customers in person!

There is no way to list the countless ways that businesses can leverage an outage. What we need is creativity and the freedom to work with the business to help them see how they can keep working as best as they can, and allow them to direct us in how we can recover in the best way for them.

Thinking about triage operations makes it evidence the importance of concepts that we have already discussed such as self-recovering systems, minimized restore sizes, and careful planning. It also highlights how important it is to have operations and other departments engaged, involved, informed of available plans, and ready to assist and coordinate when things fall apart. Good planning makes triaging better, but you cannot plan your triage operations. There are too many variables to think that we can truly plan for every contingency.

I wish that triage and disaster recovery oversight was something we could teach concretely. It is a scary situation and all we can do it make sure that the right people with triage and perception mindsets are empowered and at the ready when the time comes, have good backups, good restore processes, and as much planning as makes sense while having an organization ready to work together to minimize impact as a coherent team.

Summary

Nothing matters like backups. I feel like that is at least the fifth time that I have written that in this book, and it is certainly not enough. Today backups are more important than they have ever been. We face more disaster scenarios and more advanced data loss situations than ever before in our industry. Backups have always been and will likely always be our strongest defense against complete failure.

Backups have been changing, quite a lot, in the last several years. The assumptions as to how we would approach backups even ten years ago do not readily apply today, and yet many organizations still use legacy applications, legacy designs, and need to still use legacy backups. So, our job is a complex one and our desire for modern backups may be needed to drive towards more modern application designs so that we can protect them in a better way.

But now we understand the mechanisms underlying different approaches to backups, why we want to consider backing up in different ways, and how we can advance our backup practices into the future. Backups are probably the best place for you to set yourself apart; nothing matters more and rarely is anything as forgotten as much as backups.

In our next chapter we are going to look at how users exist and interact in Linux systems and how we can approach authentication, remote access, and security.

10
User and Access Management Strategies

In the **Linux** world it is all too easy to forget that we still have users, and they still need all of the oversight, security, and management that we would expect in the **Windows** or **macOS** worlds. Users are typically an afterthought on Linux-based operating systems as systems are often seen as just black box server workloads or bizarre appliances to which end users do not apply. This is not true, of course. Users matter on any Linux system just as they do on anything else.

In this chapter, we are going to talk about user and user access management for both servers and for end user devices. We are going to look at approaches common in the Windows world, and approaches commonly known in the UNIX world, and we are going to talk about some alternative approaches that are starting to emerge in the industry.

We will also look at remote access for Linux – that is, supporting or working from our systems remotely. Of course, all of this will be done in the context of security, as user management is fundamentally a security topic.

In this chapter, we are going to learn about the following:

- Local and remote users
- User management mechanisms
- Remote access approaches
- SSH, key management and jump boxes
- Alternative remote access approaches
- Terminal servers and VDI

Local and remote users

At the highest level, there are two basic ways to think of user accounts on any system. The first is local accounts that exist on the local system where they are being used. The second is user accounts stored remotely on some sort of server that the local system references over the network. Of course, there are hybrid methods that combine these techniques in various ways as well.

We should start by talking about the obvious benefits to both approaches. With locally managed user accounts we have lightning-fast access to our account information and no dependence on the network. This provides obvious performance advantages, better security, and protection against services failing elsewhere impacting our local systems. Local users are robust and simple, fast and easy. Until the 1990s it was rare to even consider the possibility of anything else.

Remotely managed users make up the overwhelming majority of cases today because this model allows for a *single source of truth* for users between different organizations, and in some cases, even inter-organizationally! This has many benefits, such as allowing users to make a single, complicated password that they actually have some chance of remembering and being able to use it in many places, change it once to change it everywhere, and making it simple for the support team (which might mean us as the system administrators) to reset, lock, or disable accounts.

It has become common to assume that only remote user accounts are plausible in modern businesses, but local accounts remain fully functional and viable, and are a good choice for a great number of organizations of any size. Do not simply assume that because of your size or modernity that one style or the other is appropriate for your use case. Of course, nearly all existing organizations are heavily invested in whatever architecture that they are going to use and changing is a very large undertaking. It is rare that we get the joy of implementing something like user management on any scale in a green field scenario, but it does happen from time to time.

Local user management requires less network bandwidth, but that is essentially never a concern today. Far more important are the issues around security and availability. Most networks are extremely fast today and user management uses few resources; this has led to vendors providing user management as a service allowing centralized user accounts to be provided via hosted vendors over the Internet. If the Internet can adequately provide bandwidth for these services, LAN-based versions should have no issue at all.

Security becomes a problem because having a single account that is available in many locations means that there is a single account to attack aggressively and, if compromised, will essentially provide unlimited access to all places where that account is used. If we have local user accounts and reuse the same password over and over again this mimics this problem so in many cases this becomes a moot point unless we have a scenario where our users are truly using different credentials in different locations. For traditional end users, this is not very likely to be the case. For system administrators this might be relatively easy to do. High security users with training and an understanding of the importance of their credentials may easily take advantage of the flexibility of local accounts. You have to know your users and their willingness to participate in order to really evaluate the security posture potential in this case.

The bigger issue surrounding security is the specific mechanism that supports the remote accounts, which can be attacked separately from the accounts supported by it. Whatever mechanism you chose, there must be communications over the network in order for the accounts to be able to centralize. In almost all cases, there is also some form of local fallback – such as a credential cache – that can be attacked, so having accounts exist remotely almost never eliminates the need to also have them locally as well as some mechanism. If such a mechanism is lacking, then there are readily available methods to disrupt logins by attacking network access.

Remote user management benefits from having a central store that can receive more focused security attention and that can be kept far more easily from being physically exposed to attackers. End points typically have, at most, a small number of cached user accounts or emergency access accounts on them and not the entire user list minimizing the risks of a stolen laptop or compromised desktop from impacting beyond the built-in list of accounts cached on the device.

Neither method constitutes anything close to a best practice. Both are completely viable and should be evaluated both at an organizational, as well as a workload, level. It is not uncommon for businesses to elect to use different methods for different purposes. There is generally no need to institute a single mechanism as most organizations are large enough to have varying needs and to justify implementing more than one mechanism inside of the organization. Often this hybrid approach works best as centralized accounts tend towards being most beneficial for most end users and local accounts tend towards being best for IT and other technical departments.

With that foundation under our belts, we are prepared to talk about the mechanisms that can manage these users, wherever we choose to implement them.

User management mechanisms

In the real world there are many user management mechanism implementations to consider. Some are native to UNIX or Linux, some are common in the Windows world, some are novel, and some are universal and agnostic.

It should go without saying that our first stop on any journey of investigating user management mechanisms is the Linux user system itself. Simple and universal, every Linux system of note ships with it. Of course, it can be replaced, but in practice it never is. This system carries the huge advantages of being always built in, very fast and secure, and well known by every UNIX admin anywhere. There is almost nothing to go wrong, nothing complex anywhere in the system. A few archaic components might linger on having been left over from the olden days that might be a little confusing if one finds it necessary to do manual configurations, but today almost no one manipulates these systems by hand anyway (although it is always good to know how to do so just in case.) Local users are always very easily automated by custom script, stock tools, or things like state machines.

Using automation to turn local uses into remote users

Here is one of those cases where we end up being unable to define the difference between a local user and a remote user. One important approach to local user management is to have a master user list stored elsewhere. This list might be a simple text file that we run a script against, or more likely a configuration file in a state management system used as part of an Infrastructure as Code implementation.

In this example our central management system can be used to push out user accounts, permissions, and details from the central configuration system onto each host in our network. The system might put all users on all machines, or just the users assigned to those machines. These kinds of decisions are all completely subject to your discretion during implementation.

In this example, yes, we still use the local user mechanism on each computer and there is no need to reach out over the network in order to log in at any given time, however the accounts are still centrally controlled in a way that almost completely mimics a system like Active Directory. We can centrally enforce password rules, we can centrally create or revoke accounts, we can centrally control onto which machines logins can and cannot happen, and so forth. We essentially have a central user system, not a local one.

The key difference is that with a true central user directory such as Active Directory, user activity is directed to the central system and only optionally will hit a local cache when the system is offline or in some degraded state. When possible, network activity occurs to support the login process. With a mechanism such as the one we are describing, all logins happen purely using local – that is to say, *non-networked* – resources, and the central system is only used to update the local login list and details. A carefully worded semantic difference between local accounts and locally cached accounts to be sure, but the behaviors of the two truly are different. It is a very interesting thought experiment, though. It is assumed that locally maintained accounts will have to hold all users on each device, but in practice there are many ways to limit this to just a few and sometimes even just one user, while it is assumed that none or just a few users will be cached onto a local device from a remote user management system but in some cases the entire user list might be cached there. The difference here is how the user list is created locally and knowing how that mechanism works in your case combined with knowing how the system is used will tell you the exposure of both approaches for you.

However, automation of this nature provides a pretty significant opportunity to rethink our assumptions around local user limitations. In this kind of situation, we might be able to recreate many or all of the desired features that are generally assumed to be provided exclusively by heavier and more fragile remote user systems using only local users. In a modern environment with applications that typically do not authenticate to an extension of the operating system authentication system this can potentially work beautifully, especially on a Linux-based operating system where such integrations are not common. If your network architecture is one that uses shared authentication methods, like those provided by Active Directory, to allow access network resources then this methodology will fall short of the smooth, integrated experience that one traditionally has with those models. Network resource design models that rely on those shared authentication processes are starting to lose their ground as the dominant approach today as the user experience landscape starts to change in the wake of greater work from home, user mobility, and security concerns.

Moving past local users for which there is effectively only the default mechanism we start to see a variety of legacy and more modern user management options, and more will certainly be coming in the future as this space heats up as new demands on infrastructure are shifting user management priorities for many businesses.

Traditionally in the UNIX landscape, which of course includes Linux, the standard user centralization service was the **Network Information Service** originally called YellowPages and eventually just known as **NIS**. NIS was introduced by **SUN Microsystems** in 1985 and it quickly caught on across the UNIX world and changed how people thought about users across systems. NIS was the vanguard of the movement that drove IT development in the 1990s as directory services became the hot strategic technology of the age. NIS may have not been the absolute first directory service (although maybe it was), but it was certainly the first to see widespread use or to fundamentally shift the industry's paradigm of user management.

NIS was extremely basic, but flexible and easy to manage, and lacked almost all security as it would be needed today - all things that made it effective in its time. NIS became all but universally available on all UNIX systems including commercial offerings like Solaris and AIX as well as on open source offerings like BSD and Linux-based operating systems. Learning NIS meant you could work across operating system divides easily and NIS provided interoperability between disparate UNIX flavors as well.

Given the immense age of NIS and the lack of security and scalability it might seem as though NIS would have faded into the past leaving us with nothing but stories for old timers to bring out over the campfire, but this is not the case. NIS remains in use today, especially in well entrenched, larger firms where NIS implementations might easily stretch back to the 1990s when it was the key technology still. New deployments might have all but disappeared over a decade ago, but old ones linger on. In fact, at the time of writing, every major Linux-based operating system still includes packages and support for NIS both to allow the system to act as a NIS client as well as creating new NIS servers! It has been proposed that RHEL will drop NIS in the next few years, but at this point it is still a proposal and still some ways off in the future. NIS has created quite the legacy for itself.

NIS lacked so much, especially in the areas of security and scalability, that its designers attempted to replace it after just seven years introducing NIS+ in 1992 (I told you that the 1990s were big on directory services.) NIS+ was not a direct upgrade for NIS, however, and proved to be hard to manage and was not a smooth upgrade process from NIS. NIS+ never gained strong enough traction to make it a major player and NIS actually managed to outlive it in real world utilization and in software support. SUN, who made both NIS and NIS+, announced that NIS+ was to be discontinued in 2002 and while it was still officially supported for several more years it was waning. NIS+ advanced the art of centralized user management, to be sure, but it never itself became a key technology. Its heyday was in the mid-1990s, but as so many technologies were being fielded during that era it was lost in a sea of competition from every angle - including from new players like Novell and Microsoft.

To avoid going into a history of the convoluted, and mostly forgotten, world of directory services in the 1980s and 1990s, we need only focus on one dominant technology that was introduced in 1993 and that is LDAP (the Lightweight Directory Access Protocol.) LDAP was a game changer in many ways. For one, it was vendor neutral allowing any system to implement it freely. Second it had many advanced database, protocol, and security features that allowed it to be used flexibly and security while allowing it to scale. Other technologies existed, but by 1997 when LDAPv3 was released, no other directory service was still making headlines. LDAP was seen as the clear winner and future of the LAN-based directory market.

LDAP began to replace NIS and NIS+ in the UNIX world, including on Linux-based operating systems, during the 1990s although the available implementations were complex and not well known yet. LDAP really lept forward when, in 2000, Microsoft announced that they were replacing their traditional Security Account Manager or SAM system (this is the one famous for having Primary and Backup Domain Controllers knows as PDCs and BDCs) with an LDAP implementation called Active Directory. Active Directory proved to be such a well-made and easy to manage LDAP implementation that it completely dominated the market and itself overshadowed LDAP to the point that few people are even aware that Active Directory is in essence just one of many LDAP implementations available on the market.

Linux-based operating systems can use nearly any LDAP server as a directory service source, or even Active Directory itself, which is an LDAP server but additionally has advanced requirements like Kerberos, in addition to the LDAP portion of the system, that given it expanded functionality and security over vanilla LDAP. Since the late 1990s, LDAP (in one form or another) has been the general standard for Linux systems that need to authenticate to a LAN-based directory service. Today, multiple LDAP server implementations are available that can run on Linux. There is even a SAM and Active Directory implementation for Linux to act as a server!

If using a local directory server is the chosen approach, then some implementation of LDAP is almost certainly going to be the only reasonable choice for a Linux-based operating system today. If the intent is to integrate with other systems, such as Windows or macOS, then an Active Directory flavor of LDAP with its additional *special sauce* is almost certainly the necessary option - either the version directly from Microsoft or the open-source Samba implementation available to run from almost any Linux. If implementing for an all UNIX set of devices then one of the more traditional LDAP server products is more likely to be appropriate such as OpenLDAP or 389 Directory Server.

LDAP is generally not considered to be viable for exposure to public networks (like the Internet) and is generally thought of as being LAN-centric, meaning that it relies at least partially on the network firewall to provide a safe operating space in which it can do its job. Exposing LDAP (even assuming SSL/TLS support via the upgrades LDAPS protocol) carries a bit of risk. A few companies still do this, and it can work, but it requires a lot of planning and an understanding as to the scope of exposure. Many companies accidentally expose LDAP components through a third-party proxy creating an exposure unintentionally (and highlighting the risks of the walled garden theory of LAN-centric security approaches.)

The famous RDP exposure risk

In Linux or UNIX circles, the RDP exposure risk example is not as widely known as in the Windows world, where remote access to systems by way of the Microsoft **Remote Desktop Protocol** (**RDP**) is a very common thing. However, the concept and problems often associated with RDP, such as vulnerability to brute force attacks and high visibility to potential attackers, are not actually related to Windows but to architectural design.

The issue is that exposing RDP on a publicly accessible IP address is considered to be very high risk. Yet the security on the RDP protocol is similar to that on SSH which is generally considered to be safe to exposure (within reason.) Why would two protocols of similar use and security result in two such wildly different security postures?

The secret lies either with the convention in the Windows world to use Active Directory (an LDAP implementation) ubiquitously or in the fact that Microsoft's standard multi-user RDP environment, **RDS** (for **Remote Desktop Services**), requires Active Directory. Active Directory essentially becomes a foregone conclusion when RDP is mentioned, but when using SSH it is assumed that Active Directory or some form of LDAP will not be used (at least as the external authentication method.)

Why does the underlying security method make a different when Active Directory and LDAP are highly secure on their own, and RDP is a very secure protocol? The answer is in that RDP forcibly exposes access to Active Directory in a manner very different to how it would be used on an internal LAN.

On the LAN we basically have an automatic whitelist that consists of the devices on our LAN. In many environments this will be additionally limited by VLANs that keep unnecessary devices (phones, IoT, and others) away from our end user devices. In larger environments network access controls may further limit potential exposure leaving us with a very protected environment in which our Active Directory can operate. Further, Active Directory itself generally protects itself by limiting attempts to log in to any given account locking the account for some period of time before it allows further attempts. Attacking Active Directory remotely on a LAN is generally quite difficult to do effectively.

When we add RDP and open it to the Internet at large, all of these controls drop away, completely. Of course, there are ways to limit this through IP whitelisting, VPN encapsulation, or other technique, but a stock deployment is going to expose our system broadly (if it does not do so, the purpose of exposing it is generally lost.) What is often missed is that the *lock after some number of failed attempts* mechanism that is so critical to securing Active Directory frequently offers a means of enacting denial-of-service (DoS) attacks on our users (that is, making it easy for an external user to block our users from logging in). In order to mitigate this attack risk, we have to disable this mechanism for RDP, in turn allowing essentially unlimited attacks on our accounts from the public space! Neither option is truly viable, and therefore, RDP becomes highly risky when used in a conventional way, even though both RDP and Active Directory are quite secure in their own right.

SSH and other mechanisms are most often used decoupled from central user account systems such as LDAP and Active Directory, allowing them to maintain a completely separate security mechanism and posture when compared to other user authentication methods. RDP can be used this way too, of course, but it is often assumed to have the functionality of central user account access because of Active Directory, making it difficult to treat it like a decoupled service because many in our organization many expect that because it is RDP that they will have immediate integration with Active Directory.

Today we are living in the post-LAN and generally post-LDAP world. Even LDAP's single biggest proponent, Microsoft, has moved away from it when possible and is investing probably more than any other company in LDAP alternatives, primarily, their own Azure AD service which, confusingly, keeps Active Directory in its name even though it is a completely unrelated mechanism to Active Directory (but can be tied to it to extend it.)

The biggest changes to industry accepted authentication systems is that most systems are hosted products rather than software that businesses are expected to deploy and maintain on their own; and most modern systems are Internet based allowing users to be located almost anywhere and, as long as they have an Internet connection, be able to connect to the authentication mechanisms.

These new mechanisms are coming in a continuously increasing variety of flavors from many different vendors and types of vendors. In the completely non-server, desktop only world of Chromebooks, Google themselves have an exclusive authentication mechanism and represents a very significant portion of the Linux-based end-user market today.

Are operating system logins relevant in the modern world?

This being a book on Linux administration, we are understandably talking about users at the level of the operating system, and we have to (as is best practice) question whether such a concept even matters today (or historically, for that matter) or if logging into operating systems may soon be a thing of the past entirely. Honestly, it is not an easy question to answer. The knee-jerk reaction is to jump to the conclusion that operating system users are incredibly relevant and underpin all security. But are they really?

To start, I am going to lead with saying that: Yes, generally operating system logins and user management is still important today and always has been.

How we think of user logins has changed immensely, though. The user landscape today is very different from what it was just one to two decades ago. Let's start with a little history.

During the pre-1990 era, the *archaic* computing world, very few systems used user identification mechanisms, at least at the operating system level. Systems that did, like UNIX and VMS, were special cases, big enterprise systems that were considered advanced and impressive. What most users would interact with, even using systems like Mac, Amiga, or MS-DOS, were single user, non-authenticated systems. As late as 2001, Microsoft's Windows **Millenium Edition** (**ME**) was released without true multi-user support (as it was still just a graphical shell layered on MS-DOS). In general, the idea that operating systems needed to manage multiple users was a foreign one.

In the 1990s, the shift to user security and access control as networking became generally accessible, the Internet began to come to fruition, and users needed more advanced functionality was significant. If anything, the 1990s were marked by being the era of user management (which we mentioned earlier.) Rather suddenly everyone was concerned with how we would handle many users sharing the same devices and how we would make the user experience portable across multiple devices. System administration was essentially flipped on its ear.

By the 2000s, the operating system level user experience expectations were well entrenched and user management moved from competitive advantage to commodity functionality. The last remaining operating systems of any note that did not support native multi-user functionality faded away and even casual end user products intended for entertainment user started to encourage user management and security.

When smart phones first entered the market, they were a throwback to the user-less systems of the 1980s. The holder of the device was assumed to be its universal and unnamed singular user. Even cell phones have moved to at least being a user and security centered device, even if they still focus on a single user, they do so with access controls and user identity in mind.

On the server side, user management has certainly lost its luster over the past decades. At its peak in the 1990s, Linux and its UNIX cousin systems were all the rage for direct end user logins and user management on the server was a significant portion of a system administrators' day. That trend fell away, and the idea that end users would need accounts or logins to a server at the operating system level now feels outdated, at best. Exceptions will always apply, but they are few and far between and becoming increasingly rare.

Even at its peak, the necessity that all administrators log in with individual user accounts was never universally accepted. It may never have been exactly normal for all administrators to share single accounts, but it was never exactly rare, either. Shared root (the default administration user account) account access has always been a common practice that few want to admit to having witnessed and exists commonly in Linux, Windows, and likely all other environments. The practice was (and surely still is) so widespread that techniques and tools for managing user access to the single user account through external mechanisms are even somewhat popular (such as logging into a third-party control console that grants access to root accounts on multiple servers as needed!)

As we are a book on best practices, I am going to take moment to point out that in our role as system administrators it is absolutely a best practice to maintain user identity and access control on our servers and that shared accounts are really never a good idea even when they are well managed and secured. The necessary effort to maintain more secure and auditable systems is simply not that great that we should be working to avoid it. That said, shared account access is not necessarily as risk as it sounds as there remain the potential for audit controls and access controls even when doing so and while I truly doubt that there is ever an actual justifiable use case for doing so, it is certainly possible to make shared access *secure enough* to function effectively.

More importantly, for servers (and the administrative functions of end user workstations as well) modern design techniques using state machines, infrastructure as code, application encapsulation such as application containers, **MDM** (**mobile device management**) and even **RMM** (**remote monitoring and management**) tools may entirely eliminate the need for logins at all making the entire discussion of user management moot. If we never log in, if we never create a user, we need not consider the possibility of user account access security, whether that's remote or even local.

So, it is worth considering that from a server perspective, the need for users may have been nothing more than a passing fad or fallback crutch for shops that are not able to maintain the most modern of techniques. These are administrative users, of course, but that has long been the only users expected to exist on a server at all.

On end user devices, the concepts of users have begun to change dramatically as well, but for entirely different reasons. Traditionally work on end user devices was focused on locally installed applications that themselves have no user controls and simply run under the security controls of the environment provided by the operating system. This could be any kind of application from word process, image editor, or video game. The environment was primarily defined by the accessible local storage locations. For many users, especially home users, this has almost entirely changed.

Today most applications are web applications, applications that run in the browser rather than being installed at all. Users might log into the application, but this is almost always separate from any operating system login. Even locally installed applications are starting to connect over the Internet to authenticate to a service increasingly.

As this shift occurs, the concept of the operating system user being the primary component used to define the context of security access is fading rapidly, and the idea that users need to exist at the application level instead has already become the norm. As this happens, we have to start to consider the functionality of the operating system user accounts not to be a granular access control mechanism or the end all of user management but rather little more than a security step to attempt to guarantee a safe, protected environment from which we can launch our web or networked application and sign in at the application layer where our users and access controls are far more relevant. Moving user access controls to the application layer is critical as this is where the data intelligence resides, and control can be made at a granular level rather than granting access at the level of the entire application.

As the operating system starts to lose its traditional role in user management, the benefits of strong user management and control systems begin to erode. It is far more than just the shift of user access control up the stack from operating systems to applications as the industry and software matures that is causing this. Other factors are at play. It used to be that computers were expected to user printers heavily, something that has broadly stopped being true. Printers are now an afterthought, if they even exist, rather than being a primary functionality of computers. Likewise, modern applications do not use operating system managed storage in the same ways and the need to strictly administer local storage and mapping remote storage resources has fallen away.

Even just ten years ago user management systems, Active Directory as a prime example, were used first and foremost to coordinate printing and mapped drive resources for end users. Today, between application modernization and workforce mobility and computing ubiquity, both of these things have become legacy functions. A laptop user working from home might have absolutely zero need for either mapped drives or printers, and in order to allow mobile users to function in this way, in-office workers have had to adapt to this as well. The trend in access control mechanisms is moving away from the traditional uses of operating-system-level users.

In summary, users still matter, but they do not maintain the relevance that they had in the past and the future looks like one where that value will continue to diminish.

Knowing what remote access methods are going to be most appropriate for you requires more than just an understanding of how these systems work and how their implementations play into your environment is just the first piece of the puzzle in determining what mechanism should be chosen for your organization. Learning what products are currently on the market, their current features, limitations, pricing, and other business factors are necessary. User management is rapidly becoming a market where it is more about knowing the current offerings available and much less of something that we will implement ourselves.

In the next section, we are going to move past authenticating our users and now dig into how we can provide these users with remote access to the systems that they need.

Remote access approaches

Assuming we are not using a *access-less* approach built off of state machine technology, we have a few different paths that we can popularly use to gain access to our Linux systems. In most cases with Linux based operating systems we are going to be discussing how system administrators, like you and me, are able to log in and use the operating system interactively, but any typical method that we are going to use to do this is going to be an option for end users as well. The needs of end users is generally very different from that of system administrators, but the tools that we can use are going to generally overlap.

For us, in the system administration role, access is most often defined by needing to be very quick to set up, quite temporary in its use, with the focus critically being on ensuring that the system is highly accessible and command line driven. For end users, we will expect the opposite. Administrators often have to log into many different operating systems, maybe one after another, maybe many all at the same time. A lengthy process to access them could significantly hinder our ability to be useful.

Traditionally end users will log into only a single system and remain attached to it for the duration of their work period, typically a workday or something similar. Taking more time to log in that one time, but having a more robust end user experience, is their priority. Because the needs are so different, remote access techniques may be separate. There is little need to feel compelled to merge them.

There are two main types of remote access. One is direct, meaning that we expose ports and have some protocol with as SSH, RDP, or RFB (VNC) that allows us to connect from some software client to our systems. This is the best-known type of technology and is the most straightforward to manage. It requires nothing complex and is well understood. In this approach we have a traditional service on the virtual machine in question (or potentially on the hypervisor, but the result is roughly the same) and an external client *reaches in* to access the system.

The other access methodology is indirect access where an access server is used to manage access for both the operating system to be access and for the client attempting to access it. This method requires a server that is hosted publicly (often provided as a service but can be self-hosted as well) and both the end points connect to it as clients so that nothing except for the access server need to be exposed outwardly.

Advantages of both solution types are pretty clear. Direct connections are simpler and have less to go wrong. Indirect connections require more infrastructure but reduce potential points of exposure, consolidate connections, and hide the presence of the networks making it harder to discover and attack a network based on its remote access publications.

For many reasons, there is a trend towards direct connection technologies, such as RDP and RFB (VNC), for regular end users and indirect connection technologies, such as MeshCentral, ConnectWise, and LogMeIn, for system administrators. For end users, the direct connection technologies tend to provide the most robust experience to replicate the feeling of working directly on the hardware as if it was sitting in front of you. For system administrators, the increased security and consolidation of access to systems potentially across many different physical locations is highly beneficial.

Notice, however, in all cases we keep saying *tends*. There are no hard and fast rules here, only strong tendencies. Because we are dealing with Linux, we have other considerations, however. For example, RDP tends to be less easy to manage than on Windows, while indirect access methods might be easier but not as well understood. Also, our user base may have different inherent expectations – for example, where Windows users might expect Windows-native tooling, Linux-based operating system users might be more open to less familiar access options.

Like so many things in IT, understanding base level technologies is only the first step. In our next section we will discuss ways that we can make SSH, a direct connection technology, more flexible and robust and in some ways of looking at it we will be mimicking an indirect connection using direct ones. The differences between them are not so large.

Even a rule of thumb is difficult in this situation and likely much discussion of remote access will depend upon many factors such as how that access will be used, will the infrastructure be shared with end users, what security needs may exist, do other access considerations exist such as VPNs, and is there value to creating a unified connection process that is shared with other technologies or platforms?

And then we have to consider the possibility of having multiple access methods. It is not uncommon to use more than one to ensure availability even when one has failed. One may be the convenient, yet fragile, access method while the backup is heavily secured and cumbersome to use but serves as a critical backup should all else fail.

How do I approach remote access

In a topic so devoid of strong guidance, I feel taking a moment to present my own typical access choices is valuable. I do this not to suggest that my approach is ideal, but to give some insight into a real-world decision-making process.

Most of the time when deploying Linux servers where I am overseeing their system administration and we need to have remote access to a running system instead of being managed completely by way of state machines or infrastructure as code, we opt for a two-prong approach with one direct and one indirect access method.

For indirect access we use MeshCentral which, itself, is open-source software and runs on a Linux based operating system that we host ourselves. This allows us extensive flexibility and cost savings compared to most solutions and because we are able to run it on the same operating system that we deploy internally and for customers we are able to leverage processes, tools, and skills that we are already using other places in order to maximize our efficiency. With MeshCentral, as with many indirect remote access solutions, we have remote terminal access, remote graphical desktop access if a GUI is installed on the server, and many tools for monitoring, file moving, and remote command execution.

With this indirect access the system administrators have nearly instant access to all of the servers that we maintain, across highly disparate technology stacks and physical locations. Some servers are isolated on LANs without any port forwarding available, some have public IPs, some have graphical desktops, most are command line only. No matter what they location, use case, or disposition MeshCentral gives us access to do what is needed to manage the systems.

For situations mostly involving emergency access we also maintain direct SSH access to nearly all systems. This is important as the ability to reconfigure, patch, or restart a system where indirect access methods have failed is often critical. This access would almost always be limited to access only from a local LAN on which the workload directly sits and potentially even limited further within that network scope making SSH only available from select workstations or another server designated for the purpose of highly secure remote access. In some cases, the SSH service may not even be kept running by default and only turned on by way of a state machine setting being changed; or it may be changed manually via a form of out of band management. SSH may also be useful for some forms of automation.

Having two very different forms of access, with deep security controls in place, provides the right balance of accessibility, protection, and security for us. You have to be aware that each additional method of access means another avenue of attack by malicious actors, as well, so layering on access methods that are not necessary is generally not a good option. You want to ensure reliable access without putting systems at risk.

Something often overlooked in these types of discussions is that there are some direct remote access tools that work in ways we often do not anticipate such as web based direct access tools. These tools would include such products as **Cockpit** or **WebMin** which provide a web-based interface to our systems. These tools may give access to configure our systems via a web interface and may even allow for interactive console access through the web interface allowing us to publish and secure remote access in an entirely different way.

In the Linux world, by far the most common and assumed method of accessing a computer across the room or on the other side of the world is the ubiquitous SSH protocol. Next, we will look at ways to make SSH even more powerful than it is out of the box.

SSH, key management, and jump boxes

Using SSH for remote management of Linux based operating systems is so ubiquitous that it deserves special consideration. SSH on its own is efficient and very secure, but it is well known and generally exposes such extreme functionality in our systems that it is often the target a focused attacks. We cannot be complacent in the use of SSH, especially if exposed to the Internet, as the risks are simply too high.

When using SSH we have almost a laundry list of ways that it can be secured. We will touch on several of these and how they work together to make SSH extremely difficult to compromise. SSH on Linux is provided via OpenSSH which is mature and battle tested and receives more scrutiny than almost any software package made. SSH starts as an already very harded package from most perspectives.

Our first tool for securing SSH is to consider completely removing password-based access to it in favor of using keys. Keys are fast and efficient allowing admins to access servers faster and more securely than with passwords. Keys also support passphrases which act as a form of two factor authentication. If this makes sense for your organization based on the security needs, then this requires someone to know an encrypted password and possess the private key simultaneous to attempt to breach a system through this channel. Keys have been around for remote access authentication for a very long time but have not enjoyed the popularity that they should. Too many companies take a *set up keys for yourself if you feel like it* approach, allowing far too many administrators to just not bother to take advantages of the efficiency and security that they offer.

Our second tool is account lockouts from failed attempts. The standard tool for this on Linux based operating systems is called Fail2Ban. Fail2Ban takes SSH and other services that have standard login modules, and works with them to detect suspected malicious attempts against our systems and automates our local (meaning the firewall on our Linux based operating system itself) firewall to halt traffic from the offending IP address(es) for some predetermined period of time, typically three to fifteen minutes. This approach is our most effective tool against brute force attacks.

Do you still need both a network edge firewall and an operating system firewall?

You might be shocked at just how often this question has come up over the years. What makes it such a surprising question is the context that it is always asked in.

First, some tech. In the real world, for many practical reasons, all routers are also firewalls. So functioning without a hardware firewall on the network edge is effectively impossible. In some cases, such as when installing VPS or cloud, we might be defaulted to a completely wide open firewall, but the firewall is at least still there. We have to start from an assumption that all operating system instances will always effectively be located behind a network firewall and that that firewall may be highly ineffective.

Second, some history. Operating system firewalls were extremely rare due to their performance impacts until the late 1990s. They were also not very important as the degree to which computers were networked was much lower until that era. Operating system firewalls were introduced even though network firewalls were already a universal assumption because they add granularity and automation such as we can get with Fail2Ban, and because they protect individual machines from attacks that have either breached the network firewall or, far more likely, originate from inside of the LAN.

Given the context, tech limitations, and history it seems ridiculous that someone would ever assume that both firewalls are not needed. If we had the ability to skip the network firewall, theoretically, we could do so, but there is no purpose to skipping the security at this layer; and while we can choose to not use the firewall in the operating system, this creates risks that has no other mitigation method. The operating system firewall is unique in its ability to defend against both local and remote threats, whereas the network firewall can protect against remote only. While nothing is going to be good as having both, it is the operating system firewall that we care about the most from a security perspective. Primarily we care about the network firewall only to lower the workload on the operating system firewall.

Importantly, basic firewall functionality today does not require any measurable system overhead. This was the downside to using them (around a quarter of a century ago.) Today there is no effective caveats to the use of firewalls.

Best practice here is extremely clear and quite important: both the network firewall and the operating system firewall are absolutely needed. There is no situation where one of these firewalls should be removed or disabled.

Often operating system firewalls are disabled because system administrators do not want to be bothered to have to know their networking needs or to secure their workloads properly. This is never a valid excuse. We all feel lazy at times, and we would all like to avoid having to maintain yet another point of support on our systems, but this is fundamental security and being able to properly support a system in production would require that we have all of the knowledge necessary to enable and configure the firewall regardless. Any perceived value to disabling a firewall (outside of troubleshooting, of course) should be seen as a huge warning sign that something is wrong that we need to address.

Our third tool to secure SSH, after key-based authentication and account lockouts, is network limits, generally in the form of limiting the IP addresses from which requests to connect to SSH can originate. This can come in multiple forms. Often, we would do this with the firewall on the Linux system itself, but it could also be provided by the hardware firewall on the network edge or by a network firewall from a cloud provider and so forth.

IP limits are often best set by whitelisting, when possible. Whitelisting allows us to restrict SSH traffic to just a single IP or small group of IP addresses that we believe to be known and safe such as from administrators' homes, a trusted data center, or the office IP address. This dramatically reduces the attack surface of our systems and makes them difficult to detect in the first place.

For some, though, whitelisting is not possible or, at least, not plausible. In those cases, blacklisting of IP ranges that are certainly not going to need access will still increase security. In this case, for system administration, the use of national IP ranges to block traffic from certain countries and regions can be practical. I never recommend this practice for customer facing systems where you might accidentally block a customer or business partner without realizing it and risk backlash (remember, to someone trying to use a service from a blacklisted location does not see it as blocked, but as the service having failed and being offline), but for system administration access where the locations of your administrators should be roughly known and alternative back channels for communications should always exist and where security is far more critical it can make a lot of sense.

Fourth, the use of sudo to require an additional verification before executing escalated privilege commands is very useful. With sudo we can layer on more protection over what we already can have with keys or better, keys with passphrases. If we log in using a user account that possesses sudo permissions to the root superuser account, then we have already demonstrated one or two factors of authentication just to become a standard user. To use sudo we optionally require another, separate, password for that user to gain privilege escalation. That is a lot of potential protection. Sudo also helps us avoid dangerous errors that can happen, mostly from typos, when running as the root user directly. Sudo is more likely to protect us from ourselves than it is from external actors. It's a very useful tool.

Fifth, thanks to the power of sudo mentioned above, we can disable root user logins over SSH completely keeping the most well-known, and by far the riskiest account, out of the risk equation completely. There is no need to the root user to be directly exposed when we have sudo mechanisms to protect it and to log when it is accessed already.

Does changing the default port of SSH work?

You will find many people and articles that tell you that you should always change the default port of SSH, or any access protocol, to make it harder for attackers to detect. This will then be pointed out as being a form of security through obscurity which, is generally believed, to mean no security at all.

In reality, both positions are a bit overstated. Changing the default port will truly do nothing for actual security as any real attack of any complexity or effort will find a non-standard port in a matter of seconds and likely the attacker will never even become aware that you attempted to thwart the attack by changing the address. Much like if you were to move the door to your house to somewhere along the side of the house, most thieves breaking in would not even take note of the fact that the door was in an odd location. The door is still the door and completely obvious.

Thinking of using non-standard ports as a security measure is incorrect. At best it nominally improved the security posture, at worst it may advertise that you have attempted security through obscurity and may make a good target for a more focused attack. Where non-standard ports benefit us potentially is by reducing the amount of traffic that hits our ports making it easier to store and filter logs.

Reducing log clutter can aid security, of course, simply by making a worrisome attack easier to spot or faster to diagnose, and reducing storage needs is always a nice benefit. So changing SSH ports may be beneficial, but the benefit should be kept in context.

Beyond all of these best practice security methods for SSH there are many other standard hardening options that can be used, but I would argue that outside of these that we would struggle to define any more as best practices as opposed to highly recommended or well worth considering. With a little accommodation, SSH can be ridiculously secure for almost any organizational need.

To take security in SSH to another level you can also, and quite easily, apply one-time passwords for multi-factor authentication through tools like Google Authenticator. Many third-party security enhancements exist and SSH can effectively be as secure as you want it to be.

SSH key management

It is generally assumed that if we are going to use SSH then we are going to use keys to secure the access to it. As mentioned before, they are fast and highly secure. There is little reason not to use them. It is not as simple as *just using them*, however. When we choose to use keys we then have to determine how we will manage those keys. How that is going to make sense will depend primarily on the size of our organization or, at least, on the size of our systems administration team and how many systems will be being accessed using said keys.

At its simplest SSH key management can be left up to the individual to manage on their own. If users have access to log in to an operating system already, then they are free to create and manage their own keys. For smaller organizations or those working with just a few system administrators this might make sense.

With SSH keys, of course, management includes two components, the public keys and the private keys. Typically users will manage their own private keys, keeping them safe, and public keys can be managed in nearly any fashion.

In a smaller organization looking to improve the management of keys with little other infrastructure, it can be as easy as storing the public keys for users (which includes system administrators) on a wiki or other simple documentation system where they can be easily obtained as needed. This one step alone can make a big difference in making keys very easy to use.

Keys can also be stored on the filesystem of something like a management server or workstation and pushed out through simple automation like a script that runs over SSH remotely and copies keys into place. Scripts could also pull public keys from a web page, a file share, or use something like GIT or Subversion to grab keys from a repository. Keys are simply text files and so managing them is flexible.

In a more advanced setting, state machines and infrastructure as code approaches can be used to automate key deployments through the same tooling as other automations. Keys can be just another set of files and do not need to be treated specially in any way. DevOps processes like state machines and infrastructure as code are great mechanisms to make SSH key management vastly simplified

All of that, however, is really barely worth considering. Once you are working at any scale with keys and are at a point where key management is going to become something on your system administration radar, then it is probably time to look at a **public key infrastructure (PKI)** system to manage certificates instead of keys. SSH uses TLS, the same mechanism as HTTPS and countless other secure protocols, and as such it can use the same PKI system that websites use.

Of course, in almost all cases using a publicly hosted PKI certificate system is going to be problematic for what are almost always private and internal hosts in our Linux infrastructure. So, we would be required to run our own certificate authority, known as a CA, but this is a standard practice, has extremely low cost and overhead, and while the skills to do so are not broadly available they are easily acquired. Using SSH certificates *instead* of SSH keys (instead is in quotes here because certificates contain keys, the keys always remain under the hood) gives us a mechanism to rapidly scale SSH key security for many administrators, potentially many end users.

I would not go so far as to call running your own certificate authority and building a PKI infrastructure is a best practice, it is a good rule of thumb for organizations with more than just a few users connecting to more than a few boxes over SSH. The network effect of many users to many different operating system instances can mean an explosion in SSH keys that could remain unmanaged if we do not take action. Ten administrators, twenty developers, ten testers, and one hundred virtual machines alone could create a need to monitor four thousand SSH key combinations!

With essentially all operating system supporting SSH today the benefits of a robust SSH security strategy are even larger. The easier SSH is to use securely, the more likely it is to be used over alternative technologies.

Jump boxes

Jump boxes are an important security and management tool that can simplify many aspects of system administration. As a concept they are very common, but as a term often even seasoned system administrators may not be familiar. A jump box is a system that is accessed remotely by system administrators (or regular users, but it is typically only used for technical support staff due to the cumbersome nature of the design) from which access is then granted on to the systems that will be managed.

It is called a jump box because you log into it before *jumping* to another system. It is a jumping off point for your tasks. It is common for jump boxes to be used for more than just access, but also as a central repository for tools, a temporary storage location, or a common location from which to run automation scripts.

Jump boxes are often used to provide a central point of direct access to systems to get something akin to a hybrid between the features generally associated with indirect remote access technologies and traditional direct access. Technically a jump box is just a two stage direct access system, but one that can be highly useful and avoids the need to use a network router like a VPN or complex proxies to accomplish a consolidation of access.

Through access consolidation we can more practically secure our first line of access. It is common for jump boxes to receive the most complex IP filtering, strict Fail2Ban rules, detailed logging, two factor authentication, rapid patching, and so on to tightly secure the most vulnerable point of ingress. In this fashion, system administrators might start their day by logging into their jump box and then quickly and easily attach to the systems that they administer from that point.

Because of their design, it is often easy for jump boxes to exist inside of the LAN, be hosted in colocation, or even be cloud hosted. They can be put wherever is practical and given access to resources anywhere. They can be hardened, monitored, and then connected through direct access protocols or even a VPN or multiple VPNs to systems and sites as needed.

Because a jump box is a single system it is easy to have the systems that we manage allow connections from its single IP address to allow good security even while using direct access technologies.

Jump boxes for Linux based operating systems are typically used for SSH and so may be built as a lean server. A GUI-less Linux jump box can run on one of the smallest of virtual machines, these use almost no resources making them very easy to deploy wherever needed with little cost.

Jump boxes are also built to handle other protocols, often X or RDP, for example. This is uncommon for Linux system administration as rarely is a GUI anything but an encumbrance for us and when using a jump box the resource needs and complexity of providing a central GUI source will often make us reconsider providing a GUI as well.

Jump boxes are not a best practice, but they are a common security and management tool and can be very helpful where direct access is needed to make it faster, easier, and more secure than it would likely be otherwise.

Alternative remote access approaches

Traditional remote access, at least as we tend to think of it, is all designed around the needs of end users needing to use remote sessions as a replacement for their local desktop. As system administrators, it is great for us to be able to use those tools when they make sense for us, and it is necessary that we understand those tools because they are generally components that fall to us to administer, but for our own usage they may not be the most practical.

Of course, we can include most indirect remote access technologies under the heading of alternative remote access approaches, but they are basically just traditional access that has been tweaked to be more practical for our use cases. As administrators we want to reduce our logins or interactive sessions with remote machines in the hopes of removing that access completely, at least in an ideal world.

To this end, we have other methodologies today for running commands on our servers. There are not going to replace our existing methods in all cases, but they may be a steppingstone technology to help us move from where we are to where we want to be in the future.

What makes most of our system access methods *traditional* is really that they involve complete interactivity. That means that whether we are using RDP or Splashtop or an SSH session, we assume that we are establishing a full connection to the system to be managed, complete with user-level environmental settings, and working in a manner where we have continuous input and output from the system. This is so much assumed that many applications or tools actually assume that this method is used and may require session environmental variables that are not logical or appropriate.

Short of remove access altogether as we discussed earlier when talking about infrastructure as code, our interim access step is remote command execution or using non-interactive commands. This works effective just as our other access methods do, but without the ability to become interactive. Remote command execution lets us move from manual tasks to automation more easily and is great for auditing, security, and scalability.

At its simplest, remote command execution can be handled through SSH using the same infrastructure that we would have for interactive sessions. SSH is designed to handle either method with aplomb and since it does so transparently it can be an easy tool for moving slowly from one method to the other. The methods can be mixed on a situational basis, or one administrator could use one method and a second could use the other.

With remote command execution we get the benefit that all commands are executed, and therefore can be recorded, on the originating system. Perfect for the use of something like a jump box or management server that could log all actions performed. Interactive sessions, however, even those initiated from a jump box, will log all of the important session information on the final operating system and the jump box will only know that a remote session was started - visibility into exactly what commands were run there and what their responses were will be lost to the central logging platform.

In many ways, remote command execution is a system administrator's analogue to functional programming in the software development world. Interactive sessions are more akin to procedural programming, where actions are seen as a sequence of events. Remote command execution is the use of singular remote functions to perform a task. A difficult analogy to apply if you are not familiar with these programming paradigms, but for those that have used them I think the example is valuable.

SSH may be ideal for sneaking in a casual introduction of remote command execution and is essentially always available for system administrators to use. Even in highly strict, structured, and formal process driven environments it would be highly unlikely for a policy decree to not allow administrators to use this approach whenever and wherever they want, assuming traditionally interactive SSH sessions are allowed.

Other tools, and with increasing popularity, also allow remote command execution today. This has become a standard option in most indirect remote access tools from large, commercial software as a service products to small open source products for self-hosting: almost always remote command execution is included in some fashion today. In some cases with extended features such as the ability to run the same command or set of commands against a set or list of systems simultaneously.

RMM tools are often building in remote command execution systems in the same way. This is far easier than creating custom interactive session mechanisms and can be touted as a more advanced option, while being much simpler to implement.

The most interesting place to find remote command execution, in my opinion, is in state machine systems. Just as SSH can be viewed as having remote command execution as a an optional strategy, to move away from interactive sessions and ease into something a bit less familiar, state machines can and do implement remote command execution as a way to maintain one foot in a more traditional operational mode to allow system administrators a fallback method for when state definitions are too difficult or time consuming to implement.

Remote command execution from a state machine is also a method for testing state machine access or capabilities while developing state files. State is maintained, one way or another, by running commands on the system being managed. In some cases, commands are run by an agent at arm's length and while remote commands are executed, the exact commands are not sent by the state management central system. In other systems commands may be sent directly as typed on the management system and the state machine is just an execution assistant.

Remote access, as simple as it might seem at first, is not a one size fits all solution and we do not have to use only a single solution even inside of a single organization. Consider thinking outside of the box and trying new or different approaches to make your workflows more secure, stable, and efficient.

Now that we are prepared to system administrator access and security across a broad variety of approaches, we should talk a little about technologies more applicable to end user access to systems: terminal servers and virtual desktop infrastructure.

Terminal servers and virtual desktop infrastructure (VDI)

Unlike the Windows world, remote GUI access in the world of Linux based operating systems is relatively rare. This is just not part of the Linux wheelhouse in a self-fulfilling situation where customers do not demand it, so vendors do not specialize around it, leaving customers feeling that little is available for it and the cycle continues. But that is not to say that both terminal services and **VDI** (which stands for **Virtual Desktop Infrastructure** but is more meaningful and known simply by its acronym) options cannot or do not exist for Linux based systems, they most certainly do.

Understanding terminal services and VDI conceptually

It is not uncommon for terminal servers and VDI architectures to become intertwined, this mostly has happened because of marketing departments trying to sell VDI where it does not apply and because overlapping technologies are often used. That VDI was presented as the hot, new technology as if it had not always existed added to this confusion. And Microsoft, the leader in this space, renaming their core product from **Terminal Server** to **Remote Desktop Server** (**RDS**) did not help any either. This led to the problem that many Windows administrators, let alone users, routinely confused RDP, the Remote Desktop Protocol, with RDS, Microsoft's terminal server product. One is a communications protocol used by and implemented by many different products; the other is a specific product, as well as licensing vehicle, that you have to purchase from Microsoft that may or may not utilize RDP.

So we have to start by defining these technologies. Both involve accessing a computer remotely. Both can and do use the same potential set of protocols to make this magic happen. Both serve the same basic purpose but implement it in two different ways.

A terminal server has always meant a *single server* (meaning: a *single operating system instance*) that can be accessed simultaneously by multiple remote users. In the earliest days, this was accomplished by using serial connections to dumb terminals to display text *remotely*. Later came technologies such as telnet and RSH and eventually SSH as we use today. All of these technologies advanced the state of security and accessibility, but fundamentally remote access remained a familiar command line activity resembling the original serial-based physical terminals. Decades ago, the remote use of graphical desktop environments became the norm for end users, and new protocols such as RDP, RFB, X, and NX became popular, but nothing really changed fundamentally. Many users would connect to a single operating system instance, and they would share its resources. There was only one operating system to patch, and everyone shared the same kernel and applications. A terminal server uses a *many-to-one* architecture.

VDI, which stands for virtual desktop infrastructure, refers to an alternative approach where users remotely access dedicated operating system instances that exist only for them, and which are virtualized. With VDI, each user's operating system instance could be completely different with different patch levels or even different operating systems entirely. One user might be on **Windows 11**, another on Windows XP, and another on Ubuntu. VDI means a *one-to-one* architecture.

For the most part, the concepts of and differences between these two approaches are created almost entirely out of the need to manage limitations created by Microsoft software licensing. In the Linux world, the difference between a terminal server or a VDI deployment is purely a matter of how any given system is used *in-the-moment* at the time that they are being accessed. Every Linux device is inherently multi-user already. Linux lacks the *one user at a time* framework that has always been a part of the Windows ecosystem. These concepts are often lost or confusing in a non-Windows context. For Linux administrators, any VDI system is just many terminal servers that tend to be lightly used. In the Windows world, licensing dictates every aspect of this equation in very complex ways.

In the Windows world, an end user workstation license as available for Windows 7, 10, or 11 is always a single user license, no exceptions. Remote access is always allowed, but only for the singular user never for alternative or additional users. One user sitting at the console or that user remote, but never more than one at a time. Even Windows Server licenses only allow for one user at a time (with one additional license purely for the purposes of systems administration purposes) unless RDS licensing is purchased for additional users. RDS is only available as an add-on license to Windows Server. Because of these licensing rules, the ideas of what is a server and what is an end user workstation in the Windows world is exceptionally clear and obvious.

With Linux, of course, there is no such licensing limitations. Any workstation can do the job of a server, any server can be used as a desktop or for multiple users. Any machine can be used in any way, at any time, flexibly. Terms like terminal servers and VDI refer only to how we intend to use a system or how it is being used at any given moment, but to most non-Linux administrators it is a locked-in, set in stone, expensive, licensing-driven design and decision process and so we have to be adaptable to understanding how these terms and concepts play into the consciousness of those outside of our free and flexible universe.

Of course, there is an obvious, third option that has no name. Terminal services does not consider if a system is physical (that is to say, installed directly on bare metal hardware) or virtualized, in both cases it is a terminal server, and the virtualization is irrelevant. With VDI, virtualization is part of the name, so we only consider it to be VDI when it is also virtualized for no obvious reason. If we do the same logical architecture as VDI but do so on bare metal operating system installs, then the architecture has no name, at least no well-known name. One to one remote access without virtualization is actually the most common, base approach to remote access, so much so that most people do not even think about it as an architecture. It requires essentially no planning or coordination and is often used for a variety of purposes. This unnamed design is common for any operating system Linux, Windows, macOS, or otherwise.

Linux based operating systems will support any remote access architecture that we desire and will even support protocols typically associated with other operating systems: namely RDP from the Windows ecosystem. In the Windows ecosystem, RDS is technologically bound to Active Directory causing user management to be closely coupled with remote access strategies. In Linux we have no such ties. Deploying an RDP-based terminal server can be done using any user management system that we desire.

Tools like terminal servers and VDI are far more likely to be used by end users, from office workers to developers, than by system administrators, but this is not exclusive. In my own deployment today we maintain both for use exclusively by system administration. I will use this as an example scenario to show how these technologies may be used effectively on the administration side of the fence.

On occasion, system administrators may benefit from having a graphical session for support. This may be because graphical tools like a notepad, screenshots, web interface or such make the work more efficient, or in some cases the tooling requires a graphical session. It is sadly common that many systems, rarely Linux itself but often systems required for use by Linux system administrators, will only provide a web-based GUI interface or worse, something like a Java application interface or even a native application. Having a Linux terminal server built for multi-user RDP support in our case located inside our main, trusted datacenter with a static IP allows our administration team to have access for any and all team members at any time to open a graphical session in a trusted location to perform whatever work is necessary.

Having a graphical interface is also at times highly beneficial for making documentation while working, doing training via screensharing, or similar tasks.

Terminal servers are perfect, especially with Linux based operating systems, for providing a clean, standard environment that everyone shares. VDI provides a competing approach and we, like many companies, use VDI to provide highly customized environments that individual system administrators may require such as alternative operating systems or desktop environments that would create conflicts with other users if these were implemented on a shared server environment. VDI is also better for situations where individual system administrators may require to be administrators of their own environments, often for testing, that may not be wise or plausible from a security perspective on a shared environment.

Both terminal servers and VDI, whether graphical or command line only, can be useful as platforms to be used as jump boxes, management stations, remote execution environments and so forth. We can also use them, of course, as ways to provide Linux based desktop environments to end users. There is quite simply, no reason to limit these types of technologies conceptually to Windows, Linux can shine here as well and, in many cases, excel.

Summary

Users and user access on Linux based operating systems is a complex topic, mostly because of the incredible flexibility that Linux affords us. We can approach where users exist, how we create them, how they are managed, where our source of truth resides, and how those users can access their systems in so many ways. We have ancient technologies, we have extremely modern technologies. We can use nearly any mechanism, from any era, from any ecosystem and we can have many that we build ourselves and our unique to us. We can stick to well-known traditional processes, or we can easily build our own and work in a unique way.

There is no simple best practice for user management on Linux. Instead, our best practice is, like it so often is, that we need to understand the range of technological possibilities, how different risks and benefits will apply to our unique organization and know the products that exist on the market from open source to commercial, from software to services and evaluate those needs across all axes to determine what is right for our organizations. There are no built-in assumptions. The use of local user accounts is not wrong, even when done at very large scale. Using remote users is not wrong, even at a very small scale. We do not have to maintain and run our own infrastructure for security, but we do not have to rely on third-party vendors to do it for us either. The sky is the limit – most approaches have benefits, but all of them have caveats, so they all warrant thorough investigation.

There is always a tendency to feel a need to make any system overly complex. We are taught that complex is good, as it is advanced, and it feels right. But in the end, simplicity usually provides the lowest total cost of ownership while carrying the least risk. With simple systems and simple designs there are fewer moving parts providing fewer opportunities for mistakes and this typically wins the day.

I hope that with all of this information that you are armed with the knowledge and courage to approach user management in a different light. Too many user management solutions are chosen out of a misunderstanding that industry trends should drive decisions or that common solutions are far more secure than they actually are. What is best is always what is best for your situation and what is right for you is rarely what is right for someone else that you are comparing against and only with extreme rarity has anyone that you will compare against taken the time to have truly evaluated their own needs so allowing others' decisions to overly influence us tend to be very dangerous.

In our next chapter, we are going to be tackling the incredibly complex and often confusing topic of troubleshooting.

11
Troubleshooting

Few things are as challenging in systems administration as **troubleshooting** when problems have risen. Troubleshooting is hard at the best of times, but as system administrators our job is almost always to troubleshoot a system that is either currently running in production and has to remain functional while we attempt to fix some aspect of it or is currently down and we have to get it back up and running in production as quickly as possible. The ability to work at a reasonable pace without the business losing money actively as we do so typically does not exist for us or when it does, is the exception rather than the rule. Troubleshooting is hard, critical, and stressful.

Troubleshooting involves more than just fixing an obvious technical problem, applying business logic is critical as well. We have to understand our troubleshooting in the greater context of the workload and the business and apply more than simple technical know-how. There is fixing a problem, and there is fixing the workload, and there is evaluating the needs of the workflow, and in the end there is maintaining the viability of the business.

In this chapter we are going to cover the following topics:

- The high cost of disaster avoidance
- Triage skills and staff
- Logical approaches to troubleshooting
- Investigating versus fixing

The high cost of disaster avoidance

In this chapter we are going to talk extensively about what to do after there has been a disaster. Throughout this book we consider ways to avoid disaster. Something that is easy to overlook is that there is a cost to protecting our workloads against failures and that we have to weigh that against the cost of the failure itself combined with the likeliness that that disaster will even happen.

Too often we are told, or it is implied that disasters are to be avoided at all costs. This is crazy and should never be the case. Disaster avoidance has a cost, and that cost can be quite high. The disaster itself will have a cost and while that cost might be quite high, it is not always.

The risk that we take is that the cost of avoiding a disaster is sometimes greater than the cost of the disaster itself. There was a time when it was common for companies to spend tens of thousands of dollars on fault tolerant solutions to protect workloads whose common failure scenarios would only cause a fraction of that cost in losses. The disaster was literally less of a disaster than the disaster avoidance was! And the disaster avoidance is a certain cost, a disaster is only a potential cost. If we equate both to costs we could simplify the evaluation to a simple question such as this: *Is it better to lose $50,000 today, or to maybe lose $10,000 tomorrow?* That makes it far easier than it would otherwise be and removes most emotional response.

The phrase I like to use about overspending on disaster avoidance is that it is like shooting yourself in the face today, to avoid maybe getting a headache tomorrow.

Never treat the disaster in a disaster prevention plan as a certainty, it is not. There is only a possibility that it will happen. Evaluate and use math.

In this chapter we are going to look more deeply into how support should work in a business, improving support posture for your IT organization, learn about triage needs, discuss finding the right people for your troubleshooting team, and then delve into when to investigate, and when to fix our issues.

This is some of the hardest, most ambiguous, and ultimately most important material for us to cover.

Sources of solutions

Where do we get the solutions to problems that arise when we are system administrators? I want to start this conversation with my own career anecdote, because I think that everyone gets very different perspectives on how IT support works in the broadest of senses and understanding different perspectives is important before we start to define what good looks like.

When I first started working in IT, and for the first nearly two decades, it was an assumption that any and all issues would be resolved by the IT department. Of course, situations existed where applying patches, updated, or fixes from a vendor would be part of the process, but acquiring those patches, testing them, applying them, and so forth were always completely handled by the IT staff. Even the idea that you could ask a vendor to assist, guide, or advise was foreign let alone attempting to actually do so. Reaching out to a vendor for support was assumed to be an absolutely last resort situation and reserved only for those moments when it was believed that there was an unknown and as of yet unaddressed flaw in the hardware or software that had been identified and that it would be passed over to the vendor to fix that before handing it back to IT to apply said updates or fixes, if they ever became available.

During this era IT, and especially system administrators, were expected to know systems inside and out, be able to address reasonably any issue that could arise, and quite frankly figure out what needed to be done. No ifs, no ands, and certainly no buts. If you did not know how a system worked or what might be wrong you were expected to use knowledge and logic and get to the bottom of it. A thorough understanding of how systems worked, even if specific details were sometimes lacking, and good logic essentially always would allow you to resolve an issue.

It was not until the very end of the 2000s that I first encountered IT shops that would rely on vendors and resellers for some aspect of their support. The idea that systems were being run that the company and its IT organization (internal or external, does not matter) could not install, configure, and support was totally foreign to me and to most people I had been in the industry with for years. If you required support from the vendor sometimes, how did you not need it all of the time? What purpose was the IT team fulfilling if they were not the ones who possessed the requisite knowledge to implement and operate the systems that were under their purview? You need far more knowledge to plan and consider all solution options than it takes to fix the singular one that you finally deploy. If you need your vendor for day-to-day tasks, then you obviously lack the necessary skill sets and experience for the more critical high level decision making and that is something that a vendor cannot help with. Sadly, many organizations end up simply lacking all capability around solution planning and this explains why so many horrendous solutions that should obviously have been known to not meet the business needs get deployed.

Today is seems that the world is focused on support coming from sales organizations and vendors, rather than from IT, but this creates two critical problems. First, what is IT even there to do if they are not in the critical support path. Are they even needed? And second, how is a vendor supposed to have the range of skills necessary to do internal IT needs when all they do is support products that the make? The disconnect here is significant.

There is no magic support

There tends to be this sort of unspoken belief, often amongst managers but sometimes even from IT people, that there is a group of magic companies out there, in the technological arena that can provide more or less unlimited support in a way that internal IT cannot. It is imagined, I suppose, that these companies are not comprised of human beings or perhaps it is thought that the vendors of servers, storage, and operating systems have secret manuals full of information not released to the public that include secret codes that tell misbehaving systems to start working again.

In the real world, hardware and software vendors know essentially nothing that their customers do not know, at least not when it comes to the operations of their devices. It is no different than if you called up Ford or Toyota and asked them how to drive a race car, instead of asking race car drivers. Sure, the car companies will have some people on staff with a good idea of how a car will be driven under performance conditions, but none of them can possess more knowledge than actual race car drivers and they certainly will not have as much expertise on the ground as the driver currently going around the track.

Hardware and software vendors are just groups of people, made up from and hired from the same pool of IT talent that any other firm has access to, in fact it is likely if you have worked in the industry for any length of time that someone you know will go from working in the field to working for a vendor, or vice versa, and there is a decent chance that this will happen to you yourself. I myself have been an engineer at at least half a dozen of the big ten vendors in the industry and at none of them was there any special sauce knowledge doled out in secret. All of our customer support or even internal support knowledge was acquired the same way that it was by customers. Sometimes customers actually had more access to our own documentation than we did!

If vendors knew how to make their products work so much better than they were working in the field, they would do anything that they could to get that information out there. Almost all major vendors have extensive documentation, training, certification, and other programs in the hopes that IT workers in the field will be able to do work on their own without anything going wrong at all and being able to fix it themselves whenever possible. It is not in their interest to have people say that the products do not work properly and that the vendor has to step in to fix things. The vendors desperately want IT to be able to properly deploy, configure, and maintain in the field without vendor involvement. That looks good for marketing and it generates the maximum profits. It also makes for the best relationship with IT which, by and large, is the biggest promoter and the biggest gatekeeper to hardware and software being purchased.

Perhaps there is a mistaken belief that the best of the best IT staff will automatically be gobbled up by vendors leaving the rest of the field with only those that could not cut it. That might sound plausible, but it is anything but reality. First, few vendors have the deep pockets that people imagine and there are often customers far more willing and able to pay and attract the top tier talent. Second, vendor support work is often very different from normal IT work and few people drawn into IT in general enjoy it as much simply because the technical and business aspects are different, while highly related the two jobs differ in significant ways so there is no automatic flow back and forth. Third, extremely little support from any vendor is as technical as an IT role would be. Vendor roles tend to involve sales and account management, following detailed rules and scripts, and implementing things in a pre-defined way designed not to accommodate IT processes, but sales ones. Bottom line, the best IT people rarely want to end up in vendor jobs and those that do tend to do so not because the vendor jobs are better for their IT aspirations, but because they saw an opportunity to leverage their expertise while leaving IT as a field. Many vendors have no real IT support at all.

Remember that vendors make products, they do not *do IT*. So, what a vendor is typically prepared to help support is quite different than what an IT department should be doing for their organizations. Even when they have great IT resources on staff, those resources are unlikely to be allowed to provide true IT support to a customer. Vendors typically only want to, and often only can, provide support within the very tight confines of *operating their product as intended*. It is not uncommon at all for IT organizations to know as much or more about vendor products and how to implement them best than the vendor does. The vendor simply does not have any reason to have that knowledge. A hammer maker is not likely to know as much about driving nails using their hammers as carpenters do. Engineers at Toyota are unlikely to be able to drive their own cars as well as professional, full time race car drivers. Canonical is unlikely to know the best way to deploy Ubuntu in your organization compared to your internal system administrators.

The skills of vendors, mainly hardware engineering or software engineer (or both) are different than IT skills and even the most amazing of vendors have little reason to be especially good at IT tasks. It just is not their wheelhouse, why would they have those skills, they are not a part of their business. That is our business for those of us in IT.

What vendors should be good at, and generally are, is in knowing their products. They know when there is a flaw to be fixed, they know how companies are tending to use their products in most cases, they know what changes are coming in the future (but may not be free to disclose this to customers.) The vendor is a valuable resource, but only if kept in a logical context.

Because vendors do not sell IT, but rather products, it makes no sense for a vendor to maintain highly skilled IT resources on any scale, if at all. The idea that they would even be able to do what any internal IT can do is pretty absurd, it just does not make any sense. It is most common for vendors to employ mid-career and junior resources when they do offer some amount of IT assistance because there is simply no value to having more expensive resources on staff. Customers coming to a vendor to attempt to get IT resources cannot be treating IT as a priority or logically and therefore there is no reason to provide them expensive resources that they are not prepared to leverage: the profits on selling lower cost services are simply far greater. And customers who are sophisticated enough to need high end IT resources would know not to engage a vendor for that need.

So, there is no magic. Vendors do not know things that we do not know. Generally, they know far less, at least of what is important in our environment. They have little to no access to the necessary business knowledge to make reasonable IT decisions. They are not at liberty to approach problems or solutions with the breadth of skill and products across the industry but only the products that they provide, complimentary products, and processes that encourage greater use of their products. They have no alignment with the values of IT and are financially encouraged to work in their interest, not yours. There are no shortcuts, the level of support that should be available from IT resources is second to none, no other organization has the knowledge and mandate to support your organization the way that your IT (which should almost always include external IT resources from non-vendors) does.

There are essentially three ways that we can get support for our systems. First, we have internal IT staff. Presumably given the context of this book, that is the systems administration person or team. Second, we have external IT resources from paid IT firms that provide external IT, rather than sales, resources and get paid to be an extension of the internal IT team either ad hoc or perhaps all of the time. And then third, there are vendors and value added resellers (vendor representatives.)

We have to remember that IT is not a special case and the sellers of products are not our business advisors. Just as sellers' agents in real estate cannot represent buyers, and buyers' agents cannot represent sellers we have the same conflict of interest and opposing representation in IT. The internal IT department and any external IT service providers are paid for representing the needs of the business and are legally, as well as ethically, required to do so. On the other side vendors and resellers are paid through sales profits or commissions and are financially renumerated for representing the needs of the vendor and are legally required to do so and ethically they are bound to the vendors, not to the customer. That does not mean that a vendor and reseller cannot be friendly, useful, important, or professional, of course they can be and should be. It simply means that they are representatives of product sellers and we, as IT professionals, are representatives of our constituent businesses.

We can work together, and we can do so best by understanding each others roles and obligations. In other walks of life we rarely feel that sales people or product representatives are looking out for our best interests rather than trying to promote their wares, yet in IT this is a common point of confusion.

Of course vendors as well as their reseller representatives (correctly called VARs but often presenting themselves as MSPs, but be careful not to confuse a true IT service firm with a reseller just hoping that their customers do not question the name) can be valuable allies and we should not completely discount them. They can be part of our solution process, especially when it comes to getting access to special tools, beta components, patches, release information, bug fixes, replacement hardware, and other components that come from the engineering group, rather than aspects handled by IT.

Visualizing what IT handles and what engineering handles

Even as IT professionals it is sometimes confusing to understand what falls under IT and what falls under the engineering (of a vendor or a software engineering department.) But the answers here should be quite simple.

As IT professionals, we use products made by others and assemble those products into complete solutions. We do not make Linux, we do not create a new database engine, we do not write the applications that our businesses run. We also do not form sheet metal, run chip fabs, or otherwise build computer hardware components, but we might buy parts and assemble them into a computer in rare circumstances (like how a mechanic might assemble car parts, but not actually pour metal to make them.)

Vendors are responsible for making the tools and products that IT is responsible for then using and operating. Vendors write the operating systems, we in IT install them. Vendors make the applications, we install them. Vendors are involved in making products, not solutions. IT solves organizational challenges, it does not make products.

If we were talking cars, perhaps it is more clear. Car vendors build cars. Customers ride in cars. The two obviously have an important relationship with each other, but it is pretty obvious that designing and building cars is a very different task than plotting a course and driving to a destination. We can obviously see the vendor making tools for us to use, and we are car buyers using those tools to solve transportation challenges. Apply this logic to IT and voila.

Real support, the most important support, is always going to come from our own IT team (which always includes external IT staff as well.) Our own IT staff not only has the broad range of business knowledge necessary to make key support decisions, but also has the range of potential solutions to work around vendor limitations. It is common that solutions are bigger than can be addressed by a single product vendor in isolation.

To give an analogy, if you were a logistics firm and you needed to get a shipping container from New York to Los Angeles and your truck broke down, of course you might ask the truck manufacturer for information on repairing the truck, but you would not stop there. You would look at replacement trucks, renting a truck, other truck vendors, consider the cost of using another logistics partner to ship on your behalf while you are down, or consider switching to rail or sea transport! Of all of those things, only repairing or maybe replacing your initial truck is within the potential scope of support of the truck manufacturer, and even repairing it from them is much more limited in scope than you would likely get with a mechanic. A skilled mechanic might be able to propose partial functionality, third party parts, or alternative fixes that are not possible from, or approved by, the original vendor. The original vendor has value here, important value, but only a tiny fraction of the value that the overall department would have.

It takes a much broader scope to properly deal with most disasters. Rarely do we want to just sit on our hands waiting for a vendor to determine if an issue belongs to them or not, and then decide if you have valid support or not, and then decide how they are going to deal with it. Even a great vendor, with great support has their hands dramatically tied compared to what IT staff should be doing. IT has, or should have, the scope to do whatever is necessary to protect the company. That might involve engaging the preexisting vendor or it might involve working around a vendor, or perhaps it just involves coordinating multiple vendors. Even when a vendor does need to be involved, IT should be overseeing that vendor.

IT vendor managements

More than any other IT department, system administration often has to interact with vendors and has a greater level of need to oversee them directly. At the same time, system administration is also the department most likely to not have any vendors, at least not in the traditional sense.

Vendors should not be thought of as a department, but rather more like a specific tool. A tool that needs to be overseen and used when appropriate, but at the direction and discretion of IT. A vendor on their own lack's direction and control.

Managing the vendors for the hardware, software, and services of an IT department should be an every day task of that department. The vendor relationship at that level is important as this is the level at which technical know-how should be exchanged. For us, as Linux system administrators, this means direct contact with our operating system vendor counterparts at vendors such as IBM Red Hat and Canonical who can keep us apprised of patches, upcoming changes, release dates, security alerts, and so forth and, in some cases, may be focused technical resources for us to lean on.

Systems administration may have many other types of vendors as well. We may have to work with server hardware, storage, database, and even application vendors at times. That there are so many potential vendors highlights how critical IT management of the vendors are. Without IT oversight, there is no coordination between those vendors and no mandate for them to collaborate or to work towards a common goal. The mandate to work towards the good of the business lies solely with IT in this case. It is for IT and IT alone to ensure that the vendor resources at its disposal are used for the good of the business when appropriate, rather than being sales efforts for the vendors.

The best practice is that support should come from within. Fundamentally, at the core of it all, we should see our IT team (inclusive of internal staff and external staff) as our solutions team both to design our solutions up front and to deal with them when something goes wrong. When vendors are required to be part of the solution process they should be engaged, managed, and overseen by the IT team and managed as just another resource.

Triage skills and staff

Most companies fail dramatically when disaster strikes because triage processes either do not exist or are too poor. The skills to run the business day to day are different than the decision-making processes needed in real time in a crisis: there is little time for meetings, almost no ability to consult with different parties, and planning is out of the question. When in this mode we need someone leading who is trusted, handles stress well, and is a perceiver rather than a planner - someone who thrives running with rapid decision making and does not need to have planned their events ahead of time. Planning is excellent and as much as is reasonable should be done ahead of time, but everyone involved from junior IT staff to senior executive staff should understand that true emergencies cannot be adequately planned for, and real life will involve many unknowns that have to be evaluated on the fly.

Our first process when there is a disaster is to head into triage mode. We need to know what exactly is not working, what has happened, what is the impact - basically we need to know the status, of everything. Is this something we think that we can fix in minutes? Is this going to require some investigation? How are people impacted? Are we losing money, productivity, customers?

There are so many ways that we can be impacted by a disaster. Being able to quickly get a grasp of the business effects, how does each department play into the big picture, which teams can work just fine, which teams are dead in the water, are there teams that are functional but limping, and so forth is absolutely critical and can mean all of the difference between everyone just standing around being unsure what to do and a triage manager getting things fixed right away. We need status, and we need a lot of it, very quickly.

In most cases a surprisingly small amount of our time is actually focused on solving a technical problem. This may be because it is simply a hardware failure and we just have to wait for replacement hardware to arrive. Or maybe it is complete software failure and we just have to rebuild our systems. Sometimes deep technical investigation has value, and sometimes it requires a lot of know-how to the cause of an issue, but this is not the majority case. We are much more likely to have a relatively quick fix, or at least quick in terms of the amount of administrator time is necessary to use during the process. When fixes take a long time, it is most typically because there is a third party that needs to be waited on.

Typically, we are going to mentally focus on the technical aspects of an outage. Other aspects of most outages are more important. Some organizations have operational triage experts who step in and handle these aspects of an outage allowing us as system administrators to focus purely on the technical aspects under our auspices. For most businesses, though, dealing with an IT disaster is going to require IT oversight from beginning to end. In the majority of cases, the very team that we would hope would step in to assist IT in solving issues and managing the triage of the operational environment gets in the way of finding solutions rather than being part of the solution.

I can give status, or I can fix things

Everyone who has ever dealt with any kind of outage, disaster, or what have you impacting a business knows that the expected behavior is not for executives and management to jump in and start to find ways to protect the business or to run interference to assist you in dealing with the problem that you are suddenly tasked with; but instead the very people we depend on to create an environment to minimize impact and to make us effective almost universally turn on us and begin demanding explanations, status reports, updates, estimates, promises, and miracles all of which are pointless at best and completely undermine the business at worst.

Even under ideal conditions reports, status, updates, measurements and the like have a cost and come at the expense, even if only a small degree, from productivity. During a disaster, it is rare for anyone except those in the most critical positions of attempting to mitigate and fix the disaster to have any ability to provide status. So, the most expected event is that everyone will descend on those few, critical positions and demand status updates.

There are several problems here that are all really obvious to us, working in these positions, but often lost on those who are now experiencing an impact to their productivity. We have two tools at our disposal to help with this. One, show this book to those in management and ask them to take the time to understand the situation. The second is planning. Make sure that as a part of your disaster planning and preparation process that there is training for management, and a policies and procedural plan, as to how status will be given, who may ask for it, and from whom. Consider designating and official spokesperson for IT (and other departments) who can spend all of their time giving updates as they are not involved in any other aspect of the disaster recovery efforts. Perhaps they will run a war room in person, or maintain an email messaging group, they could manage a chat room, or head a conference call that others can dial into as needed.

Then any updates that exist can be passed, proactively, to this mandated reporter(s) and they can maintain that status for the organization. The entire organization should understand the critical nature of having a mandated point of reporting so that the team actually attempting to solve the issues and get the company back to full functionality can spend their time saving the company rather than reporting on its failures. Obviously the business has some business needs to know as much as possible about what is happening. More impactful is political problems internally as managers feel that they, too, have to provide status that they cannot have and many layers of organizations will have people acting emotionally and potentially willing to cause significant financial damage in the hopes of appearing to be concerned or just to satisfy their desire to know more than anyone can really know.

When training management as to why IT (and other departments) cannot provide extensive updates we need them to understand why we, as the people attempting to remediate the issue, cannot be spending time giving status updates.

We don't have any information to give. Fundamentally, this is the biggest piece of the puzzle. While we might have a simple answer like *the replacement part is scheduled to get here tomorrow* typically we know nothing about how long something is going to take to fix. Most things in IT are fixed essentially the moment that we know what the issue is. Until it is completely fixed we normally are only working from hypothesis. Pushing us for information really just ends up being analogous to demanding that we feed intentionally false information because have nothing else to provide. It is like torturing a prisoner that does not know any information, but if you torture them they are likely to just make something up in the hopes of the torture stopping.

We are busy. If things have not already been resolved, chances are we are completely engaged in trying to get them resolved. All of the time used to give status has to come from the time working on resolving the issue. It is more than simply wasting time, it is also causing interruptions and demoralizing those trying to resolve things. It sends a huge organizational message that the issue does not matter very much and that the efforts to get it fixed are not appreciated as they should be. It makes IT wonder *if management is not prioritizing getting things fixed, why would we?*

Political Risk. In attempting to get everyone who wants to be able to *plan* a disaster information to work from those working to resolve the issue are generally in a very tough position of having to guess quite heavily as to when things will be finally resolved. Most organizations handle this uncertainty, which cannot be helped, quite negatively. Intentional bad information is often rewarded, honesty is often punished. Putting your information providers into a position of potentially having to *just tell people what they want to hear* or *providing inaccurate information for political protection* means that the business may then operate with bad data causing unnecessary additional financial loss. It is a terrible time to be pushing for bad data over no data, yet it is when it is most likely to happen.

Priorities. If the organization starts to prioritize, from management, the perception that reports, status meetings, calls, and other things that do nothing to resolve the issue to get the company back running at full speed are more important than finding a solution, this will naturally, and absolutely should, change how IT or any other department tackles the problem. If any issue is so trivial that we have meetings to discuss timelines instead of fixing the system, then overtime, rescheduling family events, even skipping lunch all become absurd IT sacrifices that obviously have no value to the company. We would never do any of those things just for some meeting, and if that meeting is more important than finding a solution, we have a relative value assessment that gives us a lot to work with.

So how should management act? Management needs to do all the opposite. If status even matters, and we understand that it generally does, then have someone that is not involved in the remediation handle those communications. Keep priorities clear. Assign teams to run interference and keep all interruptions away from the team attempting to fix broken systems. Have people bring them food, drinks, coffee, run errands, whatever is needed - show that they matter, a lot, instead of suggesting that they do not matter. Do not punish messengers for delivering bad news, reward them for honesty. Empathize and think about the best results for the business.

All things that are easy to say and hard to do when disaster has struck. Plan ahead, have these discussions, make a plan, get executive sign off before things happen. Have an action plan to put in place that says who is in charge, how things happen, and so on.

Our first stage of triage is an assessment. Do we have a plan to get back online quickly? We need to know the situation as it stands, and we need to then relay that information somehow to management. From here, things get tricky. There are so many variables that teaching triage is anything but easy. Someone who excels at triage needs to be able to take the situation as it is and gauge a range of issues from ways to fix the existing problem, potential options to work around it, and in many cases, how to modify the organization to best react to it.

This is very much a *thinking outside of the box* scenario. We need, at this point, to look at the big picture and figure out how to best keep the finances of the business running. This might seem like a management task, and again, ideally it is, but IT should play a role as we have certain types of insight that might be lacking elsewhere.

Outside of technical fixes, mitigation strategies will vary broadly based on the type of business, type of impact, and so on. Should we send everyone to get coffee? Maybe plan for a long lunch? Get everyone home on vacation now to save on insurance because it is going to be a while before anyone is productive again? Perhaps moving people to paper or from email to instant messaging? Use downtime from one type of task to focus on others. Perhaps we do a deep office cleaning while people have the time - unplug those cables and really get the place clean.

Big emergencies can present big opportunities as well. I have seen several times when companies have used catastrophic outages as chances to enact sweeping updates and changes that would require big time approval or large downtimes normally, but which could be slipped in during an outage that is happening anyway. I once even had an ISP based outage that was predicted to last so long that a team ran from New York to Washington, D.C. with a truck, put a rack of servers into the truck, and ran to a new location that was waiting and ready in New York and pulled of a datacenter migration that had been *indefinitely on hold* because of the necessary downtime and was able to bring workloads back online from New York before the ISP was able to restore service in Washington, D.C. A somewhat minor outage was turned into a huge *win* by the department. A large project that was struggling to get scheduled and approved was pulled off purely as a bonus while the team was able to simultaneously enact a significant *fix* by going to an alternative datacenter to overcome the limitations caused by the ISPs leased line at the original location.

Triage is hard because it requires that we be creative, open to alternative ideas, able to avoid panic, think broadly and outside of the box, and do so with little planning or preparation. If your organization does not have someone suited to this role, and relatively few do, then this is something you should be outsourcing, but your system administrator is one of the most likely candidates for it as the skillsets and aptitude of administration tend strongly towards triage and disaster recovery as compared to engineering and planning skillsets.

In many cases outages result in far more than a single workload being inaccessible and prioritization within the technology space is also required. Your triage person or team needs a deep understanding of how workloads interrelate to one another, which ones depend on others, what can be worked around, what can be skipped, and how all of these workloads relate to the business itself. Only by knowing the scope of the technology as well as the business can anyone provide valuable insight into what to fix, in what order, and how the business can potentially work around things.

I wish that we were able to provide concrete guidance as to how best to survive a disaster, but we cannot. Disasters come in so many shapes and sizes, and the ways that we can deal with them are more numerous still. You are best served by learning how to think, how to react, and being as prepared as possible for any eventuality.

Staffing for triage: The perceiver

I had the great benefit of once working for a large IT department where deep psychological analysis was part and parcel in the general managerial processes. This might sound like a terrible thing, but the approach was excellent and the company used proper psychoanalysis to learn how people work, how they would work together, who was expected to be strong or weak in different areas, and how best to combine people to achieve the best results.

I do not want to go in to detail in all of the ways that these techniques were or could be helpful, but one specific thing that I want to touch on is the idea of the Judger and Perceiver scale of the Myers-Briggs test. I am not a psychologist, so I recommend that you research the test and its interpretation on your own, and understand that like with all psychology it is both greatly accepted as well as heavily disputed as to its efficacy. I will not argue for or against here, but only say that understanding the judging to perceiving preference pair is highly valuable.

I generally describe the Judge as a *planner*, one who likes to organize and put things in their place before an event occurs. The Perceiver is more a *reactor* or a *responder*, someone who wants to take the world as it comes and react to it in the moment.

In our world, engineers and most managers are judgers. Their role, their value comes from thinking ahead and organizing the business or the technology to do what it needs to do. Perceivers tend to excel at being administrators, rather than engineers, and are exactly who you need to have at the ready when there is a disaster. Your perceiver personalities are your natural candidates for those who are likely to be good at handling triage operations and thinking on their feet. Humanity is naturally diverse to complement one another to handle multiple aspects of life and this is a great example of that.

There is far more to being the right person or team member for the job than just fitting an aptitude on a personality test. The Myers-Briggs assessment is simply a tool for identifying who might be strong or weak for different positions, and for explaining how people tend to think and feel. For me, discovering that I was a strong perceiver and a weak Judger was influential in helping me to understand myself and how to communicate things about me to other people. It also gave me tools to help me to understand other people in my life so that I could communicate better and set expectations better when they operate differently from me.

Whether your company uses a formal process, or you just take an online survey to learn about yourself, I recommend the Myers-Briggs and similar tools for simply helping you to understand yourself better, if nothing else. The better you know yourself, the better you can be prepared to succeed where you are strong and to ask for help where you are weak. If you are a team leader or manager, this kind of information can be useful in helping to understand your team better and how they can work together to be stronger.

Do not be tempted to read too much into psychology tools or try to apply them too broadly. By and large these tools are best when applied to yourself and when you approach them openly and honestly with a desire to learn not about your own strengths, but about your weaknesses and use them for self-improvement. Remember that a test of this nature is not about comparative results, one result is not better or worse than another; all people live on scales and neither end of the scale, nor the middle, is a good or bad result. Strong teams, however, are generally built from a variety of different combinations of aptitudes and personalities to cover many different needs.

I hope that I have just filled you with confidence and ideas that you will use to take your triage process to the next level rather than causing a panic attack about the complexities and uncertainties of dealing with disasters - that should not be the takeaway here. Use any panic that you are feeling now as motivation to immediately start your planning and documentation, and to kick off conversations with management to get stakeholder buy in now. Make it a priority and you can quickly move from being unprepared to being at the forefront of businesses ready to respond best to almost any circumstance.

Do not feel that the person who has to handle triage has to be you. Maybe you are the best person for the job, maybe you are not. Very few people have the right personality and triage is a very special aptitude to have. What is important is identifying your triage person or team, whoever they are. Everyone has a role to play, find your place, and find the right people for the roles you have to fill.

Our best practice in regards to staffing is to identify your triage people before there is a disaster and have them documented and in place to take over, and empowered to take over, when the time comes. Do not let the decision process of finding someone with a triage aptitude wait until the clock is ticking on your downtime, and do not let politics become the focus, rather than solutions, when time is of the essence.

Logical approaches to troubleshooting

Possibly the hardest thing that we have to do as system administrators is troubleshoot problems. It is one thing to be able to deploy a system initially, but a very different thing to be able to troubleshoot it when things start to go wrong. With systems administration there are so many places where things can go wrong for us; we sit at the nexus of so many technologies and so many possibilities that tracking down the source of issues can be very challenging.

Not surprisingly, experience makes this far easier than anything. The more you get experienced with maintaining and managing systems the more likely you are to be able to quickly *feel* your way around a system and often just sense what might be wrong when things get tricky. Nothing really trumps just knowing how a system will react when things are healthy and being able to sense what is wrong based on its behavior. Senior diagnosticians are often brought in for exactly this reason. With enough experience often you can just feel when an index, a cache, a disk, or lack of RAM is the issue.

Short of sheer experience, our next best tools are a deep understanding of our own systems and how they interact, deep knowledge of technology fundamentals as they apply to our situation, and logical troubleshooting.

Of course, in many situations failures are going to be quick and obvious. The power is out, a hard drive has failed, a key database has been deleted. There is nothing to track down, only things to be solved. Other times, though, we get complicated issues that could be caused by almost anything and we may have to track down something truly difficult to pinpoint.

It amazes me how often I am brought in to assist with troubleshooting only to find that the work that has already been done is haphazard and is often redundant. Of course, at times, some guesses at easy to test failure points or early tests of known common fail points can speed discovery, but we have to be careful not to lose track of what we are doing and learn systematically from what we test.

Stories of troubleshooting

A benefit of being an author is getting to regale you with tales of my own historic troubleshooting and there is no one to roll their eyes or cut me off and you cannot just walk away when you are bored. So here we go.

One time I was called in to work on a system that was used for an extremely low latency application and the team had discovered intermittent problems with the application receiving responses from other systems. The issue would arise every so many minutes that a response would be received several nanoseconds later than it was expected to have arrived. Yes, nanoseconds! Nothing broke, no results were wrong, but there was just this tiny delay, and not very often.

After much research, we finally found the issue through a combination of research and system understanding. Identifying what was happening was eventually done by hours of staring at a top monitor and looking for processes that were active around the time that the delay would occur.

Eventually the process possibilities were whittled down it was discovered that a memory garbage collection process that was soon thereafter discontinued in the kernel was using excessive system resources in its default settings and causing the system to halt for just a few nanoseconds while it processed memory for cleaning.

I was able to address the issue by setting the garbage collection process to only clean a portion of RAM on each cycle allowing it to work much faster. We ran into the issue only because we had so much physical RAM in the server that the garbage collection process took measurable time, something generally not expected.

In this case, good research and patience were certainly important. Being able to *feel* the delay in the system based on the measuring tools (no human can actually feel a delay that short), and then using logic to determine how a process doing memory garbage collection could, and would, impact a process of this nature had to come together to make troubleshooting possible. Without a deep understanding of how the system runs, it would be out of the question.

When troubleshooting anything I find that there are two key techniques that I am telling people to use over and over again. The first is to be systematic and to work from one end or the other. Avoid hopping around and testing at random. If you are testing network connectivity, for example, start at the near end of the stack and start building up a base of knowledge based on testing.

In the networking example we can start with checking if our network connection is plumbed? Does it have an IP address? Can it ping the gateway? The ISP? A public IP? Is it able to resolve DNS? Can it reach the system to which it is supposed to connect? Can it reach the right port? Does it get a proper response?

Instead of jumping around and testing different pieces, working from the nearest point and exploring helps us to understand exactly when things fail, and it tells us quickly.

The other key technique that I always teach is *work from what you know*. Basically, establish your facts. There might be many things that you do not know, but you cannot worry about those things. The unknown will always exist. There are always things that you can know, though, and these we have to establish and work from. Use the facts that we have to build up a larger body of knowledge by finding more and more things that we know for certain.

For example, if you can ping a remote server then you know that you have working plumbing, working routing, and that all of the equipment between you and that remote server is all working. Or if you know that a specific database is up and running and working properly, then you know that its operating system is also up and running, and that the hypervisor that it is on is up and running, and that the bare metal server that that hypervisor is installed on is up and running.

It is always surprising to me how often people who are troubleshooting will, take the time to establish the facts, but then question them again. In the example above, they might decide that because they are seeing an issue that maybe the hypervisor has failed and go to check it again, even knowing that they just proved that it was still working. Or in the networking example, convincing themselves that they need to check on the status of a router that they just used to prove that networking could pass through it correctly.

Going down the proverbial rabbit hole and making yourself (or your team) prove over and over again that something is or is not working that you already know the status of is a waste of time at best and can be extremely frustrating for those working with you. Once you start the pattern of reestablishing what you already solidly know, rather than trying to determine something new, you will likely continue doing so. It is very tempting to focus on those things and lose sight of growing the body of knowledge pertaining to the issue at hand.

I find that writing down what we know to be true, whether as a starting point or from investigation, is a good tool. If we feel that we have to test again something that we already proven, then we have a problem. Why do we not trust what we have already proven? Why did we feel that it was proven if we now doubt it? What is the point of testing if we are not going to trust the results of the testing?

If we test, prove, and then we are going to spin our wheels endlessly. This is a needless waste of time, time that we cannot afford during an outage. Either we need to approach what we consider to be fact differently, or we need to trust the assessment. By establishing trusted facts, we can use them to narrow down the possible issues.

Troubleshooting is hard both because it is very technical when many factors are probably unknown, and because it is emotional being done when there is stress and sometimes even panic. This is generally compounded by additional needs and pressure from our organizations as well. Keeping a clear head is key. Breath, focus, get caffeine, talk out the issue, post on technical peer review communities and forums, and maybe even engage the vendor. Have your resources written down as it is easy to forget steps when stressed.

Technical social media in problem solving

For more than two decades one of my strongest resources for dealing with serious issues has surprisingly been technical social media. I do not mean traditional social media outlets, but forums built solely for technical exchange of ideas. When faced with design challenges or, far more importantly, broken systems that need to be fixed, I have found that for me and my team that posting those problems to a forum to be absolutely invaluable.

The reasons why this is so important are not always evident. The obvious benefit is that there are many seasoned professionals happy to provide a fresh set of eyes on your problem and may easily spot something that you have missed, or they may provide insights or suggest tools of which you were not aware.

The bigger benefit, however, is in the process of requesting help. More times than not the act itself of having to write out what is wrong, the need to express it clearly in writing, and documenting the steps that I have followed will reveal to me, even before anyone has a chance to respond to me, what might be wrong. Writing down my steps encourages me to also be more methodical and to think about what obvious questions others will ask me causing me to attempt to fill in the gaps, follow good processes, and document far more than I would normally do for myself.

I use this process of posting for public review for my own team, as well; and I make others do it. Of course there are details that cannot be posted publicly, and sometimes the entire process is too sensitive to be public at all even without identifiable details, but generally at least some degree of the disaster can be reviewed publicly for assistance. Using these kinds of forums for communicating amongst my own team works wonderfully and encourages the same good behavior of thinking through what has been done, approaching how to explain the problem differently, and forces clearer documentation of the troubleshooting process between team members because even the documentation process is being reviewed publicly in real time.

This same process then provides documentation and automatic timeline of troubleshooting to use for a post mortem process. I often also invite post mortem review, generally informally, via the same mechanism. People are always happy to critique decisions. You have to be prepared to accept a bit of a brutal review.

No one is a bigger proponent of the value of public peer review in focus environments than me. I have been championing this movement since the late 1990s and spent a great deal of my career working in the public eye through these communities. It has taught me many things that I would have never been exposed to and it has forced me to work differently knowing that everything that I do will be examined, reviewed, and questioned. Being prepared to explain every decision, to defend every outcome, to resort to logic and math because your arguments for or against a decision themselves are permanently recorded for review makes you rethink what you say and, I believe, pushes you to be better at everything that you do. It is easy to make an irrational argument when you think that no one is going to question you or that no one will notice.

Troubleshooting best practices are simple: be methodical, document everything, and line up your resources before you need to rely on them.

Next, we are going to look at when it even makes sense to go through this process or if we should simply start over.

Investigating versus fixing

When we start working to deal with an outage, data loss, or other disaster, the natural inclination is to focus on finding a root cause, fixing that root cause, and then getting systems back into a working state. It makes sense, it is the obvious course of events, and it is emotionally satisfying to work through the process.

The problem with this process is that it is based on a few flawed beliefs. It is a method derived from things like getting your car or house repaired after there is damage or an accident. The underlying principle being that the object or system in question is very expensive to acquire and in relative terms, cheap to repair.

It also focuses on the value of determining why something has occurred over the value of getting systems up and running again. The assumption is that if something has happened once that it is expected to happen again and that by knowing what has failed and why that we will be able to avoid the almost inevitable recurring failures in the future.

Of course, in IT and business systems, typically the cost to build is less than the cost to fix. More importantly the cost to build is more predictable than fixing. We should, if we have planned well and documented, be able to implement a new system in a known about of time with extreme reliability. Fixing an issue may or may not happen quickly, it represents a lot of unknowns. A fix may take a very long time, and the fix may not be reliable. Root cause analysis can be time consuming and unreliable.

In most cases getting a system back up and running as quickly as possible carries great value, and while determining why a problem has occurred and finding a potential means of avoiding it in the future has little value. Business infrastructure experiences extraordinary change rates in everything from hardware to software to system design. A hardware failure that has happened today is unlikely to repeat in the same way. Software will likely be patched, updated, and modified quickly and worrying that old bugs will return is possible, but not a scenario worth a large amount of concern.

If we were dealing with a car, house, road, bridge, or other large object of this nature failures are likely to recur as the system faces small, if any, changes, over time. We need to determine the point of failure, determine the risk of recurrence, and find a way to protect against it. It is hard to separate ourselves from this mindset.

We have to evaluate the value to the business. What is the value of getting the system back up and running? What is the value to finding the root cause? We have to compare these values and, most of the time, we will find that solutions triumph over investigation.

The fix versus investigate decision gets more and more weighted towards fixing when we have more modern infrastructure with imaged systems, state machines, and infrastructure as code. The greater the quality of our automation, the faster and less costly it is to recreate systems and the lower the value to investigating an issue.

We also have to consider the possibility that with the right infrastructure that we can recreate a mirror system to use for diagnostics when it is deemed necessary. We can create an initial rebuild to get systems back up and running and build a mirror, if it makes sense, to use in order to attempt to recreate the failure and determine if there is a way or reason to protect against it in the future. Spending time attempting to identify the cause of and fixing an issue during an outage may not be the best way to accomplish that goal, even if it is deemed to be necessary.

It is all cost analysis, but one that has to be done very quickly. The unknowns are very difficult sticking points here because the time to determine the root cause is completely unknown and may take minutes or days.

This logic of just starting over applies to desktops as well as to servers. Even end user workstations have every opportunity to be interchangeable and designed in such a way to allow for rapid redeployment. If we are using images, automated software installation, and other automation it is quite standard for desktops, laptops, or whatever we are using for end users to be able to be deployed new often in a matter of minutes. A fresh deployment is more than just getting these systems back up and running with maximum efficiency, it also provides an opportunity for a completely clean installation as a bonus. Any cruft, malware, corruption or similar that might have happened on the machine will be wiped away and the end user will start as fresh as if their machine had just rolled off of the assembly line. In this way, we get a silver lining: a fresh rebuild process that we often would struggle to schedule but, ideally, would be doing on a semi-regular basis anyway.

Rebuildable systems, whether desktops, servers, or cloud instances, all mean that we only need available hardware to be able to recover and move on from most disasters. That also means, assuming that our backups are stored somewhere online or offsite, that we have the ability to walk into a new site and rebuild our entire company there. That flexible and level of comfort is a game changer - something very few companies were able to consider even just a few years ago. Knowing that starting over *from scratch* is always a possibility makes us think about everything that we do in disaster recovery completely differently.

In my own experience, even fifteen years ago, long before we had the automation and complexity of today's environments, we were moved almost entirely to *restore fast, only examine what we can after things are back online* and the ability to do so has only increased since then. Today we should, in almost all cases, be thinking of rebuilding from scratch as the default assumed starting point and we should only resort to more complicated forms of recovery when the situation demands it; and we should carefully evaluate why a situation today would demand it. That does not mean that rebuilds should be the only tool on our toolbelt, being the majority case in no way implies that they are the only correct solution, just that they are the most likely to be correct.

Traditionally there was a stigma to rebuilding, as if it meant that we had given up or were in over our heads. We have to fight this incorrect emotional response. The right way to recover an environment is whatever creates the best situation for the organization, as a whole. As with everything that we do, emotion plays no role here. This is a financial and risk calculation only. We do what is best for the company, and that is all.

Best practice:

- Evaluate each situation, but when in down err on the side of a clean rebuild. No one size fits all, but rebuilding should be the better option in most cases.

- Design systems to be able to be rebuilt quickly, easily, and automatically.

Summary

Disaster recovery, triage, proper staffing for emergencies, organizational preparedness, managerial oversight of processes during disaster situations, and every other aspect of a critical failure scenario is hard, scary, and stressful. How companies decide to handle these times often determines which companies survive, and which ones fail. We have to have the right people in place, as many organizational processes and procedures as possible, great documentation, deep knowledge of our systems, and the flexibility to do whatever it takes to make the business successful through hard times to truly succeed.

Every company struggles with these same things. These are not simple tactics that we can apply overnight. It requires buy in from organizational stakeholders, it requires professionalism and planning not just before events transpire, but maintaining those processes and professionalism during times of panic when stress causes almost anyone to act irrationally. On one side we can view this as stressful and difficult, but on the other we can recognize it for what it is: a place where nearly all organizations struggle, most fail, and our greatest opportunity shine.

Disaster planning and disaster recovery are easily your best chances to take the system administration role and grow it to something larger than the role and, often, larger than the IT organization itself. You cannot effectively isolate disaster preparedness to solely the IT department; it requires cooperation across all departments. Systems administration can lead, rather than follow, and make IT the core business unit that it always should have been.

I realize that for many, the capabilities and scope of the IT department are deeply mired in politics and cannot easily be made fungible. Challenge yourself to at least evaluate, to consider, what it would take to push your organization in new directions. No organization should force IT to wear the sales hat just to convince the organization to do what is right for itself, but here in the real world our ability to sell ourselves, our department, and our ideas is often the difference between being heralded as the savior of the company, or just ignored.

Remember that solutions come from you and not from vendors. Keep vendors and their scope in mind and remember that while disasters represent a great chance for you to save your organization, they also represent a huge opportunity for vendors to find new sales opportunities at a time when fear and emotions make well considered planning all but nonexistent. Rethink how you view vendors, make the context of the support relationship utmost in your mind. Always know who represents your interest and who is looking for where you can best serve them.

The postmortem

Putting a sidebar in the summary may seem out of place, but I think postmortems are much like a summary of a disaster themselves. So why not discuss them here?

Most companies skip the all-important task of performing a post mortem. Postmortems are not about placing blame, and if that is what your company wants to use them for then it is probably better to avoid them, but in a healthy organization they serve as a critical learning tool on many levels.

A good postmortem is going to expose mistakes in system design, documentation, planning, policies, procedures, and just about any other aspect of our systems. It should also aid us in identifying people who are strong or weak during a crisis. We should be using our postmortem processes to discover where we were weak and how we can improve, or potentially to determine not to change at all.

A postmortem should also allow us to evaluate our decision processes that led to where we are today. This is almost universally overlooked and is actually where our true value of a postmortem exists. Changing the outcomes of individual plans or decisions is good, but generally minor, but finding entire decisionmaking processes that are failing gives us an opportunity to make changes that impact all decisions going forward.

Learning how to make decisions is important and few organizations or people ever focus on the quality of the decision-making process and even fewer companies track it and attempt to improve it over time. This is a huge lost opportunity. Decision making is something that happens over and over again. Making better decisions on a regular basis is vastly more important than fixing individual decisions.

Postmortems need to dig into *why did we decide to do what we did* and then examine if that was a good decision, but we must not fall into the trap of applying current knowledge to old decisions. The *if we had only known* game is a dangerous one. We have to evaluate what we could have known at the time and determine if we researched enough, thought it through properly, applied true business goals and so forth.

It is easy to project knowledge after the fact and say *see, things failed, we lost money, that is someone's fault*. Emotionally, that feels like it must be true, but it is not. It could be true, but that is not likely a productive thing to determine. Bad things happen, risk is part of business, there is not always someone at fault causing these bad things to happen. Working in IT we deal with calculated risk every day. What we need to know in a post-mortem is if we calculated it correctly and took the right chances.

A great example of good risk is when we have to travel from New York to Los Angeles. We can take a plane or we can drive. If we look primarily at our safety during the trip, flying feels scary and driving does not seem scary at all. Yet the chances that we die in a car crash over such a long distance is many times higher than the risk of dying in a plane crash. If we took a plane and the plane did indeed crash, it would be tempting to use the new knowledge that that particular flight was going to crash to say that the right decision would have been to drive, but that is wrong. The flight was still the right decision. Both approaches had their risks and the flight was the vastly lower of the two risks. We use knowledge that that particular flight would crash because there is no way to have known that ahead of time. We were playing it safe, we made the right call; but no option is without risks and punishing people for making good choices is a terrible outcome.

People need to be rewarded for making tough choices, especially when they make the right rough choices. If we look to place blame, we risk punishing people for simply having made any decision at all, and if we do that we push them towards avoiding the decisions that protect the business to avoid being in the line of fire for false blame. Of course, if truly bad decisions were made, we want to discover that. It is just a very difficult task to maintain a focus on organizational and personal improvement rather than using postmortems to find scapegoats or deflect culpability.

Used correctly, postmortems are a powerful tool. Used incorrectly, they are a waste of time or potentially worse. Even if the organization lacks the capability of performing good postmortems, do so just within IT. If IT itself has politics that make this impossible, do so just within systems administration. Even if no one else participates, do it for yourself.

Document your post-mortems. People have a tendency to remember disasters negatively and to emotionally assign fault where none existed or where it is not deserved. Keep post-mortem documentation on hand as it is often useful for defending people, teams, or processes later. Good documentation, even after the fact, is a powerful tool.

Remember that a post-mortem does not just need to ask *could we have avoided this disaster* but also *even if we could have, should we have?* Often the cost of avoiding disaster is greater than the risk cost of having the disaster. A post-mortem should cover the decision-making process, the decision itself, as well as the response to the disaster.

I hope that the ideas and concepts in this chapter will help you to break out of the mindset of traditional roles and to tear apart *the box* and let your approach to disaster recovery reflect the best of what you and your organization can muster.

There is an obvious lack of discussion around Docker and other modern container technologies for a book on Linux. This is not an accident, it's actually by design. The reason for this is that Docker and its kin are application container technologies that leverage other technologies that we have already addressed, and their practices are their own concern. At the system administration level, application containers are simply another workload - one that happens to use Type-C virtualization and manage its own dependencies and updates. Docker or other application container management is beyond the scope of this book as well as general system administration.

In most cases, the system administrator is responsible for managing these technologies, but they're not special cases. Workloads are the same as we have discussed throughout this book. Even though they may have their own names, their own mechanisms, and their own management tools, all of these things are still governed by general case guidelines and rules that we should already know as system administrators. You will need to learn, if you're going to work with these technologies, and you very likely will if you're a system administrator today, many things that pertain specifically to the application container platform and management tools that you'll be using and apply that knowledge to what we've already learned.

Best practices focus on learning the general cases, the rules that always apply, then figuring out how different techniques, technologies, and products are covered by the general cases.

Index

X

Z

Packt.com

Subscribe to our online digital library for full access to over 7,000 books and videos, as well as industry leading tools to help you plan your personal development and advance your career. For more information, please visit our website.

Why subscribe?

- Spend less time learning and more time coding with practical eBooks and Videos from over 4,000 industry professionals

- Improve your learning with Skill Plans built especially for you

- Get a free eBook or video every month

- Fully searchable for easy access to vital information

- Copy and paste, print, and bookmark content

Did you know that Packt offers eBook versions of every book published, with PDF and ePub files available? You can upgrade to the eBook version at packt.com and as a print book customer, you are entitled to a discount on the eBook copy. Get in touch with us at customercare@packtpub.com for more details.

At www.packt.com, you can also read a collection of free technical articles, sign up for a range of free newsletters, and receive exclusive discounts and offers on Packt books and eBooks.

Other Books You May Enjoy

If you enjoyed this book, you may be interested in these other books by Packt:

Mastering Embedded Linux Programming - Third Edition

Frank Vasquez, Chris Simmonds

ISBN: 978-1-78953-038-4

- Use Buildroot and the Yocto Project to create embedded Linux systems
- Troubleshoot BitBake build failures and streamline your Yocto development workflow
- Update IoT devices securely in the field using Mender or balena
- Prototype peripheral additions by reading schematics, modifying device trees, soldering breakout boards, and probing pins with a logic analyzer
- Interact with hardware without having to write kernel device drivers
- Divide your system up into services supervised by BusyBox runit
- Debug devices remotely using GDB and measure the performance of systems using tools such as perf, ftrace, eBPF, and Callgrind

Linux System Programming Techniques

Jack-Benny Persson

ISBN: 978-1-78995-128-8

- Discover how to write programs for the Linux system using a wide variety of system calls
- Delve into the working of POSIX functions
- Understand and use key concepts such as signals, pipes, IPC, and process management
- Find out how to integrate programs with a Linux system
- Explore advanced topics such as filesystem operations, creating shared libraries, and debugging your programs
- Gain an overall understanding of how to debug your programs using Valgrind

Packt is searching for authors like you

If you're interested in becoming an author for Packt, please visit `authors.packtpub.com` and apply today. We have worked with thousands of developers and tech professionals, just like you, to help them share their insight with the global tech community. You can make a general application, apply for a specific hot topic that we are recruiting an author for, or submit your own idea.

Share Your Thoughts

Now you've finished *Linux Administration Best Practices*, we'd love to hear your thoughts! Scan the QR code below to go straight to the Amazon review page for this book and share your feedback or leave a review on the site that you purchased it from.

https://packt.link/r/1800568797

Your review is important to us and the tech community and will help us make sure we're delivering excellent quality content.

www.ingramcontent.com/pod-product-compliance
Lightning Source LLC
LaVergne TN
LVHW081330050326
832903LV00024B/1094